Developments in Environmental Modelling

Time and Methods in Environmental Interfaces Modelling: Personal Insights

Volume 29

Developments in Environmental Modelling

Developments in Environmental Modelling

Time and Methods in Environmental Interfaces Modelling: Personal Insights

Volume 29

Dragutin T. Mihailović

Igor Balaž

Darko Kapor

University of Novi Sad, Novi Sad, Serbia

ELSEVIER

AMSTERDAM • BOSTON • HEIDELBERG • LONDON
NEW YORK • OXFORD • PARIS • SAN DIEGO
SAN FRANCISCO • SINGAPORE • SYDNEY • TOKYO

Elsevier
Radarweg 29, PO Box 211, 1000 AE Amsterdam, Netherlands
The Boulevard, Langford Lane, Kidlington, Oxford OX5 1GB, United Kingdom
50 Hampshire Street, 5th Floor, Cambridge, MA 02139, United States

Notices
Knowledge and best practice in this field are constantly changing. As new research and experience broaden our understanding, changes in research methods, professional practices, or medical treatment may become necessary.

Practitioners and researchers must always rely on their own experience and knowledge in evaluating and using any information, methods, compounds, or experiments described herein. In using such information or methods they should be mindful of their own safety and the safety of others, including parties for whom they have a professional responsibility.

To the fullest extent of the law, neither the Publisher nor the authors, contributors, or editors, assume any liability for any injury and/or damage to persons or property as a matter of products liability, negligence or otherwise, or from any use or operation of any methods, products, instructions, or ideas contained in the material herein.

Library of Congress Cataloging-in-Publication Data
A catalog record for this book is available from the Library of Congress

British Library Cataloguing-in-Publication Data
A catalogue record for this book is available from the British Library

ISBN: 978-0-444-63918-9
ISSN: 0167-8892

For information on all Elsevier publications visit
our website at https://www.elsevier.com

Working together
to grow libraries in
developing countries

www.elsevier.com • www.bookaid.org

Publisher: Candice Janco
Acquisition Editor: Laura Kelleher
Editorial Project Manager: Emily Thomson
Production Project Manager: Paul Prasad Chandramohan
Cover Designer: Maria Ines Cruz

Typeset by TNQ Books and Journals

Dragutin T. Mihailović
To Gordana, Ivan, and Anja

Igor Balaž
To those who inspired it

Darko Kapor
To my family who has always supported me in many ways

Contents

PART II TIME IN ENVIRONMENTAL INTERFACES MODELLING

PART III USE OF DIFFERENT COUPLED MAPS IN THE ENVIRONMENTAL INTERFACES MODELLING

Preface

A curious reader probably scans this Preface trying to find out what the "Personal Insights" actually means. We wish to explain this term here.

All three authors are involved in the field of environmental interfaces some more formally, some less, for a very long time. During this period, we gained an insight into the field, which is definitely personal, since there are parts we like more than the others, and we do not hide it. When preparing to write this book, it was the general idea to collect our previous work and see what will come out of it. It turned out that in the meantime we have matured, and more importantly, our understanding of certain aspects of the field has matured, so we now look at the same material in a different manner. Computing techniques progressed a lot, so certain simulations became possible now. The effect is that previous research has become just a foundation for completely new constructions, and the choice of the direction follows much more personal preferences than strict research logic.

Even more importantly, we have noticed interconnections between various subjects treated separately in our previous work. This has caused a complicated structure of the book. Although we follow a certain reasonable line in presenting the material, almost everywhere where are shortcuts connecting chapters from various parts of the book. We were tempted to plot a diagram showing it, but it turned out to be a figure of large complexity, so we gave up. Actually, we think now that we had a subconscious idea about these relations, when we were choosing the next subject of research in the old days.

To illustrate this, we offer here the attitudes of three authors, which will better explain our standpoints and inspirations in writing the book.

Attitude of the first author. Seventeen years ago, I received a report from a reviewer of the *Journal of Applied Meteorology*, who ended her/his short negative report with the words … "so little science and so much imagination …" At that point, I was frustrated by the "shortsightness" that accompanied the report so I did not pay attention to the order of closing arguments and words. However, the sound emitted by those words settled deeply in my subconscious, not as an echo of criticism of my work, but rather as a message from a person who is involved in science. Finally, my brain was making a reduction of this message to the statement "imagination is behind the science," i.e., metaphorically said the science can be understood as "a train dispatcher of strictly controlled trains." It means that it is not allowed to imagination to be a locomotive of the science. This purely technological understanding of the science typical for nowadays was anticipated by many philosophers, scientists, poets, and writers. Apparently, at this moment, science is under control of "religion of the metrics," which requires rather measurable results technologically colored instead of the results that come from the world of the imagination and which push forward the frontline of science. This reasoning is metaphorically and nicely memorized by Johannes Jensen in *Madam d'Ora*: "Our little life is rounded by the sleep." Final words for describing the motive, which initiated this book I found in a short story *The Landscape of Mountain* (*Meifu sansuizu*) by Shumon Miura. The main character of the story Chen was trying needlessly to paint the mountain landscape better than it had done by the nature, and finally he said, "Ultimately, we are all apprentices of time. If it wishes, a worthless paper will be left to future generations as a masterpiece while an original masterpiece will be displaced aside. I do not want to be an apprentice. From that reason, I am ready to become an artist who does not take brush in hand. Like the Time itself. It is paradoxically, but…"

My own attitude about science and its role is based on a clear distinction between discovery (science) and invention (technology and technique). The discovery could be defined as a final step in finding "something that is not still discovered but it exists as a truth," while invention means "something that has been found in the field of the scientifically established discoveries" (Darko Kapor, personal communication). In fact, this intuitive understanding of discovery and mission of science is one possible reading of the Gödel's first incompleteness theorem (and the second incompleteness theorem as an extension of the first), which is important both in mathematical logic and in the philosophy of mathematics. The formal theorem is written in highly technical language. It may be paraphrased in English as, "Any effectively generated theory capable of expressing elementary arithmetic cannot be both consistent and complete. In particular, for any consistent, effectively generated formal theory that proves certain basic arithmetic truths, there is an arithmetical statement that is true, but not provable in the theory" (Kleene, S.C. 1967 Mathematical Logic. John Wiley, pp 250.). There exists always synchronization between science and technology either on the lower or higher levels. Undoubtedly, the science provides the "field of truth" pushing forward the technology, while in return the technology provides advanced products for challenge of science (Darko Kapor, personal communication).

To make step forward, of any size, in science we must have (1) a dream and (2) a deep belief in that dream. (1) The dream including imagination about something is a *condition* sine *qua non* of science. It provides voyaging through the field of the hidden truths waiting to be seen. The power of that dream is warmly and lyrically described in the poetical movie *Do You Remember Dolly Bell* by Emir Kusturica. A metaphysical philanthropist father Fahro is passing away, while his son Dino is reading aloud to him a newspaper article and says, " If the Earth's axis could be moved just a little and Indian Ocean dried, it would be so much wheat that all raya ('people' in Bosnian slang) in the world would not be hungry." He is finishing his life with the words, "I will not join to raya but my dream came true." (2) The dream without belief would be incomplete. The belief leads the dream toward the discovery. The strength of the belief in its epical beauty I found at the funeral ceremony in case one of the local bohemians from my hometown Čačak (Serbia) when his friend, in farewell speech, said, "You are the only one who believed that all the rivers, seas, and oceans were originating from the Morava River." Such deepness and strength of the belief in something is almost metaphysical. That is my understanding of the scientific work. In that sense, all my reflections including potential misapprehension in the last 15 years are incorporated in this book (Dragu-tin T. Mihailović).

Attitude of the second author. The concept of environmental interfaces shares some similarities with the concept of complexity. For both of them, it can be argued that there are more notions used to describe a particular subset of natural phenomena than precisely defined concepts shared among the majority of scientist. Some may say that they are too general to have much content. Wide breadth of topics covered in this book accompanied by a diverse set of mathematical tools could support that view. Here, I will not argue against it. Instead, I will offer an alternative reading.

So-called systems thinking was born almost 70 years ago as a search for common properties shared by all organized systems. Initial enthusiasm was supported by some of the finest scientific minds of that time. Two strong disciplines emerged from that wave: cybernetics and general systems theory (GST). They helped establishing precise understanding of the role of feedbacks in organizing systems and introduced *self-organization* and *emergence* into modern science. Over time, their grandiose approach was toned down, and the whole discipline seemingly disappeared from the scene. From today's perspective, it is debatable to what extent GST and cybernetics changed scientific landscape, but it was a fertile movement that left a deep mark in several thriving disciplines such as control theory, systems biology (with genomics and other omics), and systems engineering. In numerous other fields, systemic approach became a norm. It demonstrates how changing perspective of scientific inquiry can open up a vast field of new insights. The main topic of this book, environmental interfaces, arises as an offspring of that, so-called, systems thinking. In that light, I hope that some of the ideas presented here could be inspiring enough to open up new avenues in our striving to understand organization of environment (Igor Balaž).

Attitude of the third author. After many years spent working on environmental problems (among other things), I started to believe that most of the concepts could be transferred to the humanities too. So, the human society or, in particular, the circle of people around us can be treated as an environment. However, each person within this environment is a microcosm of its own, so its contact with the everyone else can be treated as being realized through an environmental interface. This interface has two aspects: there is a physical one, bordering our body and consisting of skin and senses and a psychological one, probably a real or virtual aura encapsulating us. While we know a lot about the events happening on the physical environmental interface, it is the psychological one that determines our behavior. Actually, to a person who decides to reflect a lot about its environment, most of the events happen on this, other environmental interface. Many events, occurring in the interaction with Prof. Mihailović and Dr Balaž resulted in joining the forces to prepare this manuscript through which we tried to present our ideas about many things invading in this way, auras of people who would care to read it (Darko Kapor).

To visualize our attitudes, we choose for the cover page of the book, the photo of the architecture of Antonio Gaudi inside the Park Güell (Barcelona, Spain), which is a public park system composed of gardens. This ingenious Catalan artist gave birth to this park in such a way that no one before him did do it, nor will do it so after him, since "his works acquire a structural richness of forms and volumes, free of the rational rigidity or any sort of classic premises." ("Park Güell", Wikipedia, The Free Encyclopedia) Selected landscape can be seen as a sort of interface that relies on another space, while the corridor symbolizes the space of solution, which is almost at hand distance, but still unapproachable.

This book has three formal authors, yet, it is based on the research performed in collaboration with many people, all of them also dear friends. It is essential to mention them here and thank them for a long and fruitful cooperation. Department of Mathematics and Informatics of the Faculty of Sciences, University of Novi Sad, was the source of information and help. Prof. Mirko Budinčević shared with us his knowledge on nonlinear dynamics and difference equation, while Prof. Siniša Crvenković introduced us to the category theory and the formal concept analysis. They were of great help, even though permanently wondering why people of our profile need such knowledge. Prof. Vladimir Kostić helped us learn about the new fields of spectra and pseudospectra, i.e., the behavior of nonnormal matrices and operators.

Teaching process was a permanent source of inspiration for research. The interaction with students—undergraduates and graduates, as well as with fellow teachers, was a fountain of new ideas and different approaches. It is here that we wish to thank all the students and colleagues we have met during many years of teaching at the Faculty of Agriculture, Department of Physics, Faculty of Sciences, and the Center for Meteorology and Environmental Modelling, all at the University of Novi Sad. Many names come to our minds, but in this way, we also recognize the merits of the very institutions, not just the particular people.

This highly interdisciplinary book deals with mathematical methods in modelling of environmental interfaces from nanotubes and cell to planetary scale. On the other hand, the exposition is accompanied by personal insights of the authors based on their long-lasting activity in the fields covered by the book. In this way, the reader is provoked to establish his own standpoint which might or might not agree with the one of the authors. Many numerical simulations offered, and extensive list of cited literature will provide solid basis for this. Finally, let us mention that we used various synonyms equivalently, exploiting the rich structure of English language. The book is divided into 7 parts containing 26 chapters.

Part I contains an introductory material and starts with a chapter where we give a definition of the environmental interface, which broadly covers the unavoidable multidisciplinary approach in environmental sciences and also includes the traditional approaches in environmental modelling. The interface between two different environments itself is considered as a complex system itself, in the sense that "a complex system cannot be decomposed nontrivially into a set of part for which it is the logical sum." (Rosen R. 1991 Life itself. Columbia University Press) In Chapter 2, we review advanced theoretician's tools in the modelling of the environmental interface systems. An extensive discussion of various aspects of modelling is offered in Chapter 3 with an illustration through the solution of the energy balance equation for the ground surface, which is often used in environmental modelling. We state our opinion about dilemma whether the environmental interface systems models should be built in the form of differential or difference equations, i.e., whether we should either deal with the continuous-time or discrete-time, where time is considered as a continuous or discrete variable, respectively. We end this part with a chapter on the use of formal complex analysis in solving the environmental problems.

Part II is devoted to the role of time in environmental interface modelling since with the progress in this field, the question of the concept of time becomes more authentic. We first elaborate understanding the time in physics and philosophy in Chapter 5, going over to Chapter 6 dealing with time in biology. It is formalized in Chapter 7 by the introduction of functional time in generalized functional systems. By the notion of the functional system, we cover all systems where processes unfold following a set of known rules and which exhibit repetitive pattern. Using mathematical formalism, we show on several examples how the functional time is formed as a result of consistent change of concrete material object states. Examples are: (1) the response of the functional system on a stimulus (mollusk time reflex formation); (2) the response of the functional system on a cognitive level (prisoner time formation in the cell), and (3) the process of substance exchange on the cellular level (time formation in process of biochemical substance exchange between cells).

Part III is an very important one since it introduces the material necessary for understanding the rest of the book. It considers the use of different logistic maps in the coupling in the environmental interfaces. In Chapter 8, we consider coupled logistic maps, through their diffusive, linear, and combined coupling. We give an example of diffusive coupling through interaction of two environmental interfaces on the Earth's surface. We analyze the stability of this dynamical system using the Lyapunov exponent. Chapter 9 is devoted to the logistic difference equation on the extended domain. We extend the domain [0,4] in which the logistic parameter of the classical logistic equation is defined to the domain [−2,4], and we discuss and analyze properties of the parameter of difference equation, which is ranged in this domain, using bifurcation diagram, Lyapunov exponent, sample and permutation entropies. As the next step, in Chapter 10, we introduce the logistic equation with affinity, and then, we demonstrate its use in modelling turbulent fluxes over the heterogeneous environmental interfaces. First, we give a mathematical background of a map with cell affinity in the form of a generalized logistic map. Second, analyzing the model outputs and observed data, we summarize uncertainties that occur in modelling the turbulent energy exchange over the heterogeneous environmental interfaces, with setting an accent on the Schmidt's paradox.

Chapter 11 deals with the maps serving the different coupling in the environmental interfaces modelling. First, we consider behavior of a logistic map driven by fluctuations. We give an overview of literature about logistic map driven by periodic signal, quasi-periodic signal or noise. Second, we analyze the behavior of the coupled maps serving the combined coupling in the presence of dynamical noise. In the case of uncoupled nonlinear oscillators, we demonstrate that the addition of parametric fluctuations has a pronounced effect on the dynamics of such systems. Finally, we consider the behavior of the coupled maps serving the combined coupling when we introduce a parametric noise in their all parameters.

Part IV is devoted to the concepts of heterarchy and observational heterarchy and their relation to the exchange processes between the environmental interfaces. The concept of heterarchy in environmental modelling is introduced and some ecological examples are given in Chapter 12. This concept is then applied to biochemical substance exchange in a diffusively coupled ring of cells in Chapter 13. We first consider the observational heterarchy consisting of two sets of intralayer maps, called Intent and Extent perspectives, and interlayer operations using the formalism of the category theory. Looking from the intent and extent perspective in a cell, we address the synchronization of the passive and active coupling for two cells using the generalized logistic equation with the affinity. We perform simulations of active coupling in a multicell system. Finally, in Chapter 14, we study the heterarchical aspect of the albedo over heterogeneous environmental interfaces.

Many results in environmental studies are presented in the form of measured or modeled time series for certain important quantities, since it is essential to know how to study the complexity of the environmental system, based on this series. This is the subject of Part V. We first introduce the concept of Kolmogorov complexity and other complexity measures based on it in Chapter 15. Number of example follows. In Chapter 16, we first perform a complexity analysis of ^{222}Rn concentration variation in a cave. Second, we use complexity analysis in analyzing the dependence of ^{222}Rn concentration time series on indoor air temperature and humidity. Finally, we apply the Kolmogorov complexity and use its spectrum in analysis of the UV-B radiation time series. In Chapter 17, we deal with complexity analysis of the environmental flow time series. First, we use it to quantify the randomness degree in river flow time series of two mountain rivers in Bosnia and Herzegovina, representing the turbulent environmental fluid. Next, we analyze the experimental data from a turbulent flow collected in a laboratory channel with bed roughness elements of different densities and variable bed slope. Finally, we use the Kolmogorov complexities and the Kolmogorov complexity spectrum to quantify the randomness degree in river flow time series of seven rivers with different regimes in Bosnia and Herzegovina, representing their different type of courses. Since climate is a typical example of the complex system, we discuss various approaches to its complexity in Chapter 18. Thus, we use complexity measures to analyze spatial and temporal distribution of air temperature and the observed precipitation time series. Finally, we give an example of comparison between complexities of a global and regional model.

In Part VI, we address the problem of the chaotic phenomena in computing the environmental interface variables. Such a study must begin (Chapter 19) with the analysis of the relations between mathematics and environmental sciences. In that sense, we consider: (1) the role of mathematics in environmental sciences and (2) difference equations and occurrence of chaos in modelling of phenomena in the environmental world. First, in Chapter 20, we consider the climate predictability and climate models through: (1) giving the short survey on the predictability and (2) gathering current issues in modelling the global climate system. Second, we give an example of the application of the regional climate models with an overview of its outputs. The outputs were obtained by dynamic downscaling of climate simulations conducted with the ECHAM5 GCM (General Circulation Model) coupled with the Max Planck Institute Ocean Model. The downscaling of the GCM climate simulations was performed with the coupled regional climate model EBU-POM (Eta Belgrade University - Princeton Ocean Model). In Chapter 21, we deal with occurrence of chaos in exchange of vertical turbulent fluxes over environmental interfaces in climate models, concentrating on the occurrence of the chaos in computing the environmental interface temperature. We have derived criterion for choice of the time step used in environmental models for numerical solving of the energy balance equation to avoid situation when the environmental interface cannot oppose an enormous amount of energy suddenly entering system. We also perform a dynamic analysis of solutions for the environmental interface and deeper soil layer temperatures represented by the coupled difference equations to find regions where solutions show chaotic behavior. We consider synchronization and stability of horizontal energy exchange between environmental interfaces in climate models in Chapter 22, by considering it as a diffusion-like process described by the dynamics of driven coupled oscillators enhancing the conditions when the process of exchange is synchronized. Then, we consider asymptotic stability of horizontal energy exchange between environmental interfaces introducing a dynamical system approach that provides more realistic results in modelling of energy exchange over the heterogeneous grid-box than the flux aggregation methods.

The last part of the book (VII) includes the following topics: environmental interfaces and their stability in biological systems, synchronization of the biochemical substance exchange between cells, complexity, and asymptotic stability in the process of biochemical substance exchange in multicell systems and use of pseudospectra in analyzing the influence of intercellular nanotubes on cell-to-cell communication integrity. Chapter 23 is devoted to the biological environmental interfaces and their ability to perceive the changes in the environment. Going further, in Chapter 24, we consider synchronization of the biochemical substance exchange between cells mathematically modeled as a system of difference equations of coupled logistic equations. Then, we add the fluctuations of environmental parameters to the model. In Chapter 25, we deal with the issue of complexity and asymptotic stability in the process of the biochemical substance exchange in multicell system, using the model described in the previous chapter. After calculating the Kolmogorov complexity measures, we focus on the asymptotic stability of the intercellular biochemical substance exchange. In Chapter 26, we examine how the biochemical substance exchange through tunneling nanotubes (TNT) (besides common exchange through gap junctions (GJ)) affects the functional stability of the multicellular system. We answer whether TNT can destabilize the intercellular communication through GJ and how to determine the threshold at which the destabilization occurs. One way to answer is the application of the concept of pseudospectra.

This Preface might look too long, but we find it important to state our positions in advance and, in this way, prepare the reader for an adventure that expects him/her. We felt the writing of this book as a great adventure and we do hope that the readers will feel at least some of the excitement we did.

PART

I

Introduction

Environmental interface: definition and introductory comments

1

Complex systems science has contributed to our understanding of environmental issues in many areas from small to large temporal and spatial scales (from the cell behavior to global climate and its change). Environmental systems by themselves are both complicated and complex. Complicated, in that many agents act upon them; complex, in that there are feedback loops connecting the state of the system back to the agents, and connecting the actions of the agents to one another. Complex systems have complex dynamics usually characterized by the so-called tipping points, abrupt changes in the state of the system caused by seemingly gradual change in its drivers (Gladwell, 2000). For example, a climate tipping point is a somewhat ill-defined concept of a point when global climate changes from one stable state to another stable state. After the tipping point has been passed, a transition to a new state occurs. Many scientists now use the power of computer models to advance their subjects. But there is a choice: to simplify complex systems or to include more details (Paola and Leeder, 2011). Further advances in these areas will be necessary before complex systems science can be widely applied to understand the dynamics of environmental systems. In this book we will consider *environmental interfaces* as complex systems through their main features. There are many contemporary researches that deal with specific aspects of the environmental interface. However, in this book we will consider the temporal aspect, various recent approaches to it and complexity in environmental interfaces modelling through our personal insights.

Definition of environmental interface: Technically speaking, the *interface* is a space at which independent systems or components meet and act or communicate with each other. Interfaces can appear between system elements and they can also exist between a system element and the system's environment. In the latter case we speak about *environmental interface*. It can be specifically defined depending on the science where it is used (ecology (Sizykh, 2007), ecological economy (Lehtonen, 2004), social sciences (Rasmussen and Arler, 2010), programming languages, and simulations support systems (Banks et al., 2009), etc.). We define the *environmental interface* as an interface between two abiotic or biotic environments that may be in relative motion and exchange energy, matter (substance), or information through physical, biological, or chemical processes, fluctuating temporally and spatially regardless of the space and time scale. It is slightly different from its formulation in Mihailovic and Balaž (2007) and Mihailović et al. (2012). This definition

Developments in Environmental Modelling, Volume 29, ISSN 0167-8892, http://dx.doi.org/10.1016/B978-0-444-63918-9.00001-6

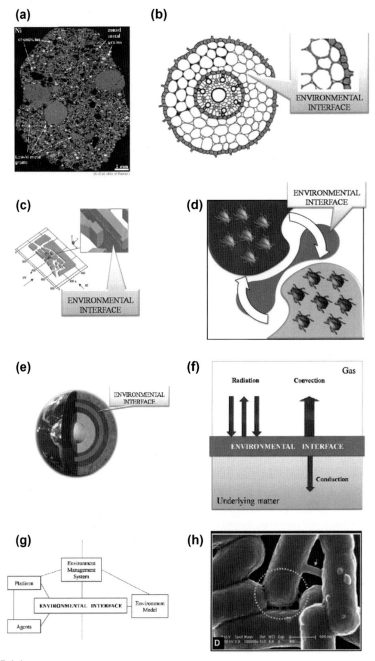

FIGURE 1.1

broadly covers the unavoidable multidisciplinary approach in environmental sciences and also includes the traditional approaches in environmental modelling. For example, such interfaces can be (1) placed in between different environments and (2) extended from micro to planetary scales. Through these interfaces environments exchange energy, matter, and information (Fig. 1.1). For example, those processes are (a) ions exchange in metals (Krot et al., 2001), (b) intercellular exchange of biochemical substances (Mihailović et al., 2011a), (c) exchange of air volumes in a macroscale of urban conditions (Neofytou et al., 2006), (d) periodic migrations between populations (Lloyd, 1995), (e) heat exchange in Earth's interior consisting of central core, a mantle surrounding the core and lithosphere, (f) energy exchange between solid matter and gas in natural conditions (Mihailović et al., 2011b), and (g) information exchange in a specific environment model combined with the environment interface describing their interactions (Behrens, 2009).

The interface between two media is a complex system itself. We use the term complex system in Rosen's sense (1991) as it was explicated in the comment by Collier (2003) as follows: "In Rosen's sense a complex system cannot be decomposed nontrivially into a set of part [sic!] for which it is the logical sum. Rosen's modelling relation requires this. Other notions of modelling would allow complete models of Rosen style complex systems, but the models would have to be what Rosen calls analytic, that is, they would have to be a logical product. Autonomous systems must be complex. Other types of systems may be complex, and some may go in and out of complex phases." Also, we will explain in which sense the term complexity will be used in further text. Usually, that is an ambiguous term, sometimes used to refer to systems that cannot be modeled precisely in all respects

---

Examples of environmental interfaces: (a) ions exchange in metals. The space highlighted by a *dashed red line* (dark gray in print versions) indicates an interface; (b) intercellular exchange of biochemical substances (Mihailović et al., 2011a); (c) exchange of air volumes in urban conditions; (d) migration of insects (Lloyd, 1995); (e) heat exchange in Earth's interior consisting of central core, a mantle surrounding the core and lithosphere; (f) energy exchange between solid matter and gas in natural conditions (Mihailović et al., 2011b); (g) information exchange in a specific environment model combined with the environment interface describing their interactions (Behrens et al., 2009) Intercellular TNTs between neighboring cells, and (h) a field of cells exchanging the substance with a cluster of smaller tunneling nanotube TNTs (highlighted by a *dashed circle*) and a more pronounced larger tube (indicated by an *arrow*).

(a) Reprinted with permission from Krot, A.N., Meibom, A., Russell, S.S., Conel, M.A., Jeffries, T.E., Keil, K., 2001. A new astrophysical setting for chondrule formation. Science 291, 1776–1779. (c) Reprinted with permission from Neofytou, P., Venetsanos, A.G., Vlachogiannis, D., Bartzis, J.G., Scaperdas, A., 2006. CFD simulations of the wind environment around an airport terminal building. Environ. Model. Softw. 21, 520–524. (h) Reprinted with permission from Dubey, G., Ben-Yehuda, S., 2011. Intercellular nanotubes mediate bacterial communication. Cell 144, 590–600.

(Rosen, 1991). However, following Arshinov and Fuchs (2003) the term "complexity" has three levels of meaning. (1) There is self-organization and emergence in complex systems (Edmonds, 1999). (2) Complex systems are not organized centrally but in a distributed manner; there are many connections between the system's parts (Edmonds, 1999; Kauffman, 1993). (3) It is difficult to model complex systems and to predict their behavior even if one knows to a large extent the parts of such systems and the connections between the parts (Edmonds, 1999; Heylighen, 1997). The complexity of a system depends on the number of its elements and connections between the elements (the system's structure). According to this assumption, Kauffman (1993) defines complexity as the "number of conflicting constraints" in a system; Heylighen (1996) says that complexity can be characterized by a lack of symmetry (symmetry breaking) which means that "no part or aspect of a complex entity can provide sufficient information to actually or statistically predict the properties of the others parts." Edmonds (1996) defines complexity as "that property of a language expression which makes it difficult to formulate its overall behavior, even when given almost complete information about its atomic components and their inter-relations." Aspects of complexity are things, people, number of elements, number of relations, nonlinearity, broken symmetry, nonholonic constraints, hierarchy, and emergence (Flood and Carson, 1993). Note, that the interactions between parts of the complex environmental interface systems are nonlinear, while their interactions with the surrounding environments are noisy that is mathematically well elaborated in Liu and Ma (2005), Serletis and Shahmoradi (2006), Savi (2007), Serletis et al. (2007a), (2007b), Mihailović et al. (2012), among others.

In this introductory part we cannot avoid a short overview of some epistemological points from the 20th century onward. Until recently, discussions about scientific truth were filled with numerous metaphysical assumptions. They usually converged to one question (more or less explicitly stated): "How can we reach objective truth about natural processes?" However, during the 20th century, this question first became less important and then gradually disappeared from the epistemological scene as a relic from the age of naive realism. Now, in contemporary epistemology of science, it is well established that there is a fundamental difference between phenomenon and noumenon. Therefore, the object of scientific analysis cannot be the nature by itself, but only highly constructivistic products, i.e., conceptually embedded sets of observer's experiences. Accordingly, scientific theories are now understood as logical instruments of organization of human thought, through which we can interpret and organize experimental laws (Nagel, 1961). Also, since they have constructivistic character, their relation to nature should not be considered through the vocabulary of logic; they are not truth statements and they are not logical derivatives of observed facts but only sets of rules and guiding principles for analysis of empirical facts (Nagel, 1961). Therefore, in the development of a scientific theory, it is not a problem to make approximations that can never reach reality. It is inevitable. But believing that relations of abstractions are exactly the same as relations in nature can be very problematic. Firstly, it can usually become a source of unfruitful debates about the "true" nature of nature. Secondly, from such a perspective it is impossible to see and analyze the consequences of the interface perspective, where the observer is within the universe he observes. A clear example of both mentioned problems can be found in the development of the contemporary physics. At the very beginning of the 20th century, Pierre Duhem (1906) asserted that physical theories are not simple reflexions of natural processes, but rigorous logical systems, which operate with abstract symbols and which are connected with nature through system of measurements and scales. Such approaches put forward the process of encoding of natural processes into the domain of formal systems, as the first and crucial step in the development of a physical theory. However, in his opinion, a pattern of encoding depends almost entirely on the previously accepted theories. Therefore, empirical observations cannot be separated from the current state of affairs in a given scientific discipline, since theoretical assumptions determine what will be observed, how it will be observed, and how results will be interpreted. Although Duhem's approach can be characterized as conventionalism, his contribution to the general trend of development of thought in theoretical physics remains immense.

Few decades later, the explosive growth of quantum mechanics raised some fundamental questions about the status of observation in physics, and how our measurement procedures can affect the observed physical properties ("measurement problem"). In short, Einstein, opposing the Copenhagen interpretation of physical properties of quantum systems, claimed that under ideal conditions, observations reflect the objective physical reality. On the other hand, Bohr asserts that in quantum mechanics the measured quantum system and the measuring macroscopic apparatus cannot be considered as separate within a scope of scientific consideration. In other words, the physical properties of quantum systems are essentially dependent on the applied experimental apparatus. One of the most famous moments of the debate is now well known as Einstein–Podolsky–Rosen paradox (EPR) (Einstein et al., 1935). In the short paper they showed, that if the quantum mechanics description of reality is complete, then the noncommutable operators corresponding to two physical quantities can have simultaneous reality. In other words, quantum mechanics is inconsistent with the reduction of the wave-packet postulate. Later, Bell (1964) revealed that the EPR paradox stands only under the set of supplementary assumptions, among which there is the assumption of locality. Moreover, within quantum mechanics there is no need to accept them all. Although it can look like a closing chapter in the debate on "measurement problem," this question evolved from the limited scope of quantum mechanics and took a more general form: "how the observations are affected by the fact that the observer is within the universe he observes?" This is certainly not a new question in the history of human thought, but (until recent partial attempts) in the natural sciences it never gets a formal explanation. In developmental psychology, Piaget (1973a,b) clearly demonstrated that elementary categories of human thought are construed during one's development, and how externality of cognitive entities is restructured in accordance with its functional purposes through the process of assimilation of external changes with the operative schematism of that entity, and finally, in the world of logic and formal systems, Gödel shook the scientific community with his proof of incompleteness of formal systems (for extensive discussion see Rosen, 1991 and Nagel and Newman, 1958). Now, in the natural sciences, this problem is finally recognized and dispersed attempts of its formal treatment fall under the umbrella of discipline called endophysics. This term was originally suggested by David Finkelstein in personal communication with Otto Rössler. Later, it was comprehensively elaborated in detail by Otto Rössler (1998).

Finally, although the question of time in the modelling relation is the theme of the next part, here we will make some comments about time in the context of teleological as well as causal dynamics (the term, "causal" is used in the broader sense of "governed by influences from the past"). A usual approach in physics is that the present state is strictly a result of its evolution from the past. However, it has been shown that some phenomena in the real world can be explained, if we accept that the present state of a system is defined by its *past*, in the sense that the past determines the possible states that are to be considered, and by its *future*, in the sense that the selection of a possible future state determines the effective *present state*. Namely, past and future measurements, taken together, provide complete information about a quantum system. Pioneering step about this subject has been done by Watanabe (1955) whose work was later experienced again by Aharonov et al. (1964), who later renamed it the Two-State Vector Formalism (Aharonov et al., 1964; Aharonov and Vaidman, 1997, 2008). This, a time-symmetrized approach in quantum theory is particularly helpful for the analysis of experiments performed on preselected and postselected ensembles (Aharonov and Vaidman 2008; Brodutch, 2014). The two-state quantum dynamics is used for designing the phenomenological model of the reionization process, when this dynamics is adopted for the vicinity of the potential barrier top (Aharonov et al., 1964). This is just a short reminder for the environmental modelling community that, in the modelling of complex environmental interface processes, we should bear in mind a possibility of using two-state formalism in the modelling procedure (Nedeljkovic and Nedeljkovic, 2003).

REFERENCES

Aharonov, Y., Vaidman, L., 1997. Protective measurements of two-state vectors. In: Cohen, R.S., Horne, M., Stachel, J.J. (Eds.), Potentiality, Entanglement and Passion-at-a-distance, Quantum Mechanical Studies for A. M. Shimony, Volume Two, pp. 1−8.

Aharonov, Y., Vaidman, L., 2008. The two-state vector formalism: an updated review. Lect. Notes Phys. 734, 399−447.

Aharonov, Y., Bergmann, P.G., Lebowitz, J.L., 1964. Time symmetry in the quantum process of measurement. Phys. Rev. B 134, 1410−1416.

Arshinov, V., Fuchs, C., 2003. Preface. In: Arshinov, V., Fuchs, C. (Eds.), Causality, Emergence, Self-Organisation. NIAPriroda, Moscow, Russia, pp. 1−18.

Banks, J., Carson, J.S., Nelson, B.L., Nicol, D.M., 2009. Discrete-Event System Simulation. Prentice Hall, Upper Saddle River, NJ.

Behrens, T.M., Dix, J., Hindriks, K.V., 2009. Towards an Environment Interface Standard for Agent-Oriented Programming. Technical Report IfI-09-09. Clausthal University of Technology.

Bell, J.S., 1964. On the Einstein Podolsky Rosen paradox. Physics 1, 195−200.

Brodutch, A., 2014. Weak Measurements and the Two State Vector Formalism. https://uwaterloo.ca/institute-for-quantum-computing/sites/ca.institute-for-quantum-computing/files/uploads/files/ab_11.pdf.

Collier, J.D., 2003. Fundamental properties of self-organisation. In: Arshinov, V., Fuchs, C. (Eds.), Causality, Emergence, Self-organisation. NIA-Priroda, Moscow, Russia, pp. 150−166.

Dubey, G., Ben-Yehuda, S., 2011. Intercellular nanotubes mediate bacterial communication. Cell 144, 590−600.

Duhem, P., 1906. The Aim and Structure of Physical Theory (in Serbian). Izdavacka knjizarnica Zorana Stojanovica, Novi Sad, 2003.

Edmonds, B., 1996. Pragmatic Holism. CPM Report 96−08. MMU.

Edmonds, B., 1999. What is complexity? The philosophy of complexity per se with application to some examples in evolution. In: Heylighen, F., Aerts, D. (Eds.), The Evolution of Complexity. Kluwer, Dordrecht, The Netherlands, pp. 1−18.

Einstein, E., Podolsky, B., Rosen, N., 1935. Can quantum-mechanical description of physical reality be considered complete? Phys. Rev. 47, 777−780.

Flood, R.L., Carson, E.R., 1993. Dealing With Complexity: An Introduction to the Theory and Application of Systems Science. Plenum Press, New York, NY, USA.

Gladwell, M., 2000. The Tipping Point: How Little Things Can Make a Big Difference, first ed. Little Brown, London.

Heylighen, F., 1996. What is complexity? In: Heylighenand, F., Aerts, D. (Eds.), The Evolution of Complexity. Kluwer, Dordrecht, The Netherlands.

Heylighen, F., 1997. The growth of structural and functional complexity. In: Heylighen, F. (Ed.), The Evolution of Complexity. Kluwer, Dordrecht, The Netherlands.

Kauffman, S., 1993. The Origins of Order. Oxford University Press, Oxford, UK.

Krot, A.N., Meibom, A., Russell, S.S., Conel, M.A., Jeffries, T.E., Keil, K., 2001. A new astrophysical setting for chondrule formation. Science 291, 1776−1779.

Lehtonen, M., 2004. The environmental−social interface of sustainable development: capabilities, social capital, institution. Ecol. Econ. 49, 199−214.

Liu, Z., Ma, W., 2005. Noise induced destruction of zero Lyapunov exponent in coupled chaotic systems. Phys. Lett. A 343, 300−305.

Lloyd, A.L., 1995. The coupled logistic map: a simple model for effects of spatial heterogeneity on population dynamics. J. Theor. Biol. 173, 217−230.

Mihailovic, D.T., Balaž, I., 2007. An essay about modeling problems of complex systems in environmental fluid mechanics. Idojaras 111, 209−220.

Mihailović, D.T., Budincevic, M., Balaž, I., Mihailović, A., 2011a. Stability of intercellular exchange of biochemical substances affected by variability of environmental parameters. Mod. Phys. Lett. B 25, 2407−2417.

Mihailović, D.T., Budinčević, M., Kapor, D., Balaž, I., Perišić, D., 2011b. A numerical study of coupled maps representing energy exchange processes between two environmental interfaces regarded as biophysical complex systems. Nat. Sci. 1, 75−84.

Mihailović, D.T., Budinčević, M., Perišić, D., Balaž, I., 2012. Maps serving the combined coupling for use in environmental models and their behaviour in the presence of dynamical noise. Chaos Solitons Fractals 45, 156−165.

Nagel, E., 1961. The Structure of Science: Problems in the Logic of Scientific Explanation. Harcourt, Brace World, Inc, New York.

Nagel, E., Newman, J.R., 1958. Gödel's Proof. New York University Press, New York.

Nedeljković, L.D., Nedeljković, N.N., 2003. Rydberg-state reionization of multiply charged ions escaping from solid surfaces. Phys. Rev. A 67, 032709.

Neofytou, P., Venetsanos, A.G., Vlachogiannis, D., Bartzis, J.G., Scaperdas, A., 2006. CFD simulations of the wind environment around an airport terminal building. Environ. Model. Softw. 21, 520–524.

Paola, C., Leeder, M., 2011. Environmental dynamics: simplicity versus complexity. Nature 469, 38–39.

Piaget, J., 1973a. Introduction to Genetic Epistemology. 1) Mathematical Thought (in Serbian). Izdavacka knjizarnica Zorana Stojanovica, Novi Sad, 1994.

Piaget, J., 1973b. Introduction to Genetic Epistemology. 2) Physical Thought (In Serbian). Izdavacka knjizarnica Zorana Stojanovica, Novi Sad, 1996.

Rasmussen, K., Arler, F., 2010. Interdisciplinarity at the human-environment interface. Dan. J. Geogr. 110, 37–45.

Rosen, R., 1991. Life Itself, a Comprehensive Inquiry Into the Nature, Origin, and Fabrication of Life. Columbia University Press.

Rössler, O.E., 1998. Endophysics: The World as an Interface. World Scientific Publishing Co. Pte. Ltd., Singapore.

Savi, M.A., 2007. Effects of randomness on chaos and order of coupled maps. Phys. Lett. A 364, 389–395.

Serletis, A., Shahmoradi, A., 2006. Comment on "Singularity Bifurcations" by Yijun He and William A. Barnett. J. Macroecon. 28, 23–26.

Serletis, A., Shahmoradi, A., Serletis, D., 2007a. Effect of noise on the bifurcation behavior of nonlinear dynamical systems. Chaos Solitons Fractals 33, 914–921.

Serletis, A., Shahmoradi, A., Serletis, D., 2007b. Effect of noise on estimation Lyapunov exponents from a time series. Chaos Solitons Fractals 32, 883–887.

Sizykh, A.P., 2007. Plant communities of environmental interfaces as a problem of ecology and biogeography. Biol. Bull. 34, 292–296.

Watanabe, S., 1955. Symmetry of physical laws. Part III. Prediction and retrodiction. Rev. Mod. Phys. 27, 179.

Advanced theoretician's tools in the modelling of the environmental interface systems

The environmental interfaces are formed in a space that is rich with complex systems. Each such system, as an open one, interacts with other systems in a coherent way, producing new structures and building cohesion and new structural boundaries. It undergoes emergence and self-organization. Thus, in the modelling of the environmental interfaces, it is necessary to consider the following points: (1) the need for new modelling architecture and (2) usage of new mathematical tools like Category Theory (whose first use for research in ecology was originally proposed by Rosen (1985, 1991)), Mathematical Theory of General Systems (Mesarovic and Takahara, 1975), Formal Concept Analysis (FCA) (Wille, 1982; Ganter and Wille, 1997), and Nonlinear Dynamics and Chaos (Mihailovic and Balaž, 2007).

2.1 MODELLING ARCHITECTURE

Modelers of environmental interface systems in numerically oriented studies base their calculations on mathematical models for the simulation and prediction of different processes, which are exclusively nonlinear in describing relevant environmental quantities (Rosen, 1991). A theoretical description of any environmental interface system includes at least two important aspects. First, one should construct a concrete mathematical model of both the admissible states of the system and the transitions between these states. Second, one should establish the rules of selecting among many theoretically admissible states of the system only those states that are realized in nature under the given external conditions (Flood and Carson, 1993).

In the modelling community dealing with complex systems, Rosen's diagram (1991) is a recognizable guide. Fig. 2.1 is a slightly modified Rosen's diagram and schematically depicts a modelling relation when a natural system (N) and a formal system (F) are given. As above, two arrows represent the respective entailment structures: inference in formalism (F) and causality in a natural system (N). Now, the two established dictionaries provide encoding the phenomena of N into the propositions of F and another for decoding the propositions of F back to the phenomena in N. As mentioned above, there are two paths in diagram (1) and (2) + (3) + (4). According to Rosen (1991), the first of them (path (1)) represents

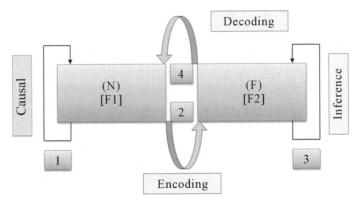

FIGURE 2.1

Schematic diagram representing both (i) the comparison of two formalisms F1 and F2 and (ii) modelling relation when we have given a natural system (N) and a formal system (F) (Rosen, 1991). Here, 1 represents causal entailment within the natural system (N); 2 represents encoding, where the observer's propositions about N are used as hypotheses in constructing formal system (F); 3 is the generation of theorems in F, which function as a model of N; and 4 is decoding, where the theorems of F are applied back to N in the form of predictions (Mihailovic et al., 2012).

the causal entailment within N (what an observer will see by simply sitting and watching what is happening). Arrow (2) encodes the phenomena in N into the propositions in F. In this route, we must use these propositions as hypotheses based on which the inferential machinery of the formal system F may operate (denoted by arrow (3)); it generates theorems in F, entailed precisely by the encoded hypotheses. Finally, we have to decode these theorems back into the phenomena of N, via arrow (4). At this point, the theorems become *predictions* about N. Then the formal system F is called a *model* of the natural system N if we always get the same answer regardless of the fact whether we follow path (1) or path (2) + (3) + (4). The process of modelling complex systems is a very comprehensive one. A system is to be treated as a complex structure, as for instance in Peter Checkland's definition: "A system is a model of a whole entity; when applied to human, the model is characterized fundamentally in terms of hierarchical structure, emergent properties, communication and control"(Levich and Solovyov, 1999, p. 318). The major components of complexity are openness and freeness, but the distinctive characteristic is "natural activity" like self-organization, and still of great importance as intraactivity but now joined by the phenomena of anticipation (Checkland, 1981) and interactivity between systems to be found in global interoperability. The transition from connectivity to activity involves a type change and therefore requires a formal system with an inbuilt facility to cross between the levels. Thus, intraconnectivity between the components cannot give rise to interactivity between those components without some nonlocal integrity coming into play (Klir, 2002). The nonlocality is a principle that is, among some others, specific for a certain object area such as the inorganic, living, or human realm. This principle means certain interaction between the elements of the system that is treated as the transmission of information at infinite speed (see, for instance, Bell (1964)).

2.2 BASICS OF CATEGORY THEORY

Category Theory, a discipline developed by Mac Lane (1971) and recommended by Rosen (1991) as a modern tool for complex (and living) systems, is found to have a formal expressive power for exploring the fundamental nonlocal concept of adjointness needed to understand complex systems. The arrow of Category Theory does not have just a formal meaning. According to Rossiter and Heather (2005), it formalizes the principle of *constancy* (originally introduced by Heraclites and Parmenides) that is provided by a common source and target. Such an arrow refers to the situation in which a source and target are indistinguishable. In a defined system, the collection of entities can be identified as objects, while operations between them are defined by arrows. Fig. 2.2 shows that there may exist many possible arrows between objects. However, Category Theory holds that a unique limiting arrow may exist for all of these possible arrows that represent the resulting *intraconnectivity* of a local system. There is an order between the two entities established by the directions of arrows (Manes and Arbib, 1975). This means that the arrow limit between two entities is also a limit of all possible paths. Because of the existence of limits and all possible connectivity, this is classified by axiomatic categories as a Cartesian closed category. Moving up one level, there is a grand limiting arrow for all of the aforementioned limits, existing as an identity functor characterizing the type and therefore the system as a category (Fig. 2.3). A system as a category may then be drawn as a circular arrow, which is the identity functor that identifies the type of a system (Manes and Arbib, 1975; Rossiter and Heather, 2005). Therefore, the system can be represented as an arrow, i.e., a process in which the internal arrows are simply the components of one arrow. This then leads to *interconnectivity* between the systems. Also, the functor between two categories is conceptually the same as internal arrows between the arrows

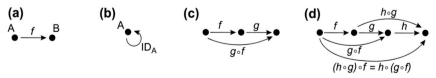

FIGURE 2.2

Schematic diagram representing the Category Theory essentials: (a) morphism (*arrows*), objects, domain, and codomain, (b) identity morphism, (c) composite morphism, and (d) identity composition and associativity (Mihailovic et al., 2012).

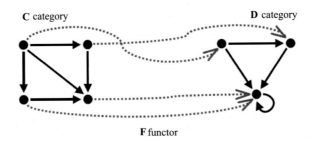

FIGURE 2.3

Schematic diagram representing the functor "action" (*dashed line*) (Mihailovic et al., 2012).

above. Within the framework of this theory, it is possible to repeat the abstraction to one level higher, to the so-called natural transformations. This level is the level of *interactivity.* It is important to note that the self-organization of a Category Theory system (intraactivity) arises when the category-system pair is indistinguishable. Finally, Category Theory is a very useful tool when we meet difficult problems in some areas of mathematics, ecology, physics, computer sciences, biological nano-engineering, and the self-organization of cell function in living systems (Wolkenhauer and Hofmeyr, 2007), among many others. They can be translated into (easier) problems in other areas (e.g., by using functors, which map one category to another).

Let us briefly expose essentials of Category Theory in a condensed form. *A category* is a quadruple $\mathscr{A}(O, hom, id, \circ)$, where

1. O is a class of $\mathscr{A}$-objects,
2. for each pair (A, B) of $\mathscr{A}$-objects, a set $hom(A, B)$ is the set of $\mathscr{A}$-morphisms, from A to B [$f \in hom(A, B)$ is expressed as $f: A \to B$ or $A \xrightarrow{f} B$],
3. for each $\mathscr{A}$-object A, a morphism $A \xrightarrow{id} A$ is called the $\mathscr{A}$-identity of A,
4. a *composition law* associating each $\mathscr{A}$-morphism $A \xrightarrow{f} B$ and each $\mathscr{A}$-morphism $B \xrightarrow{g} C$ is an $\mathscr{A}$-morphism $A \xrightarrow{g \circ f} C$, called the *composite* of f and g,

Impose the following conditions:

1. composition is associative, i.e., for morphism $A \xrightarrow{f} B$, $B \xrightarrow{g} C$, and $C \xrightarrow{h} D$, the equation $h \circ (g \circ f) = (h \circ g) \circ f$ holds,
2. $\mathscr{A}$-identities act as identities with respect to composition, i.e., for $\mathscr{A}$-morphism $A \xrightarrow{f} B$, we have $id_B \circ f = f$ and $f \circ id_A = f$,
3. the sets $hom(A, B)$ are pairwise disjoint.

If $\mathscr{A} = (O, hom, id, \circ)$ is a category, then

1. the class O of $\mathscr{A}$-objects is denoted by $Ob(\mathscr{A})$.
2. the class of all $\mathscr{A}$-morphisms is denoted by $Mor(\mathscr{A})$ (or $Hom(\mathscr{A})$) is defined to be the union of all the sets $hom(A, B)$ in $\mathscr{A}$.
3. if $A \xrightarrow{f} B$ is an $\mathscr{A}$-morphism, we call A the *domain* of f [$A = dom(f)$] and call B the *codomain* of f [$A = cod(f)$].

4. the composition ∘, is a partial binary operation on the class $Mor(\mathscr{A})$. For a pair (f, g) of morphisms, $f \circ g$ is defined if and only if the domain of f and codomain of g coincide.

If we have $h = g \circ f$, then sometimes it is denoted by $A \xrightarrow{f} B \xrightarrow{g} C$ or by saying that the following triangle commutes

$$
\begin{array}{ccc}
A & \xrightarrow{f} & B \\
& h \searrow & \downarrow g \\
& & C
\end{array}
$$

Similarly, the statement that the square commutes means that $g \circ f = k \circ h$. A morphism $A \xrightarrow{f} B$ in a category is called an *isomorphism* provided that there exists a morphism $g\colon A \to B$ with $g \circ f = id_A$ and $f \circ g = id_B$. Objects A and B in category are said to be *isomorphic* provided that there is an *isomorphism* $f\colon A \to B$. If $\mathscr{A}$ and $\mathscr{B}$ are categories, then a *functor* from $\mathscr{A}$ to $\mathscr{B}$ is a function that assigns to each $\mathscr{A}$-object A a $\mathscr{B}$-object $F(A)$ and to each $\mathscr{A}$-morphism $A \xrightarrow{f} B$ a $\mathscr{B}$-morphism $F(A) \xrightarrow{F(f)} F(B)$ such that

$$
\begin{array}{ccc}
A & \xrightarrow{f} & B \\
h \downarrow & & \downarrow g \\
C & \xrightarrow{k} & D
\end{array}
$$

1. F preserves composition, i.e., $F(f \circ g) = F(f) \circ F(g)$ whenever $f \circ g$ is defined;
2. F preserves identity morphisms, i.e., $F(id_A) = id_{F(A)}$ for each $\mathscr{A}$-object A.

A functor $F : \mathscr{A} \to \mathscr{B}$ is called an *isomorphism* provided that there is a functor $G : \mathscr{B} \to \mathscr{A}$ such that $G \circ F = id_{\mathscr{A}}$ and $F \circ G = id_{\mathscr{B}}$. The categories $\mathscr{A}$ and $\mathscr{B}$ are said to be isomorphic if there is an isomorphism $F : \mathscr{A} \to \mathscr{B}$.

2.3 BASICS OF MATHEMATICAL THEORY OF GENERAL SYSTEMS

Following Mesarovic's Mathematical Theory of General Systems (Mesarovic and Takahara, 1972), if we observe interactions of agents with their surrounding environment, such a system can be defined as a set of interacting objects $S \subseteq O_1 \times O_2 \times O_3 \times \ldots \times O_n$. If we denote the population of agents under consideration as $p = \{p_1, p_2, p_3, \ldots, p_n\}$ and a set of external influences as $E = \{e_1, e_2, e_3, \ldots e_n\}$ (these influences can be either other agents or extrasystemic influences), then the state of such formed systems at any particular moment in time can be defined as the Cartesian product $s \subseteq P \times E$. Because our system is a dynamical network of interactions where at each moment the hierarchical status of network elements can vary significantly, we have to define state of the population P as a mapping $\omega : e \rightarrow p$, $e \in E$, $p \in P$. Both e and p are defined as temporal sequences of events such that $E = \{e: T \rightarrow I\}$ and $P = \{e: T \rightarrow R\}$, where T is a set of time points t, I is a set of external stimuli on a particular agent such that at each time system receives stimulus $i(t)$ and R is a set of responses, $r(t)$. Furthermore, both P and E are formal systems. Therefore, the occurrence of p and the occurrence of e at some particular time point t are governed not only by mapping ω but also by the internal rules of these systems, which are partially independent. Thus, it is obvious that changes in an environment induce appropriate responses in agents through the model of coupled input/output pairs. In real systems, the reverse situation is also possible such that some external changes can be influenced by the activity of organisms. It is clear that a critical factor in building an evolvable model as described above is choosing the appropriate structure for the mapping $I \rightarrow R$. When dealing with models usually developed as prediction tools, it is sufficient to assume the attitude of analyzing a "black box." Therefore, we can propose a function that should summarize all available experimental data and obtain a set of more or less accurate predictions for various initial conditions. However, in such a case we will neglect the real meaning of the nature of mappings within E and P. Taking a slightly closer look at these relations, we can see that a somewhat hidden problem is that of how I is generated from the wholeness of external changes and what is the connection between generating I with a constitution of the corresponding R. Although this connection can be efficiently represented using the FCA (Ganter and Wille, 1997), its evolvability demands a more advanced formal treatment to be fully comprehended.

2.4 FORMAL CONCEPT ANALYSIS IN MODELLING THE INTERACTION OF LIVING SYSTEMS AND THEIR ENVIRONMENTS

To establish a more accurate estimation of the pattern of interactions of biological systems with their environment, as well as interactions among living systems, it is necessary to take into account the manner in which they "observe" the environment, separate it into different functional patches, and associate the patches with an internal functional schematism. The formal representation of such processes can be elegantly and efficiently performed by using the FCA, which is a branch of the applied lattice theory. FCA was introduced by Wille (1982) and defines a concept as a unit of two parts: extension and intension. The extension covers all objects belonging to a particular concept, and the intension comprises all attributes valid for all of those objects. Both attributes and objects are united by a triple (G, M, I), which is called a formal context if G and M are sets and $I \subseteq G \times M$ is a binary relation between G and M. Also, between these two closure systems (G and M), a dual isomorphism is established. The ordered set of all formal concepts of (G, M, I) forms the concept lattice of (G, M, I), which is always complete. The

FCA has been academically and commercially applied over a wide range of domains, such as medicine, biology, psychology, musicology, archeology, law, civil and industrial engineering, library and information science, computer science, and mathematics. In these applications, the main achievements of concept lattices are due to the support of general tasks such as exploring, searching, recognizing, identifying, analyzing, investigating, and deciding (Ganter and Wille, 1997). The mathematization of concepts may be understood as a first step in mathematizing the traditional philosophical logic, understood as a doctrine of the forms and functions of thinking based on concepts, judgments, and conclusions, which leads to contextual logic (Wille, 2000; Crvenković et al., 2009, 2012). However, one of the most important characteristics of interaction with the environment in organisms is versatility. Therefore, an FCA-based formalism can be used within a much broader framework that allows for the comparison of substantially different structures. Here, we consider concept lattices and we give a description of the fundamental theorem on concept lattices.

Concept lattices. An *order* (or *partial order*) on a set P is a binary relation $\leq$ on P such that, for all $x, y, x \in P$,

1. $x \leq x$,
2. $x \leq y$ and $y \leq x$ imply $x = y$,
3. $x \leq y$ and $y \leq z$ imply $x \leq z$.

A set P equipped with an order relation $\leq$ is said to be an *ordered set*.

Let P be an ordered set and let $S \subseteq P$. An element $x \in P$ is an upper bound of S if $s \leq x$ for all $s \in S$. A *lower bound* is dually defined. The set of all upper bounds of S is denoted by S^u and the set of lower bounds by S^l

$$S^u := \{x \in P | (\forall s \in S) \; s \leq x\}$$
$$S^l := \{x \in P | (\forall s \in S) \; s \geq x\}.$$

If S^u has a least element, x, then x is called the *least upper bound* of S. Equivalently, x is the least upper bound of S if

1. x is an upper bound of S *and*
2. $x \leq y$ for all upper bounds y of S.

Dually, if S^l has the largest element x, then x is called the *greatest lower bound of S*. Least elements and greatest elements are unique, so least upper bounds and greatest lower bounds are unique when they exist. The least upper bound of S is also called the *supremum* of S and is denoted by sup S; the greatest lower bound of S is also called the *infimum* of S and is denoted by inf S.

We write $x \vee y$ in place of sup $\{x, y\}$ when it exists and $x \wedge y$ in place of inf $\{x, y\}$ when it exists. Similarly we write $\vee S$ instead of sup S and $\wedge S$ instead of inf S when these exist. Let P be a nonempty ordered set.

1. If $x \vee y$ and $x \wedge y$ exist for all $x, y \in P$, then P is called a *lattice*.
2. If $\vee S$ and $\wedge S$ exists for all $S \subseteq P$, then P is called a *complete lattice*.

The theory of ordered sets and lattices provides a natural setting in which we can discuss and analyze hierarchies occurring within mathematics and in the "real" world.

A *concept* is considered to be determined by its extent and intent: the *extent* consists of all objects belonging to the concept, while the *intent* is the collection of all attributes shared by the objects. It is often difficult to list all the objects belonging to a concept and usually impossible to list all its attributes; therefore, it is natural to work with a specific *context* in which the objects and attributes are fixed. A *context* is a triple (G, M, I) in which G and M are sets and $I \subseteq G \times M$. The elements of G and M are called *objects* and *attributes*, respectively. $(g, m) \in I$ means "the object g has attribute m."

For $A \subseteq G$ and $B \subseteq M$, we define

$$A' = \{m \in M | (\forall g \in A)(g, m) \in I\},$$
$$B' = \{g \in G | (\forall m \in B)(g, m) \in I\}.$$

Therefore, A' is the set of attributes common to all objects in A, and B' is the set of objects possessing the attributes of B. The *concept* of the context (G, M, I) is a pair (A, B), in which $A \subseteq G$, $B \subseteq M$, $A' = B$, and $B' = A$. The *extent* of the concept (A, B) is A, while the intent is B. A subset $A \subseteq G$ is the extent of some concept if and only if $A'' = (A')' = A$, in which case the unique concept of which A is an extent is (A, A'). The corresponding statement applies to these subsets B of M, which are the intents of some concept.

The set of all concepts of the context (G, M, I) is denoted in the literature by $\mathscr{B}(G, M, I)$. For concepts (A_1, B_1) and (A_2, B_2) in $\mathscr{B}(G, M, I)$, we write $(A_1, B_1) \leq (A_2, B_2)$ and state that (A_1, B_1) is a subconcept of (A_2, B_2) or that (A_2, B_2) is a superconcept of (A_1, B_1) if $A_1 \subseteq A_2$ (which is equivalent to $B_1 \supseteq B_2$). Assume that (G, M, I) is a context, and let $A, A_j \subseteq G$ and $B, B_j \subseteq M$, for $j \in J$. Then

(i) $A \subseteq A''$,	(i)' $B \subseteq B''$,
(ii) $A_1 \subseteq A_2 \Rightarrow A_1' \supseteq A_2'$,	(i)' $B_1 \subseteq B_2 \Rightarrow B_1' \supseteq B_2'$,
(iii) $A' \subseteq A'''$,	(iii)' $B' \subseteq B'''$,
(iv) $\left(\bigcup\limits_{j \in J} A_j\right)' = \bigcap\limits_{j \in J} A_j'$	(iv)' $\left(\bigcup\limits_{j \in J} B_j\right)' = \bigcap\limits_{j \in J} B_j'$

Let Q be an ordered set, and $P \subseteq Q$. Then, P is *join-dense* in Q if for every element $s \in Q$ there is a subset A of P such that s is the supremum of A in Q, i.e., $s = V_Q A$. The dual of join-dense is *meet-dense*.

The fundamental theorem of concept lattices. Let (G, M, I) be a context. Then, $(B(G, M, I); \leq)$ is a complete lattice in which join and meet are given by

$$\bigvee\limits_{j \in J} (A_j, B_j) = \left(\left(\bigcup\limits_{j \in J} A_j\right)'', \bigcap\limits_{j \in J} B_j\right)$$

$$\bigwedge\limits_{j \in J} (A_j, B_j) = \left(\bigcap\limits_{j \in J} A_j, \left(\bigcup\limits_{j \in J} B_j\right)''\right)$$

Conversely, if L is a complete lattice, then L is isomorphic to $\mathscr{B}(G, M, I)$ if and only if there are mappings $\gamma: G \to L$ and $\mu: M \to L$ such that $\gamma(G)$ is join-dense in L, $\mu(M)$ is meet-dense in L, and $(g, m) \in I$ is equivalent to $\gamma(g) \le \mu(m)$ for each $g \in G$ and $m \in M$. In particular, L is isomorphic to $\mathscr{B}(L, L, \le)$ for every complete lattice L. The proofs of the previous assertion can be found in Davey and Priestley (1990). Therefore, the class of complete lattices coincides with the class of concept lattices.

If we are interested in algebraic laws satisfied by complete lattices, it is obvious that the class of all complete lattices could not be defined by a finite set of lattice identities. This is a consequence of the fact that inf and sup are basically infinitary operations. A lattice $\mathscr{L}(L, \wedge, \vee)$ is said to be *distributive* if the following law holds for $\mathscr{L}$:

$$x \wedge (y \vee z) = (x \wedge y) \vee (x \wedge z).$$

An empirical experience seems to show that contexts arising from concrete problems in real life rarely happen to have distributive concept lattices. The following may be found in Erne (1993). Define for any $A \subseteq G$ the *conditional incidence relation* $_A I$

$j_A I m$ if every consequence of m that holds for all objects in A is valid for j.
Dually, for any set $B \subseteq M$ we define
$j I^B m$ if every specialization of j possessing all attributes in B has property m.
We call the pair (A, B) *discriminating* if

$$I = {}_A I \cap I^B.$$

A concept lattice $(\mathscr{B}(G, M, I); \le)$ is distributive if and only if each concept of the context (G, M, I) or, equivalently, each pair (A, B) with $A \times B \subseteq I$ is discriminating.

The *arrow relations* of context (G, M, I) are defined as follows: for $h \in G, m \in M$ let

$$g \swarrow m : \Leftrightarrow \begin{cases} (g, m) \notin I, & \text{and} \\ \text{if } g' \subseteq h' \text{ and } g' \ne h', \text{ then } (h, m) \in I \end{cases}$$

$$g \nearrow m : \Leftrightarrow \begin{cases} (g, m) \notin I, & \text{and} \\ \text{if } m' \subseteq n' \text{ and } m' \ne n', \text{ then } (g, n) \in I \end{cases}$$

$$g \leftrightarrow m : \Leftrightarrow g \swarrow m \text{ and } g \nearrow m.$$

All the lattices in our examples are finite. We have the following: a finite lattice with standard context (G, M, I) is distributive if and only if

$$g \leftrightarrow m, g \nearrow n \Rightarrow m = n.$$

A subrelation $J \subseteq I$ is *closed* if every concept of (G, M, I) is also a concept of (G, M, I). S is a complete sublattice of $(\mathscr{B}(G, M, I); \le)$ if and only if $S = (\mathscr{B}(G, M, I); \le)$ for some closed subrelation $J \subseteq I$ (Davey and Priestley, 1990).

A class $\mathscr{K}$ of lattices is a *variety* if $\mathscr{K}$ is defined by a set of lattice identities. It is well known, from universal algebra, that a class $\mathscr{K}$ is closed under H (homomorphic images), S (sublattices), and P (direct product).

$$Var(\mathscr{K}) = HSP(\mathscr{K}).$$

A variety is finitely based if it can be defined by a finite set of identities. The following result is provided by McKenzie (1970): for any finite lattice L, the variety $Var(L)$ is finitely based.

Also, because we are dealing with finite lattices and every finite lattice is a complete lattice, it is good to know the following. The variety $\mathscr{L}$ of all lattices is generated by its finite numbers. Therefore, the class of finite concept lattices generates the whole variety of lattices. What is the meaning of this assertion? Think of any nontrivial lattice identity. There is a finite concept (G, M, I) such that $(\mathscr{B}(G, M, I); \leq)$ does not satisfy this identity. The theory of lattices and techniques of universal algebra provide a powerful tool for the identification of the lattice identities of a given lattice.

2.5 BASIC CONCEPTS OF THE CHAOS THEORY

As it was said, the environmental interfaces are formed between complex systems which are *per definitionem* nonlinear ones. Dynamics of those systems can be described by the sets of differential equations, which cannot be solved analytically, even in the case if a complex system is described by system of equations that "completely" captures the whole system. Often, when working with such systems, scientists rely on tools developed for simpler, linear systems. Their idea about solving this problem is quite similar to one attributed to King Christian IV of Denmark from 17th century, who standing face to face with the "unsolvable problem," always asked his own advisor what he would to do. "Your Majesty, the same as you always do in these situations. Make a simplification of the complex problem to the simplest one. It always works!" (Tremain, 2001). He, as many others, didn't make a clear distinction between problems that can be successfully divided into parts, and those where such an approach is not advisable. Similarly, in physics, only *linear systems can be broken into parts* (Strogatz, 2007). Pursuing this way then without many obstacles we get the answer. First we solve each part separately and secondly, by their recombination, we reach that a linear system is accurately equal to the sum of its parts. This idea offers to scientists a great simplification of complex problems transferred for the world of nonlinear physics to the linear one. However, things and processes in nature definitely do not follow this way. For example, when somebody at the same time listens to two favorite piano concerts (what is a complex event in any sense), then the listener does not get double pleasure since the principle of superposition fails dramatically in "nonlinear" life. The "linear sight" on the "nonlinear" world works well up to some level of accuracy and some range for the input values, but some interesting phenomena such as *chaos* and singularities (Frisch and Morf, 1981) are hidden by linearization. It follows that some aspects of the behavior of a nonlinear system appear commonly to be chaotic, unpredictable, or counterintuitive.

Chaos theory is a field of study in mathematics that has applications in several disciplines including meteorology, physics, technique, economics, and environmental sciences. It studies the behavior of dynamical systems that are highly sensitive to initial conditions. In layperson language, when we talk about "chaos" we mean "a state of disorder." However, in deterministic chaos theory, this term is defined more precisely. Although there is no universally accepted mathematical definition of chaos, a commonly used definition says that, for a dynamical system to be classified as chaotic, it must have the following properties (Hasselblatt and Katok, 2003): (1) it must be sensitive to initial conditions; (2) it must be topologically mixing; and (3) its periodic orbits must be dense. Apart from the chaotic, there exist other types of nonlinear behaviors which are (1) *multistability*—that alternates between two or more exclusive states; (2) *aperiodic oscillations*—functions that do not repeat values after some period (otherwise known as chaotic oscillations or chaos); and (3) *solitons*—self-reinforcing solitary waves (Khalil, 2001). Since most chapters in this book deal with different aspects of nonlinear dynamics, here we will not go into details of mathematical formalisms. They will be explained separately for each specific example.

REFERENCES

Bell, J.S., 1964. On the Einstein Podolsky Rosen paradox. Physics 1, 195–200.

Checkland, P.B., 1981. Systems Thinking, Systems Practice. Wiley, New York.

Crvenković, S., Mihailović, D.T., Balaž, I., 2009. Use of formal concept analysis for construction of subjective interface between biological systems and their environment. In: Mihailovic, D.T., VojnovićMiloradov, M. (Eds.), Environmental, Health and Humanity Issues in the Down Danubian Region: Multidisciplinary Approaches. World Scientific Publishing Co., Singapoore.

Crvenković, S., Mihailović, D.T., Balaž, I., 2012. Formal concept analysis and category theory in modeling interaction of living systems and their environments. In: Essays of Fundamental and Applied Environmental Topics. Nova Science Publisher Inc., New York, pp. 23–44.

Davey, B.A., Priestley, H.A., 1990. Introduction to Lattices and Order. Cambridge Mathematical Textbooks.

Erne, M., 1993. Distributive laws for concept lattices. Algebra Univ. 30, pp. 538–580.

Flood, R.L., Carson, E.R., 1993. Dealing With Complexity: An Introduction to the Theory and Application of Systems Science. Plenum Press, New York, NY, USA.

Frisch, U., Morf, R., 1981. Intermittency in nonlinear dynamics and singularities at complex times. Phys. Rev. A 23, 2673.

Ganter, B., Wille, R., 1997. Formal Concept Analysis: Mathematical Foundations. Springer-Verlag, Berlin.

Hasselblatt, B., Katok, A., 2003. A First Course in Dynamics: with a Panorama of Recent Developments. Cambridge University Press, NY.

Khalil, H.K., 2001. Nonlinear Systems. Prentice Hall, NJ.

Klir, G.J., 2002. The role of anticipation in intelligent systems. In: Dubois, D.M. (Ed.), Computing Anticipatory Systems (CASYS'01), vol. 627, pp. 37–46.

Levich, A.P., Solovyov, A.V., 1999. Category-functor modeling of natural systems. Cybern. Sys. 30, 571–585.

Mac Lane, S., 1971. Categories for the Working Mathematician. Springer-Verlag.

Manes, E.G., Arbib, M.A., 1975. Arrows, Structures and Functors, the Categorical Imperative. Academic Press.

McKenzie, R.N., 1970. Math. Scand. 27, 118.

Mesarovic, M., Takahara, Y., 1972. General Systems Theory: Mathematical Foundations. Academic Press, Inc., London.

Mesarovic, M., Takahara, Y., 1975. General Systems Theory: Mathematical Foundations. Academic Press, Inc., London.

Mihailovic, D.T., Balaž, I., 2007. An essay about modeling problems of complex systems in environmental fluid mechanics. Idojaras 111, 209–220.

Mihailovic, D., Budincevic, M., Balaz, I., Crvenkovic, S., Arsenic, I., 2012. Coupled maps serving the exchange processes on the environmental interfaces regarded as complex systems. Nat. Sci. 4, 569–580.

Rosen, R., 1985. Anticipatory Systems: Philosophical, Mathematical and Methodological Foundations. Pergamon Press.

Rosen, R., 1991. Life itself, A Comprehensive Inquiry into the Nature, Origin, and Fabrication of Life. Columbia University Press.

Rossiter, N., Heather, M., 2005. In: Conditions for Interoperability, 7th International Conference of Enterprise Information Systems (ICEIS), Florida, USA, p. 92.

Strogatz, S.H., 2007. Nonlinear Dynamics and Chaos: With Applications to Physics, Biology, Chemistry, and Engineering (Studies in Nonlinearity). Perseus Books, Cambridge.

Tremain, R., 2001. Music & Silence. 1999. Washington Square Press, USA.

Wille, R., 2000. Boolean Concept Logic. ICCS, pp. 317–331.

Wille, R., 1982. Restructuring lattice theory: an approach based on hierarchies of concepts. In: Rival, I. (Ed.), Ordered Sets: Proceedings. NATO Advanced Studies Institute, vol. 83. Reidel, Dordrecht, pp. 445–470.

Wolkenhauer, O., Hofmeyr, J.-H., 2007. An abstract cell model that describes the self-organization of cell function in living systems. J. Theor. Biol. 246, 461–476.

Approaches and meaning of time in the modelling of the environmental interface systems

3

Designing the models and their use in computer simulation in the environmental sciences has opened many epistemological questions (Heymann, 2010; Tolk, 2013). Although, practical considerations often overruled the problems of epistemology (Heymann, 2010) sometimes it is necessary to make basic epistemological choices, especially in modelling. In the previous chapter, we highlighted the importance of being nonlinear. If we decide to linearize "the object of modelling" then we use linear equations where the variables and their derivatives must always appear as a simple first power. The theory for solving linear equations is very well developed because linear equations are simple enough to be solvable. The shortcoming of this approach is the fact that many things and phenomena, even important ones, remain hidden. However, if we decide to follow as much as possible the existing nonlinearities in the object that we model, we have to consider the following key points: (1) model choice; (2) continuous time versus discrete time in building the model; and (3) time in building the model.

3.1 MODEL CHOICE

Ceteris paribus is a Latin phrase meaning "if all other relevant things, factors, or elements remain unaltered" or "all or other things being equal or held constant". *Ceteris paribus* laws are defined as natural laws that are accurate in expected conditions but can have exceptions. Whereas physics has a tendency to state universal laws that hold true in "normal conditions," in other sciences, like biology, psychology, or economics, laws usually have exceptions, the so-called *ceteris paribus* laws (Reutlinger, 2014). The laws of nature involve more formal hidden assumptions, about which we have no awareness. Those laws are expressed through mathematical equations or formulae that include mathematical premises, which were unknown at the time when the law was formulated. For example, physics for a long time functioned on an assumption that the equation of motion in classical mechanics is a strictly deterministic equation which provides a complete prediction of the future. What is *ceteris paribus* condition in this case? The answer to that can vary greatly (Earman et al., 2002). One line of thought is that all physical laws are true and universal claims. So, the Newton's Second Law of Motion is always true. The other opinion is that even the basic laws of physics contain (perhaps implicit) *ceteris paribus* clauses. So, to the basic equation, we would need to add a special constraining condition that the equation holds so long as everything that can affect the targeted effect is describable in the theory (Cartwright, 2002). An illustration for the *ceteris paribus* is the energy balance equation for the ground surface, which is often used in boundary layer and numerical weather modelling (Bhumralkar, 1975). This is a typical example of an environmental interface in nature, where exist all three mechanisms of energy transfer: incoming and outgoing radiation, convection of heat and moisture into the atmosphere, and conduction of heat into deeper soil layers of ground. This partial differential equation can be easily solved numerically by stepping either forward or backward in time from a known initial condition after it is written in the form of a difference equation. Under some conditions and expected conditions in atmosphere the energy balance equation can be written in the form (Mihailovic and Mimic, 2012)

$$X_{n+1} = A_n X_n - B_n X_n^2 \tag{3.1}$$

where X is the dimensionless environmental interface temperature, while dimensionless coefficients A_n and B_n include an inverse form of resistance in calculating the turbulent fluxes, which change periodically during a day (Pielke, 2002). This equation is a nonlinear autonomous difference equation that represents time changes of the dimensionless environmental interface temperature response to the radiative forcing (Stull, 1988). Its solution can exhibit chaotic fluctuations in the considered system because the environmental interface cannot oppose an enormous radiative forcing, suddenly reaching the interface. Therefore, it raises the question whether we can find either domain or domains where physically meaningful solutions exist. Fig. 3.1 depicts (a) chaotic fluctuations of solution in Eq. (3.1) and regions of stable and (b) unstable solutions of this equation determined by the values of Lyapunov exponent as a function the coefficients $A \in (0, 2)$ and $B \in (0, 0.5)$.

In choosing the model, scientists often apply a *heuristic* technique that could be defined as any approach to problem solving that makes use of a practical method not guaranteed to be optimal or perfect but sufficient either for the immediate goals or until a better approach is reached. We meet this approach in many sciences, in particular technical and environmental, when some phenomena cannot be expressed through time-dependent equations, whereas they have to be parameterized. This approach in modelling the turbulence inside the canopy turbulent is plastically described by Sellers et al. (1986). He said: "We have mentioned before that use of 'K-theory' within the canopy may be physically unrealistic, but because it yields reasonable results we shall use this method until suitable second-order closure can be applied to the problem."

To illustrate this situation we use as example the differential equation describing the wind profile within such a canopy architecture where the canopy is considered to be a block of constant-density porous material "sandwiched" between two heights, canopy height H and canopy bottom height h (Mihailovic and Kallos, 1997;

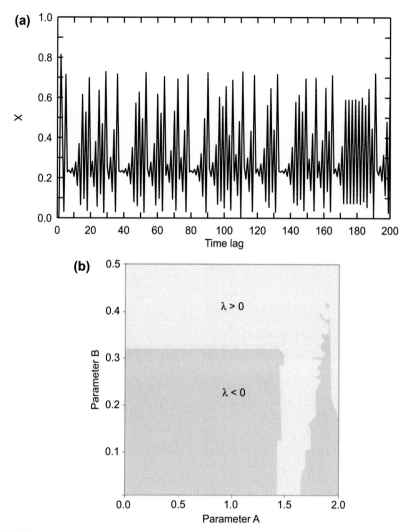

FIGURE 3.1

(a) Chaotic fluctuations of environmental interface temperature (X) in Eq. (3.1); (b) Regions of stable and unstable solutions of Eq. (3.1) determined by the values of Lyapunov exponent (λ) in dependence of the coefficients $A \in (0, 2)$ and $B \in (0, 0.5)$ (Mihailovic and Mimić, 2012).

Mihailovic et al., 2004). Within this architecture the equation can be written in the form

$$\frac{d}{dz}\left(\frac{d}{dz}K_s\right) = \frac{C_d L_d (H - h)}{H} u^2 \tag{3.2}$$

where K_s is the turbulent transfer coefficient within the canopy, C_d is the leaf drag coefficient, L_d is the canopy density, and u is the wind speed within the canopy. To solve this equation, we have to know how K_s depends on parameters that represent the canopy's aerodynamic and morphological features. Mihailovic et al. (2004) used an approach in which K_s is proportional to wind speed u, i.e., $K_s = \sigma u$ were the scaling length σ is an arbitrary, unknown constant. With this assumption we solve Eq. (3.2) to get the wind speed profile. Although this approach is not physically unrealistic, from Fig. 3.2 it is seen that the profile obtained from Eq. (3.2) suitably agrees with observed data.

3.2 CONTINUOUS TIME VERSUS DISCRETE TIME IN BUILDING THE MODEL

Many mathematical models, more or less sophisticated, of environmental interface systems have been built and will be built in the form of differential or difference

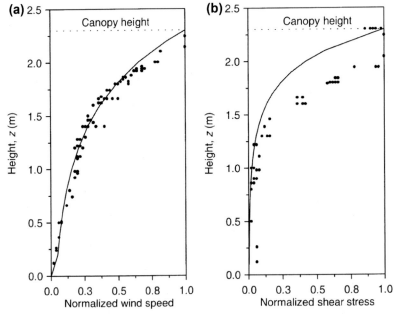

FIGURE 3.2

Profiles of (a) wind speed and (b) shear stress inside a maize crop. The *black circles* are observations (Wilson et al., 1982) and the *solid lines* are plotted using calculated values. The wind speed u and shear stress τ are normalized by their values $u(H)$ and $\tau(H)$ at the canopy top height (Mihailovic et al., 2004).

equations or systems of such equations. It means that we confront the choice whether we will deal either with the continuous-time or discrete-time environmental interface systems, where time is considered as a continuous or discrete variable, respectively. The dilemma about this choice is yet to be solved. For example, the qualitative models (describing qualitative relations between the observed variables), which seem heuristically close to the continuous-time models, exhibit drastically different behaviors when they are designed in the discrete-time interpretation. Therefore, it "could be naively to believe that continuous-time and discrete-time models have the same qualitative characteristics" (Istas, 2005).

Many modelers in this area use mathematical techniques with an idea to replace the given differential equations by apposite difference equations. It opens the question "How to choose suitable difference equations whose solutions are 'good' approximations to the solutions of the given differential equation?" (van der Vaart, 1973). So a huge effort has been invested into choice of appropriate difference equations. This question includes a requirement for better understanding of the fundamental problem: interrelations between *classical continuum* mathematics and reality in different sciences. For many environmental interface phenomena the "continuum" type of thinking, that is, at the basis of any differential equation, is not natural to the phenomenon but rather constitutes an approximation to a basically discrete situation. In many papers dealing with this approach, the "infinitesimal step lengths" handled in the reasoning which lead us to the differential equation are not really thought of as infinitesimally small but as finite. However, in the last stage of such reasoning, where the differential equation rises from the differentials, these "infinitesimal" step lengths go to zero, that is, where the above-mentioned approximation comes in. Under this kind of circumstances, it seems more natural to build the model as a discrete difference equation from the start, without going through the painful, doubly approximative process of first, during the modelling stage, finding a differential equation to approximate a basically discrete situation and then, for numerical computing purposes, approximating that differential equation by a difference scheme. In modelling procedure we meet three problems (Mihailovic et al., 2012). The first problem is this: (1) environmental scientists (also physicists and biologists among them) come to us with a theory in the form of differential equation including the mathematical concept of the first derivative; (2) this is done in spite of the fact that this concept is not a fairly suitable reflection of many environmental phenomena as a difference equation would be. The second problem is the possible way for a given differential equation to construct a difference equation with exactly the "same" collection of solutions. The third problem is defined conversely to the second one: whether we in any way for a given difference equation can construct a differential equation with exactly the same solutions? In this book we will give the advantage to discrete-time approach in building the models describing the environmental interface phenomena.

It is worth mentioning that the traditional mathematical analysis of physical and other dynamical systems tacitly assumes that integers and all real numbers, no matter how large or how small, are physically possible and all mathematically possible trajectories are physically admissible (Kreinovich, 2003). Traditionally, this approach has worked well in physics, biology, and in engineering, but it does not lead to a very good understanding of chaotic systems, which, as it is now known, are extremely important in the study of real-world phenomena ranging from weather to biological and environmental interface systems (Mihailovic, 2012).

3.3 TIME IN MODEL BUILDING

In classical physics, the time is an objective continuous function. Traditionally, time has been modeled as a basic variable taking its values from an interval on a real axis. The pervasiveness of this concept was largely due to the success of the models it supported, in particular to the expression of physical laws by differential equations which ultimately relied on the limiting process, inherent in the notion of a (total or partial) derivative. Despite this success at the computational level, it has long been clear that the truly ramified nature of time cannot be captured by what amounts to a mathematical convention (Smith, 2003), although, all of the fundamental theories of physics are symmetric with respect to time reversal. The only fundamental theory that picks out a preferred direction of time is the second law of thermodynamics, which asserts that the entropy of the Universe increases as time flows toward the future, providing an orientation, or arrow of time, and it is generally believed that all other time asymmetries, such as our sense that future and past are different, are a direct consequence of this thermodynamic arrow (Eddington, 1928; Feng and Crooks, 2008).

In contrast with classical physics in biology, the concept of "time's cycle" is commonly applied as a metaphor (Günther and Morgado, 2004). These two and other notions of time often present in environmental complex systems, we will consider in Part II.

REFERENCES

Bhumralkar, C.M., 1975. Numerical experiments on the computation of ground surface temperature in an atmospheric general circulation model. J. Appl. Meteorol. 14, 1246.

Cartwright, N., 2002. In favor of laws that are not ceteris paribus after all. Erkenntnis 57, 425–439.

Earman, J., Glymour, C., Mitchell, S., 2002. Editorial. Erkenntnis 57, 277–280.

Eddington, A.S., 1928. The Nature of the Physical World. Cambridge University Press, Cambridge.

Feng, E.H., Crooks, G.E., 2008. Length of time's arrow. Phys. Rev. Lett. 101, 330 090602.

Günther, B., Morgado, E., 2004. Time in physics and biology. Biol. Res. 336, 759–765.

Heymann, M., 2010. The evolution of climate ideas and knowledge. WIREs Clim. Change 1, 581–597.

Istas, J., 2005. Mathematical Modeling for the Life Sciences. Springer-Verlag, Netherland.

Kreinovich, V., 2003. Kolmogorov complexity and chaotic phenomena. Int. J. Eng. Sci. 41, 483–493.

Mihailović, D.T., 2012. Preface. In: Mihailović, D.T. (Ed.), Essays on Fundamental and Applied Environmental Topics. Nova Science Publishers, New York.

Mihailovic, D.T., Alapaty, K., Lalic, B., Arsenic, I., Rajkovic, B., Malinovic, S., 2004. Turbulent transfer coefficient and calculation of air temperature inside the tall grass canopies in coupled land-atmosphere scheme for environmental modelling. J. Appl. Meteorol. 43, 1498–1512.

Mihailovic, D.T., Kallos, G., 1997. A sensitivity study of a coupled-vegetation boundary-layer scheme for use in atmospheric modelling. Bound. Layer Meteorol. 82, 283–315.

Mihailovic, D.T., Mimić, G., 2012. Kolmogorov complexity and chaotic phenomenon in computing the environmental interface temperature. Mod. Phys. Lett. B 26 (27).

Pielke Sr., R.A., 2002. Mesoscale Meteorological Modelling. Academic Press, San Diego.

Reutlinger, A., Schurz, G., Hüttemann, A., 2014. Ceteris paribus laws. In: Zalta, E.N. (Ed.). The Stanford Encyclopedia of Philosophy.

Sellers, P., Mintz, Y., Sud, Y.C., Dachler, A., 1986. A simple biosphere model (SiB) for use within general circulation models. J. Atmos. Sci. 43, 505–531.

Smith, J.D.H., 2003. Time in biology and physics. In: Buccheri, R., Saniga, M., Stuckey, W.M. (Eds.), The Nature of Time: Geometry, Physics and Perception. Kluwer, Dordrecht.

Stull, R.B., 1988. An Introduction to Boundary Layer Meteorology. Kluwer Academic Publisher, Dordrecht.

Tolk, A., 2013. Ontology, Epistemology, and Teleology for Modeling and Simulation: Philosophical Foundations for Intelligent M&S Applications. Springer-Verlag Berlin Heidelberg.

van der Vaart, H.R., 1973. A comparative investigation of certain difference equations and related differential equations: implications for model building. Bull. Math. Biol. 35, 195–211.

Wilson, J.D., Ward, D.P., Thurtell, G.W., Kidd, G.E., 1982. Statistics of atmospheric turbulence within and above a corn canopy. Bound. Layer Meteorol. 24, 495–519.

Examples of use of the formal complex analysis

4

In this chapter, we present two simplified but illustrative examples of formal complex analysis (FCA): (1) making of the conceptual hierarchy of animals based on their attributes and (2) construction of the subjective interface between biological systems and their environments. These examples are based on reasoning from Crvenkovic et al. (2009, 2012) and Wolff (1994).

4.1 USE OF FORMAL COMPLEX ANALYSIS IN THE CONTEXT OF ANIMALS: AN EXAMPLE

The following example is adapted from Wolff (1994). Table 4.1 describes which of the mentioned attributes some animals have. This is indicated by crosses. An empty cell indicates that the corresponding animal does not have the corresponding attribute. To explain the notion of a formal concept of a context, we look at the attributes of the FINCH and look for all other animals, within the same context that share the same set of attributes. Hence, we obtain sets A = {FINCH, EAGLE} and B = {flying, bird}. A is the set of all objects having all of the attributes of B, and B is the set of all attributes that are valid for all of the objects of A. Each such pair (A, B) is called a formal concept. Between the concepts of a given context there is a natural hierarchical order, the "subconcept—superconcept" relation. For example, the preying, flying birds describe a subconcept of the concept of the flying birds. The extent of this subconcept consists only of the EAGLE, and the intent consists

Table 4.1 Object Versus Attributes in the Context of Animals

Attribute Animals	Preying	Flying	Bird	Mammal
Lion	X			X
Finch		X	X	
Eagle	X	X	X	
Hare				X
Ostrich			X	

Reprinted from Crvenković, S., Mihailović, D.T., Balaž, I., 2012. Formal concept analysis and category theory in modeling interaction of living systems and their environments. In: Essays of Fundamental and Applied Environmental Topics. Nova Publishers, Nova Science Publisher Inc., New York, pp. 23–44; with permission from Nova Science Publishers, Inc.

Developments in Environmental Modelling, Volume 29, ISSN 0167-8892, http://dx.doi.org/10.1016/B978-0-444-63918-9.00004-1

of the three attributes preying, flying, and bird. In the example we started with, the extent is the set A, while the intent is the set B. In the following line diagram, we represent the conceptual hierarchy of all concepts of the context ANIMAL (Fig. 4.1).

Following the reading rule for conceptual hierarchy, we can recognize from the line diagram that the lion has the attributes preying and mammal. Using the reading rule, we can easily understand from the line diagram the extent and the intent of each concept by collecting all of the objects below each respective attribute above the circle of the given concept. Hence the object concept "finch" has the extent finch and eagle and the intent flying and bird. For this example, the extent of the top concept is the set of all objects, while the intent of it does not contain any attribute. However, in other contexts, the intent of the top concept may not be empty, e.g., if we add to the given context the attribute "animal" with crosses in each row, then the top concept would be the attribute concept of "animal" and the intent of the top concept would contain only the attribute "animal."

If we extend the set of objects, we must change the lattice of conceptual hierarchy. For example, the new object bee has an attribute "flying" but it is not a bird. Thus, we have to separate "flying" and "bird" to meet finch.

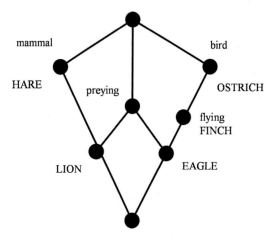

FIGURE 4.1

Conceptual hierarchy of all concepts of the context animal according to Table 4.1. A line diagram consists of circles, lines, objects (written in capital letters), and attributes (small letters). The relation between concepts and attributes can be read from the line diagram by the following simple reading rule: an object g has an attribute m if and only if there is an upward leading path from the "g" circle to the "m" circle.

Reprinted from Crvenković, S., Mihailović, D.T., Balaž, I., 2012. Formal concept analysis and category theory in modeling interaction of living systems and their environments. In: Essays of Fundamental and Applied Environmental Topics. Nova Publishers, Nova Science Publisher Inc., New York, pp. 23–44; with permission from Nova Science Publishers, Inc.

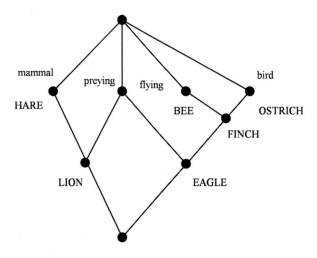

FIGURE 4.2

Modification of conceptual hierarchy to incorporate flying animals that are not birds.

Reprinted from Crvenković, S., Mihailović, D.T., Balaž, I., 2012. Formal concept analysis and category theory in
modeling interaction of living systems and their environments. In: Essays of Fundamental and Applied Envi-
ronmental Topics. Nova Publishers, Nova Science Publisher Inc., New York, pp. 23–44; with permission from
Nova Science Publishers, Inc.

According to the second diagram (Fig. 4.2), "mammals" and "flying" imply "preying" and "bird", i.e., they meet at the bottom of the lattice, and "preying" and "bird" are above. However, it does not correspond well with our world, since bats are flying mammals and not birds. Thus, we have to extend the lattice by adding a new vertex bat. This is shown in the third diagram (Fig. 4.3).

4.2 USE OF FORMAL COMPLEX ANALYSIS IN CONSTRUCTING THE SUBJECTIVE INTERFACE BETWEEN BIOLOGICAL SYSTEMS AND THEIR ENVIRONMENTS

In this example, we present a simplified but illustrative example of an abstract living system's interaction with the environment. Our goal is to represent how subjective processing of the environment is achieved, what are the functional consequences and how this process influences the dynamics of the organization of living systems. This example is based on reasoning from Crvenkovic et al. (2009, 2012), with additional explanations and clarifications.

Let us define a set of environmental objects $O = \{o_1, o_2, o_3, ..., m_n\}$. From the perspective of living systems, only objects that have some recognizable attributes could be perceived. Therefore, we can define a set of attributes $M = \{m_1, m_2, m_3, ..., m_n\}$. Within FCA, objects and attributes are mutually defined,

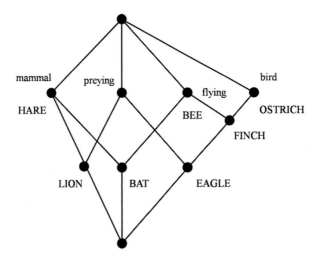

FIGURE 4.3

Further modification of conceptual hierarchy to separate flying birds from flying mammals.

Reprinted from Crvenković, S., Mihailović, D.T., Balaž, I., 2012. Formal concept analysis and category theory in modeling interaction of living systems and their environments. In: Essays of Fundamental and Applied Environmental Topics. Nova Publishers, Nova Science Publisher Inc., New York, pp. 23–44; with permission from Nova Science Publishers, Inc.

so we automatically obtain set $G \subseteq O : \{\forall g \in G | (g, m) \in I\}$, where I is a binary relation between M and G. In other words, every object that does not form a binary relation with a corresponding attribute and vice versa, simply does not exist from the perspective of a particular formal context. Thus, the concept of the "whole" environment changes its epistemological status and is reduced to an observable environment, i.e., to a set of objects characterized by defined attributes. After recognizing an object, the living system categorizes it into functional subsets. So, the supposed object can be categorized, for example, as a source of food, place for shelter, or threat. In terms of the FCA, the environment is separated into a set of concepts within a context defined by the triple (G, M, I). In simple organisms, the attributes used in separating environmental factors from one another can be divided into two main categories: physical influences (various types of radiation or temperature) and chemical influences (various types of molecules or ions). However, these attributes are further divided into a set of subattributes, which can be represented as a scale of values within a given attribute. In this way, a segment of the environment is encircled by a certain attribute and is further divided according to the given scale within that attribute, thus establishing a many-valued context (G, M, W, I) where G is a set of objects, M is a set of many-valued attributes, W is a set of attribute values, and I is a ternary relation $I \subseteq G \times M \times W$.

To illustrate that, we will use the very simple model of a typical photosynthetic organism. Organisms interact with their environment using evolved receptors.

At any given time, the set of active receptors reflects metabolic state of the organism. In our example, we can divide receptors into two groups: into photoreceptors and receptors for various types of external molecules. Therefore, in our conceptual scheme we postulate the existence of two main attributes: radiation and molecule (Table 4.2). The attribute "molecule" is further divided into two subclasses based on their size, denoted as L and S. All molecules that can spontaneously enter or leave cells without interacting with receptors are in the S group. These can be water molecules or small water-soluble ions. Consequently, all other molecules are in the L group. This group is further divided into several types of recognizable molecular structures (Table 4.2). Similarly, "Radiation" attribute is divided into several subattributes. Finally, we introduce one more attribute, defined as a global regulator. To preserve the simplicity of the model, this attribute does not have any subattributes. In living organisms, the role of global regulators is to coordinate metabolic response of the cell as a response to change of external conditions. As has already been mentioned, each attribute defined here corresponds to the receptive ability of an organism to perform the following chain of actions: (1) recognition of some stimulus (to form concept), (2) assimilation of stimulus and/or changing its own configuration, and (3) activation (indirectly or directly) of some other molecules to process the received information.

Because FCA is strictly lattice based, it is unable to depict the sequential dynamics of some process. Therefore, we will construct several different formal contexts and corresponding lattices, which will demonstrate how the internal network responds to external stimuli.

We will first define formal context that shows the interaction of the generic photosynthetic organism with the environment. According to the previously described process, the organism has a limited set of active receptors, and through them, it determines which kind of external influences it can observe and react to them. In this example (Table 4.3), the organism can initially sense water and small ions (denoted as Mol. smaller than[1]), three kinds of organic molecular structures (denoted as X, Y, and Z), and three kinds of radiation sources (denoted as 700 nm, 450 nm, and UV radiation).

Table 4.2 List of All Attributes Used in the Model. In Further Text, the Term Structure (X...Q) Will Be Denoted by the Strings Str(X...Q)

Attribute	Subattributes						
Molecule	L	S	Structure X	Structure Y	Structure Z	Structure W	Structure Q
Radiation	700 nm	450 nm	UV				

Reprinted from Crvenković, S., Mihailović, D.T., Balaž, I., 2012. Formal concept analysis and category theory in modeling interaction of living systems and their environments. In: Essays of Fundamental and Applied Environmental Topics. Nova Publishers, Nova Science Publisher Inc., New York, pp. 23—44; with permission from Nova Science Publishers, Inc.

Table 4.3 Formal Context Which Represents Initial Steps in Interaction With the Environment

	Molecule	Radiation	Mol. Larger Than[1]	Mol. Smaller Than[2]	Mol. StrX	Mol. StrY	Mol. StrZ	Rad. 700 nm	Rad. 450 nm	Rad. UV
Object 0	X	X						X		
Object 1	X	X							X	
Object 2	X	X								X
Object 3	X			X						
Object 4	X			X						
Object 5	X			X						
Object 6	X		X		X					
Object 7	X		X			X				
Object 8	X		X				X			

Reprinted from Crvenković, S., Mihailović, D.T., Balaž, I., 2012. Formal concept analysis and category theory in modeling interaction of living systems and their environments. In: Essays of Fundamental and Applied Environmental Topics. Nova Publishers, Nova Science Publisher Inc., New York, pp. 23–44; with permission from Nova Science Publishers, Inc.

Fig. 4.4a shows that the complete environment is divided into two main concepts, which further diverge into several more specific concepts. Because each formed concept in this model is connected with a previously defined chain of events, Fig. 4.4b shows how a simple feedback-governed metabolic chain would look like from an FCA perspective. The association of molecules with structures Z (MolstrZ) and corresponding receptors (concept <10>) activates the production of the next molecule (<12>), which in turn generates the product (<13>). Generation of the final product inhibits the production of strZ receptors, the intermediate product spontaneously degrades, and some other metabolic chain utilizes this final product (<13>). Because (<13>) is no longer available, the production of strZ receptors can start again.

In the second formal context, we introduced the activation of the supposed global regulator (Fig. 4.5). Since the global regulator is able to coordinate different metabolic processes, their hierarchical configuration is now different. Hierarchy is not formed only as a result of direct relations between metabolic processes but also as a result of global regulations. It is very important to emphasize that the activation of global regulators in organisms is the inevitable consequence of the assimilation of almost any external stimulus. More generally, an interaction with the environment

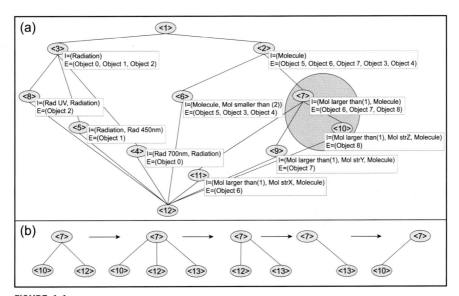

FIGURE 4.4

Lattice of the formal context given in Table 4.2 (a) Branching of the complete environment into two concepts and (b) Example of a simple metabolic-governed metabolic change. (lattice generation was performed using open-source software package "Galicia (2001)").

Reprinted from Crvenković, S., Mihailović, D.T., Balaž, I., 2012. Formal concept analysis and category theory in modeling interaction of living systems and their environments. In: Essays of Fundamental and Applied Environmental Topics. Nova Publishers, Nova Science Publisher Inc., New York, pp. 23–44; with permission from Nova Science Publishers, Inc.

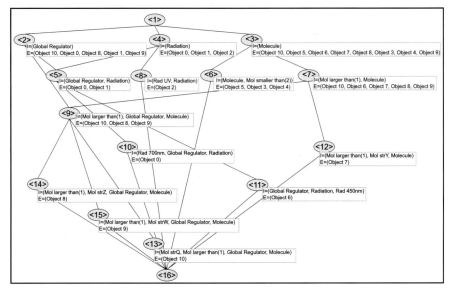

FIGURE 4.5

Lattice of the formal context for the same living system when global regulator is active.

Reprinted from Crvenković, S., Mihailović, D.T., Balaž, I., 2012. Formal concept analysis and category theory in modeling interaction of living systems and their environments. In: Essays of Fundamental and Applied Environmental Topics. Nova Publishers, Nova Science Publisher Inc., New York, pp. 23–44; with permission from Nova Science Publishers, Inc.

through the construction of a subjective perspective for living organisms is connected with the temporality of the internal conceptual lattice. This means that finalizing the formation of one configuration simultaneously causes its degradation and reconfiguration.

This model is an oversimplified example of a generic photosynthetic organism, but it can still reveal some important characteristics of functioning of living systems. First, because they are able to form a subjective perspective, living systems operate only within a limited scope of an entire environment. Moreover, they shape the environment according to their ability to functionally process some external segments, thus forming the subjective environment. This subjective construction is highly variable because it is defined not as a reflection of the complete ability of a living system to sense its environment but only as a reflection of currently active receptors. Second, the formation of an internal conceptual lattice is inevitably connected with the destruction of the previous one. At the molecular level, formation of each conceptual lattice corresponds with the activation of different molecules, receptors, and/or regulators. Each of them has strictly defined scope of action: what could be its input and what are possible outputs. Therefore, change of composition of activated receptors or regulators change the current functional context and lead to reconfiguration of the conceptual lattice.

REFERENCES

Crvenković, S., Mihailović, D.T., Balaž, I., 2012. Formal concept analysis and category theory in modeling interaction of living systems and their environments. In: Essays of Fundamental and Applied Environmental Topics. Nova Publishers, Nova Science Publisher Inc, New York, pp. 23–44.

Crvenković, S., Mihailović, D.T., Balaž, I., 2009. Use of formal concept analysis for construction of subjective interface between biological systems and their environment. In: Mihailovic, D.T., Vojnović-Miloradov, M. (Eds.), Environmental, Health and Humanity Issues in the Down Danubian Region: Multidisciplinary Approaches. World Scientific Publishing Co, Singapore.

Galicia, 2001. Galicia: Galois Lattice Interactive Constructor. Available at: http://www.iro.umontreal.ca/~galicia/.

Wolff, K.E., 1994. A first course in formal concept analysis — how to understand line diagrams. In: Faulbaum, F. (Ed.), SoftStat '93, Advances in Statistical Software 4. Fisher Verlag, Stuttgart, pp. 429–438.

Time in environmental interfaces modelling

II

Time in philosophy and physics

5

Let us suppose that we ask a community of scientists dealing with the problems of environmental modelling, the following question: "How do you understand the time?" Our anticipation is that the most common answer could be synthesized by the following sentences: (1) "That is clear!"; (2) "The time flows from the past to the future"; and (3) "The time quite well 'pursues' equations, which we use in our models." Are these simple answers even close to be satisfactory? Certainly not. We accept, continuous, linear notions of the time almost instinctively since it looks natural. However, of all of the theories of the Universe, the one of time is the most enigmatic and enchanting. Because of the multitude of usages of the concept of time in various disciplines, the often evoked question is "What is the meaning of time?". Our idea of time we can possibly have in mind is essentially connected to the events. Therefore, time makes sense only when it is in relation with something. If not, in emptiness we cannot have time because there would be nothing to relate it to. In psychology and neuroscience, time perception is a well-developed field dealing with subjective experience of time. Although origins of perception of time are not completely understood, it would not be a surprise that our sense of time entirely is influenced by the nature of events themselves, since we measure time by the events that mark it. With the progress in modelling of environmental interfaces, the question of the concept of time becomes more authentic. In this chapter we shortly outline understanding of the time in philosophy and physics.

5.1 TIME IN PHILOSOPHY

Scientists in some scientific communities, especially technical and technological, have the perception that many problems can be solved by using "common sense" (sound and prudent judgment based on a simple perception of the situation or facts). However, this concept is not the prevailing one in all scientific disciplines, especially fundamental ones. Thus, if the problem requires logic or language that is not "under the control of common sense," they usually naively but honestly say: "It is a matter for philosophy!" By this statement they imply that philosophy is a discipline in which "we can ask any question." Insights originated from philosophy can became an integral part of a scientific discourse, as it was the case with Mach's principle (Gürsey, 1963) or the problem of defining species (Boas, 1951). These new insights became integrated into the body of scientific knowledge and, as such, the basis for further development of science. One of the key philosophical issues is the question of time. In everyday life, the time is duration measured by clock. Despite a long investigation of the nature of time, many issues about it are unresolved. Dowden (2015) summarized some of the most important issues that are under discussion in philosophical community regarding time as follows: (1) What should a philosophical theory of time do? (2) How is time related to mind? (3) What is time? (the variety of answers, time vs. "time," linear and circular time) Does time has a beginning or end? Does time emerge from something more basic? (4) What does science require of time? (5) What kinds of time travel are possible? (6) Does time require change? Does time flow? (McTaggart's A series and B series—subjective flow and objective flow (McTaggart, 1908)) (7) What are the differences among the past, present, and future? (presentism, the growing-past, eternalism, and the block-universe; Is the present, the now, objectively real?) (8) Are there essentially tensed facts (using a tensed verb is a grammatical way of locating an event in time)? (9) What gives time its direction or arrow? (time without an arrow? What needs to be explained? Explanations or theories of the arrow, multiple arrows) and (10) What is temporal logic? Here, we will consider just one issue upon which philosophers are deeply divided: What is the ontological difference between, the present, the past, and the future? Our analysis will largely follow the one laid out previously (Dowden, 2015), but we will keep it as concise as possible. For more elaborated analysis, the interested reader should check Dowden's article (2015).

Philosophers of time could be divided into two broad groups in relation to this question. Philosophers from the "A group" see time through the following points: (1) events are always changing; (2) the now is objectively real and so is time's flow; (3) ontologically we should accept either presentism or the growing-past theory (the philosophical doctrine that only events and entities—and, in some versions of presentism, timeless objects or ideas like numbers and sets—that occur in the present exist); (4) predictions are not true or false at the time they are expressed; (5) tenses are semantically basic; and (6) the ontologically fundamental entities are three-dimensional objects. In contrast to them, members from the "B group" say that: (1) events are never changing; (2) the now is not objectively real and neither is time's flow; (3) ontologically we should accept eternalism and the block-time or block-universe theory (this would mean that time is just another dimension, that future events are "already there" and that there is no objective flow of time); (4) predictions are true or false at the time they are uttered; (5) tenses are not semantically basic; and (6) the fundamental entities are four-dimensional events or processes (space and time are merged into a space-time unchanging four-dimensional "block") (Dainton, 2010). This separation into groups is done following the basic idea dating back at least to philosopher McTaggart's B Theory of time (first published in 1908, only three years after the first paper on relativity) (McTaggart, 1908), who proposed two ways of linearly ordering all events in time by placing them into a series according to the times at which they occur, where this ordering can be created in two ways, an A way and a B way.

Here, we shortly describe McTaggart's A and B series: Let us consider two past events *a* and *b*, in which *b* is the most recent of the two (Fig. 5.1). In McTaggart's B series, event *a* happens before event b in the series because the time of occurrence of event *a* is less than the time of occurrence of event *b*. But when ordering the same events into McTaggart's A series, event *a* happens before event *b* for a different reason, because event *a* is more in the past than event *b*. Both series produce exactly the same ordering of events. Fig. 5.1 graphically depicts the ordering where *c* is an event that happens after *a* and *b*. Obviously, there are many other events that are placed within the series at the location of event *a*, namely all events simultaneous with event *a*. If we were to consider an instant of time to be a set of simultaneous events, then instants of time are also linearly ordered into an A series and a B series. However, McTaggart (1908) himself believed the A series is paradoxical, but he also believed the A properties such as being past are essential to our current concept of time, so for this reason he believed our current concept of time is not coherent. Now, let us include event *c* to occur in our present after events *a* and *b*. The information that *c* occurs in the present is not contained within either the A series or the B series. However, the information that *c* is in the present is used to create the A-series; it is what tells us to place *c* to the right of *b*. In contrast to that, this information is not used to create the B series. In metaphysic community, philosophers dispute whether the A theory or, the B theory is the correct theory of reality. The A theory includes two theses, each of which is contrary to the B theory: (1) time is constituted by an A series in which any event's being in the past (or in the present or in the future) is an intrinsic, objective, monadic property of the event itself. It is not merely a subjective relation between the event and us who exist. (2) The second thesis of the A theory is that events change, as explained by McTaggart (1908): "Take any event—the death of Queen Anne, for example—and consider what change can take place in its characteristics. That it is a death, that it is the death of Anne Stuart, that it has such causes, that it has such effects—every characteristic of this sort never changes. […] But in one respect it does change. It began by being a future event. It became every moment an event in the nearer future. At last it was present. Then it became past, and will always remain so, though every moment it becomes further and further past." Here, we will not deal with the question of time in biology and physics, as seen through the optics of the philosophy. Instead of that, in the rest of this chapter we will shortly describe understanding the time in physics, while in the next two chapters we will be devoted to time in biology and functional time.

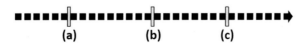

(a)   (b)   (c)

FIGURE 5.1

McTaggart's A series and B series (McTaggart, 1908).

5.2 TIME IN PHYSICS

In physics, the concept of time was changing and it was closely related to qualitative leaps in physical science. However, regardless of this fact the concept of a single underlying time dimension parameterized by a real interval remained. In our opinion there are two reasons why this concept is preserved. One is our intuitive sense of time, which flows from the past to the future, while the second one is the practical success of models based on such notions. Namely, the expression defining the physical laws is given by differential equations which, ultimately rely on the limiting process inherent in the notion of a derivative that can be either total or partial (van der Vaart, 1973). In our statement, we use the word "intuitive" in the sense that during early cognitive development, all children develop temporal concepts such as "before" and "after" (Hoerl and Savitt, 2011). According to Immanuel Kant, time and space are just forms that the mind projects upon the external things-in-themselves (Gardner, 1999). Further, he claimed that our mind structures our perceptions in a way that space always has an Euclidean geometry, while time is like the structure of the mathematical line. This Kant's idea of time as a form of apprehending phenomena suggests that we have the ability to experience things and events in time, i.e., we have no direct perception of time. Let us make now a short walk through the history of the physical concept of time.

To measure time, people recorded the number of events of some periodic phenomenon. The regular repetition, of the seasons and the motions of the celestial objects, were noted and recorded for millennia, before the physical laws were formulated. The Sun, the Moon, and the Stars were main natural timekeepers, while oil lamps, candle clocks, and water clocks were earliest man-made inventions for measuring time. In the 14th century, a mechanical clock was built, and then it became miniaturized enough for personal, standard, and scientific use.

In 1583, Galileo Galilei discovered, by observing the oscillation of a votive lamp at the cathedral of Pisa, that a pendulum's harmonic motion has a constant period. Half century later (1638) in his "Two New Sciences" (Galileo, 1954) he described an experiment with a water clock, which was used to measure the time by which a bronze ball rolls a known distance down an elevated plane. It was engineered to preserve laminar flow of the water during the experiments, thus providing a constant flow of water for the durations of the experiments. In this experiment, literally said Galileo measured the flow of time to describe the motion of a ball. Note that the Galilean transformations assume that time is the same for all reference frames.

Isaac Newton (1999) in his "Philosophiae Naturalis Principia Mathematica" introduced the concepts of absolute time and space providing a theoretical basis for the Newtonian mechanics. According to Newton, absolute time and space are independent parts of objective reality: "Absolute, true and mathematical time, of itself, and from its own nature flows evenly regardless of anything external, remains always similar and immovable." Namely, absolute time exists independently of any observer and goes alone at a constant pace throughout the Universe. According to

Newton, humans are only capable of observing relative time, which is a measure of observable objects in motion which is commonly used instead of true time. Newton believed that absolute time, in contrast to relative time, was imperceptible and could only be understandable through mathematics. Since Newton spoke about linear flow of time (what he called mathematical time), time could be considered to be a parameter, which linearly varies. In Newtonian mechanics and in corresponding form in quantum mechanics, Lagrangian's (and their Legendre transformation, i.e., Hamilton's equations) bespeak a conception of reversible time (Lagrange, 1796).

The beginning and almost the entire 19th century were distinguished by thermodynamics. The nature of the phenomena that have been studied and their encoding in the laws of physics have set new requirements regarding the concept of time. This issue is open through the Loschmidt's paradox also known as the reversibility paradox, irreversibility paradox, or Umkehreinwand, first published in 1874 by William Thomson (Lord Kelvin). The paradox claims that from time-symmetric dynamics, it should not be possible to deduce an irreversible process. Therefore, there is a conflict between time symmetry of fundamental physical processes and broken time symmetry of macroscopic systems governed by the second law of thermodynamics. Correspondingly it opened a question of the arrow of time. It refers to processes going in a particular direction in such a way that any state in that progressing cannot spontaneously be reformed in any point that has passed. For example, eggs may break, but they never spontaneously reform. In 1927, British astronomer and physicist Arthur Eddington introduced the term "arrow of time" in his book "The Nature of the Physical World," connecting it to the one-way direction of increasing entropy required by the second law of thermodynamics. This arrow is also now known as the "thermodynamic arrow." Note that the arrow of time cannot be identified by time itself. Symbolically speaking arrow of time looks like a vector having a direction and undefined magnitude indicating to the way how the Universe and its contents evolve. Besides the thermodynamic arrow of time, which is distinguished by the growth of entropy, Stephen Hawking (1996) assumed two more arrows of time: (1) psychological as our perception of an inevitable flow and (2) cosmological introduced because of the universe expansion in a single direction from the initial state of Big Bang. Time in biological and other complex systems is often represented using logarithmic scale, to better deal with large fluctuations in their internal parameters. Another example is the thermodynamic time defined by Ilya Prigogine (1961) as the time scale with respect to which the rate of entropy production in the system was constant.

Modern conception of time in physics started to emerge in 1864, when James Maxwell presented a combined theory of electricity and magnetism combining all known laws related to these two phenomena into four equations (Maxwell, 1865). These equations are known as Maxwell's equations for electromagnetism, which allow the solutions in the form of electromagnetic waves, propagating at the frequency of the electric charge which generates those fields. However, they came into conflict with Galilean transformations. According to Maxwell's equations, the light is independent of inertial reference frame and has constant speed, which is in stark contrast with Galilean relativity. There were three possible solutions to this situation: (1) that Maxwell's equations are not correct, (2) to introduce the concept of the luminiferous aether and then the propagation of waves in a vacuum would become the propagation of the acoustic waves in the air, and (3) existence of a third principle of relativity, which is valid for the mechanics and electrodynamics but not based on Galileo's principles.

Some physicists were inclined either to the first or second solution, while Einstein was slanted to the third solution. From his thinking has been created one of the greatest theory in science, i.e., the special relativity, which is based on two postulates: (1) the laws of physics are the same in all inertial frames of reference (principle of relativity) and (2) the speed of light in free space has the same value c in all inertial frames of reference (invariance of c). To fulfill the first postulate, transformations for transition from one to the other inertial system had to be modified. At that time there were known transformations which left Maxwell's equations invariant (the Lorentz transformations discovered by Hendrik Lorenz in 1875), and a consequence of the spatial part of these transformations (Fitz-Gerald Lorentz contraction) were also known. However, the interpretation of these terms remained in the domain of electrodynamics, or just moving within imaginary luminiferous aether, while the temporal part of these transformations nobody understood, although the Lorentz transformations, except space contraction, predicted time dilatation. Albert Einstein directly in 1905 interpreted the Lorentz transformations from his postulates and, in fact, set them to the level of postulates, i.e., all laws of physics must be invariant with respect to the Lorentz transformations.

According to the Lorentz transformations, the time does not remain invariant and this makes a crucial difference to the Galilean transformations. Each observer has its own time and its own spatial coordinates but space and time of another observer in another inertial frame are linear combinations of time and the coordinates of the first frame. It is clear that the Lorentz transformation put an end to the gap between the temporal and spatial dimensions.

In Chapter 1, we briefly mentioned that some quantum phenomena in the real world can be explained, if we accept that the present state of a system is defined by its past, in the sense that the past determines the possible states that are to be considered, and by its future, in the sense that the selection of a possible future state determines the effective present state. Here, we will describe that in more detail. Namely, "[t]he concept of a quantum state is time-asymmetric: it is defined by the results of measurements in the past. This fact by itself is not enough for the asymmetry: in classical physics, the state of a system at time defined by the results of the complete set of measurements in the past is not different from the state defined by the complete measurements in the future. [...] In quantum mechanics this is not so: the results of measurements in the future are only partially constrained by the results of measurements in the past. Thus, the concept of a quantum state is genuinely time-asymmetric" (Aharonov and Vaidman, 2008). This asymmetry is removed by introducing the two-state vector formalism of quantum mechanics (TSVF) originated in a seminal work of Aharonov et al. (1964) which later was extended in (Aharonov and

Vaidman, 1997, 2008), providing a time-symmetric formulation of quantum mechanics. Shortly, a system at a given time t is described completely by a *two-state vector* $\langle\Psi||\Phi\rangle$, which consists of a quantum state $\langle\Psi|$ defined by the results of measurements performed on the system in the past relative to the time t and of a backward evolving quantum state $|\Phi\rangle$ defined by the results of measurements performed on this system after the time t. Note, that the status of the two-state vector might be interpreted differently but a noncontroversial fact is that it yields maximal information about how this system can affect other systems interacting with it at time t (Aharonov and Vaidman, 2008).

REFERENCES

Aharonov, Y., Bergmann, P.G., Lebowitz, J.L., 1964. Time symmetry in the quantum process of measurement. Phys. Rev. B 134, 1410–1416.

Aharonov, Y., Vaidman, L., 1997. Protective measurements of two-state vectors. In: Cohen, R.S., Horne, M., Stachel, J.J. (Eds.), Potentiality, Entanglement and Passion-at-a-Distance, Quantum Mechanical Studies for A. M. Shimony, Volume Two, pp. 1–8.

Aharonov, Y., Vaidman, L., 2008. The two-state vector formalism: an updated review. Lect. Notes Phys. 734, 399–447.

Boas, G., 1951. The influence of philosophy on the sciences. Proc. Am. Philos. Soc. 95, 528–537.

Dainton, B., 2010. Time and Space, second ed. McGill-Queens University Press, Ithaca.

Dowden, B., 2015. Time. The Internet Encyclopedia of Philosophy. ISSN: 2161-0002. http://www.iep.utm.edu/.

Einstein, A., 1905. Zur Elektrodynamik bewegter Körper (On the electrodynamics of moving bodies). In: The Principle of Relativity: Original Papers by A. Einstein and H. Minkowski. (1920). University of Calcutta, pp. 1–34.

Galileo, 1954. Dialogues Concerning Two New Sciences. Dover Publications, New York.

Gardner, S., 1999. Kant and the "Critique of Pure Reason". Routledge, London.

Gürsey, F., 1963. Reformulation of general relativity in accordance with Mach's principle. Ann. Phys. 24, 211–242.

Hawking, S., 1996. The Illustrated Brief History of Time: Updated and Expanded Edition. Bantam Book, New York.

Hoerl, C., Savitt, S., 2011. Time in cognitive development. In: Callender, C. (Ed.), The Oxford Handbook of Philosophy of Time. Oxford University Press, pp. 439–459.

Lagrange, 1796. Dynamics is a four-dimensional geometry. In: The End of Certainty by Prigogine I. (1997). Free Press, New York.

Maxwell, J.C., 1865. A dynamical theory of the electromagnetic field. Philos. Trans. R. Soc. London 155, 459–512.

McTaggart, J.M.E., 1908. The unreality of time. Mind 17, 457–474.

Newton, I., 1999. The Principia: Mathematical Principles of Natural Philosophy. University of California Press, Los Angeles.

Prigogine, I., 1961. Introduction to Thermodynamics of Irreversible Processes, second ed. Interscience, New York.

Thomson, W., 1874. The kinetic theory of the dissipation of energy. In: Brush, S. (Ed.), Kinetic Theory. (1966), vol. 2. Pergamon Press, pp. 176–187.

van der Vaart, H.R., 1973. A comparative investigation of certain difference equations and related differential equations. Bull. Math. Biol. 35, 195–211.

Time in biology

6

The problem of understanding the time flow in biological systems has a long tradition and is often connected with inextricable problems derived from undeveloped conceptual differentiations. In the beginning of our analysis we should therefore clearly distinguish duration as an objective property of matter and time flow (chronology) as the subjective construction of an observer which cannot be equated with duration or changes in duration (for example, see Deleuze, 1990; Husserl, 1964; Merleau-Ponty, 1945) Therefore, time in biological systems cannot be considered as an independent flow, inert to changes in itself nor as a simple line of successive infinitesimal quantities, but as duration filtered through perception (Balaz, 2005). Therefore, changes in duration are only a basis upon which every perceptive entity can construct its own time flow. The structure of that flow is not at issue in this chapter), but rather how constructions of subjective structure(s) influence the organization and functioning of living systems. First, systemic time flow cannot be established by perceptivity itself, since the prerequisite for establishing systemic time relations is the ability to compare different systemic states. As long as perceptivity is not incorporated into the network of systemic relations, the system will remain in a state of independent linear flows. Only by developing such relations, the simple succession of states can be manipulated and arranged. In other words, systemic time can be established. Therefore, as a first step toward further analysis, we should take a brief look at the generation of processes.

According to Luhmann (2012), we can define a process as a mutually connected succession of events where the scope of selections is reestablished at each stage. A specific characteristic of processes understood in that manner is their anticipatory structure—because such a succession of events is also inherently an accumulation toward less and less probable states (less probable from the perspective of the beginning of the process) which are a necessary (structurally but not logically) consequence of previous stages. But, there is a very important difference between elementary processes (e.g., separate enzymatic transformations) and processes developed from them (e.g., metabolic pathways). The first ones can be identified as (temporally) irreversible transformations toward a determined state (in an ideal case) or a group of very similar states. However, in living systems, that kind of almost indispensable flow enclosed in a rigid structure is only a basis for the further development of functionality. These irreversible sequences can be rearranged into higher order structures (so-called metabolic pathways) thus gradually relativizing indispensability of stages within the process, with every superposed level of

constructed functionality. In such a structure, individual events are no longer necessary for continuing the line of transformations (which is the case with elementary
processes), but become "one of" the possible realizations of functionality. In spite
of such relativization, what remains inherently connected for processes is their
necessary differentiation along the former/latter axis. In elementary processes, the
structurally (materially) defined flow of transformations from a currently actual state
into a necessary prospective state is the primary form of temporal differentiation.
However, that primacy is erased at the systemic level and is transformed into sets
of probable states by which *intrasystemic* time is relieved from the uniformity of uni-
lateral flow.

Through this relativization of necessities, the system becomes able to construct
anticipatory structures which in one available now (or more precisely: in a perceptively constructed present) choose indicators that are in correlation with changes in
the future, associating them with adequate systems of transformations and therefore
preparing themselves for the following events. From such a perspective, it is obvious
that mutual interactions of anticipatory structures are not only based on the possibility of perceiving signals, but also on the possibility of anticipating future states (e.g.,
establishment of regulations based on feedbacks). Such anticipations are inherently
connected with systemic expectations in which realization or nonrealization becomes a powerful *intrasystemic* regulative factor. Therefore, it is not only important
to accomplish some function but equally important is temporal compatibility with
other, parallel processes. And only a combination of these two factors, the possibility
of anticipation and regulation by anticipation, can create a basis for the construction
of an autonomous, systemic time flow, as a generalization of the validity of partial
intrasystemic time horizons across functional elements and (organizational) structures within certain subsystems. In other words, during interactions—along with
perceptive normativity—subsystems also impute time (their own construction of
time) to each other. The organizational dynamics of mutual influences results purely
from such imputations. However, to functionalize the *intersubjective* temporal field
which has been formed, all interacting elements must share the same normative
rules, because simultaneity of (functionally meaningful) interactions cannot be
achieved in two different times with no points of contact. In this way, *intersubjec-
tivity* is achieved only within purposeful situations.

Although such relativizations make organizational manipulations available, still
we cannot talk about systemic comparisons (since an abstract measuring of empty
intervals is not possible in such systems) and hence, about systemic time. The final
precondition is to establish composite repetitions and the successive process of
transformations. In other words, by parallel and multiple repetitions of transformations which are successively superpositioned, the system enters a state of former/
latter processes which may be equivalent or different, where the consequences of
such transformations are transferred to the cycles of successive processes. In this
way, the dynamics of cycles is not self-sustained but is always constituted with reference to previous operations; i.e., it is constituted by a rudimentary "comparison" of
intervals. Only then do processes became autonomous axes for establishing systemic

time relations and the following construction of organizational regulations and controls. Before proceeding further, it should be emphasized that there is no such thing as a universality of time intervals within living systems in the first place, because no internal meta-systemic observer exists which would be able to grasp the systemic wholeness and impose a perspective of uniform relations. Therefore, in living systems we cannot talk about the objective metrics of time relations but only about partial relations where the norms are defined in accordance with the actual context.

One of the most important ways in which living systems are able to manipulate within the temporal dimension are regulations of physical duration of constitutive elements (i.e., temporalized reproduction of elements). That idea is certainly not new. It is very well known in empirical investigations (e.g., Eden et al., 2011), but its theoretical treatment is surprisingly ignored. Although the theory of autopoietic systems made obvious that systemic elements should always be self-reproduced (Zeleny, 1981), models of global regulations in organisms mainly neglect this fact and its consequences. Therefore, it is necessary to pay particular attention to the development of functionality based on cyclic degradations and reconstructions of intrasystemic material structures. Before that, it should be borne in mind that temporalized reproduction is not a mere repetitive circle producing sameness, but rather produces with variations whose roots are in the foregoing but also deviate from it. And deviation is not only change in the sense of small structural alterations or achievement of higher or lesser efficiency (compared with previous state) but is always potential transformation into some other framework. It is production based on one's own needs, which are being constantly surveyed, and constantly changed.

The continuous decomposition of segments of processes compels them to be in constant reconstruction, thus making space available in the organizational structure for different insertions, divergences, and reroutings without the need to construct specific mechanisms (in the form of localized regulators) for each specific case. Also, through *temporalization*, the system purposefully eliminates groups of elements, concordantly eliminating them from the possibility of direct reaction (regarding other elements, subsystems, etc.). In this way the system's internal structure perpetually reconstitutes the causal basis for its own processes and the past is not merely a fixed set of preceding events which linearly vanished but is rather a dynamic accumulation where, according to its relevancy to the current state (a relevancy constantly updated with causal reconstitution), some elements and structures can be summoned while others disappear without any further functional influences. Thus, the primacy of successive stepwise regulation is greatly diminished and the structure itself becomes a major determinant in the regulation of reproductive periodicity. In this manner, after introducing the idea of temporality, the organizational model of living systems becomes clearly different from the (first-order) cybernetic perspective which is usually applied. It is legitimate to use such models when dealing with short-time segments of processes. In that case, analysis is focused only on those elements whose life span exceeds the duration of process itself. However, by this approach we get only a naive sketch of living systems. To move forward from this rudimentary understanding, we need to change the paradigm and postulate a continuous instability of mere elements by which fundamentally new types for achieving functional flexibility are established. As a universal consequence, the system is obliged to continuously self-adapt. Since each subsystem has different dynamics of degradation with each cycle of reconstitution, the structure and organization of subsystems are in continuous adaptation to the (internal) environment: starting from differences in protein folding (static disorder), through continuous changes in protein composition (qualitatively and quantitatively), and on to hierarchical and communicative variations caused by reproductive cycles. Through such cycles, the system is forced not to be self-adapted but rather in self-adaptation, because material realizations based on previous informational context are constantly decomposed and, to maintain functionality, the system must constantly deal with external signals.

Finally, since time in biological systems (as a perceptive construct) is only a reflection of some aspects of duration, it is obviously liable to different strategies of manipulation. By perceptive deconstruction of the continuity of external changes, alteration of a previous state is not only a variation but appears as a functional novelty—and is processed as such. It allows functionally meaningful time manipulations, since the dynamic of perceived changes (faster/slower) is least connected with objective changes of duration. What really generates the rapidity of time flow are perceptive scopes, i.e., the distribution of boundaries between different absolutes of perception. These boundaries may be distributed homogeneously, thus generating the illusion of general acceleration or deceleration of external changes, or they can build a heterogeneous construction displaying acceleration/deceleration of externality in accordance with functional context. By such strategies, systemic time can be distributed in accordance with needs and coordinated with external pressures which are not liable to direct manipulations.

Before closing this chapter, it is necessary to answer one more question: What is the main precondition for temporalization of elements without destroying systemic functionality? Since an organism's survival is inherently connected with undisturbed metabolism run, temporalization should not influence the continuity of metabolic processes. Is it justified then to assume the existence of elementary functional units which cannot be, and should not be, perturbed? If we analyze metabolism as a whole, at each situation we can identify some segments which are essential for survival of the organism and which will be safeguarded from the possibility of internal violations (e.g., by excessive synthesis of groups of enzymes coupled with a decrease in the level of specific chaperone). However, here we should bear in mind that the determination of such "elementary units" is highly dependent on context: both materially (e.g., availability of certain nutrients, the constellation of environmental factor), as well as functionally (e.g., hierarchical variations regarding actual distribution of subsystems). Therefore, what is usually considered as a main quality of units, namely their perseverance through different contexts, is lost. However, if we transfer our focus down to the level of concrete, material transformations, we can see that every single step in the processes of metabolic transformations is performed by enzymes whose actions are not liable to cutting or dividing into independent phases. In this manner, single enzymatic transformation can be considered as an elementary event, an atom of functionality. Only through the enclosure of single occurrences in a web of functionally meaningful events does it become possible for systems to base their functionality on recursive, reflexive reproduction of elements. The rise of elementary events from the level of discrete, meaningless occurrences to the level of finished processes, lays the groundwork for a situation where any kind of interruption (in the sense of physical elimination of functional elements) or rearrangement cannot violate the fundamentality of such units. Without that kind of organization, temporalizing constitutive elements into systems would be destructive for them.

REFERENCES

Balaz, I., 2005. Construction of endo-time and its manipulation in autopoietic systems. In: Buccheri, R., et al. (Eds.), Endophysics, Time, Quantum and the Subjective. World Scientific, Singapore, pp. 139−151.

Deleuze, G., 1990. Bergsonism. Zone Books, New York.

Eden, E., Geva-Zatorsky, N., Issaeva, I., Cohen, A., Dekel, E., Danon, T., Cohen, L., Mayo, A., Alon, U., 2011. Proteome half-life dynamics in living human cells. Science 331, 764−768.

Husserl, E., 1964. Phenomenology of Internal Time Consciousness. Indiana University Press, Bloomington.

Luhmann, N., 2012. Introduction to Systems Theory. Polity Press, Cambridge, UK.

Merleau-Ponty, M., 1945. Phenomenology of Perception. Routledge Classics, New York.

Zeleny, M., 1981. Autopoiesis: A Theory of Living Organization. North Holland, New York.

Functional time: definition and examples

In the previous chapters we covered several notions of time. First, we started with the nature of time in philosophy and physics, and then we continued with the phenomenology of time in biological organisms (previous chapter). In this chapter we will keep our focus on the phenomenology of time, but from the perspective of functional systems. By the notion of the functional system we will cover all systems where processes unfold following a set of known rules and which exhibit repetitive pattern. In such systems, we can measure time in several ways. Most obvious would be external clock-time. In that way, as a result of observing systemic events we will obtain a sequence of intervals, where each interval has its own duration. On the other hand, we can shift our perspective and move from the position of the external objective observer, to the process itself. How is time formed from the perspective of the functional system? And what is the structure of time flow from that perspective? These questions have a long history (Whitehead, 1978; Luhmann, 2012; Lolaev, 1995, 1996, 1998; Mihailović and Balaž, 2012b). Despite numerous differences in existing approaches, common denominator is the view that the structure of time in functional systems (*functional time*) should be considered as different from the universal, abstract time. Instead of being always synchronized with the global timekeeper, functional time is derived from the concrete, material systems and their processes. Therefore, functional time ultimately depends on the quality changes within systems and processes forming them. As a result, in contrast to the notion of time in classical physics, functional time is not always linear but takes a shape of interaction of processes that constitute the functional system.

In complex biological systems, the formation of functional time strongly depends on the state of the system. For the sake of further consideration in this section, we will introduce definitions of some terms.

Definition 1: For the sake of simplicity, we can define functional system R as the system which has the following properties: (1) partial decomposability to subsystems, and processes r_1, r_2,... depending on each other, (2) strong hierarchical ordering, and (3) complexity. We call a set $R = \{r_1, r_2,..., r_N\}$, where N is number of elements in the set R a complex functional system (hereafter, functional system).

Definition 2: Set $S = \{S_1, S_2,...\}$ represents the states during time evolution of the functional system R which passes through those states.

Comment: The term state is differently described in physics, mathematics, biology, chemistry, computing, sociology, etc. In further text we use its meaning in the common sense, i.e., that it refers to the present condition of a system or entity.

Developments in Environmental Modelling, Volume 29, ISSN 0167-8892, http://dx.doi.org/10.1016/B978-0-444-63918-9.00007-7

Definition 3: To each element of the set S we assign value 0 or 1. If the system is unstable or unsynchronized then $m = 0$, otherwise $m = 1$. The set of all values of m we call the measure $M = \{m_1, m_2,...\}$ of the set of states S that indicates the status of functional system R.

Comment: The status of the functional system R is described: (1) as either stable or unstable, where the term stable we use in the sense of the structural stability, i.e., that small perturbations to the system do not determine emergence of the qualitatively new features (Jen, 2003) and (3) as either synchronized or unsynchronized. Let us note that all elements from the set $R = \{r_1, r_2,..., r_N\}$ have particular measures corresponding to elements from that set. The status of the functional system R is measurable either through the observation or computationally. Thus, the stability or synchronization in the functional system can be established by measurements or by some methods from the archive of nonlinear dynamics, as, for example Lyapunov exponent (Pikovsky et al., 2001).

Definition 4: Subsystems, phenomena and processes of the functional system R, emit the set of signals $I = \{i_1, i,..., i_I\}$ either directly or indirectly indicating that the system remains functional until any signal reaches the location of an observer, where I is the maximal number of signals that the system emits.

Definition 5: Subsystems, and processes of the functional system R, we call cardinal ones, when they lose their functionality determining the termination of that system.

Definition 6: If the functional system R emits a noncardinal signal (i.e., a signal which does not come from the cardinal subsystem), it means (1) that the system is in function sending other signals but without signal from that subsystem or (2) that subsystem waits to be synchronized with other subsystems in order to send the signal.

The fact that the functional time is formed as a result of consistent change of concrete material object states we illustrate using the following examples: (1) response of the functional system on a stimulus (mollusk time reflex formation); (2) response of the functional system on a cognitive level (prisoner time formation); and (3) process of substance exchange on the cellular level (time formation in process of biochemical substance exchange between cells).

7.1 MOLLUSK TIME REFLEX FORMATION

Here, as an example of formation of the functional time we will analyze the mollusk reflex formation. Loalev described the experiments as follows: "[T]he mollusk receives shocks with low-power current every 5 min. After shock it hides in a shell for a short while and then continues its motion. After the shocks stop the mollusk continues to hide in a shell every 5 min. It proves the availability of time system. In this connection we remark first of all that this example is not a proof of the mollusk's astronomic time counting system, as there is no such time in nature. The mollusk hides in a shell every 5 min, not due to the availability of counting systems of postulated nonexistent time in nature but because every 5 min consistent change of definite, strictly identical number of states takes place in the mollusk's organism. As a result the own time of mollusk is formed in which it lives, exists." (Lolaev, 1996).

We formalize this example of functional time formation in the following way. Here, the functional system is a mollusk (Fig. 7.1(a)), which receives external, electrical shocks every 5 min forming the reflex to hide in a shell upon receiving the shock. Symbolically, we have $R = \{r_1, r_2\}$ where $r_1 = $ *mollusk*—the mollusk as the one element of the set R and $r_2 = $ *reflex* the phenomenon as the second one. The mollusk is passing through states $S = \{s_1,..., s_{20}, s_{21},..., s_{30}, s_{31}, ..., s_{50}, ...\}$ where $s_1, s_2,..., s_{20} = $ *out of the shell* corresponding to stable states (the mollusk does normal life routines, having a communication with the surrounding environment), and $s_{21}, s_{22},..., s_{30} = $ *in the shell* including unstable ones (the mollusk is hiding within the shell, having no communication with surrounding environment), and so forth. The measure of the state in this example is $M = \{\,1,...,1,...,1\,,0,...,0,...,0,\,1,...,1,...,1,\,...\}$, where 0 and 1 correspond

$$\underbrace{}_{1-20}\;\underbrace{}_{21-30}\;\underbrace{}_{31-50}$$

to unstable and stable states, respectively. Thus, an observer can receive signals $I = \{i_1, i_2,...\}$, for example visualized on a display, like bars as in Fig. 7.1(b). In this example, functional time forms at the level of reflexes, without any cognitive influences. We would like to emphasize that this example is not a proof of the mollusk's clock-time counting system, as there is no such time in nature. Here, a mollusk does not react on passed 5 min of astronomical time but on strictly definite number of states that consistently changed in its organism during these symbolic 5 min.

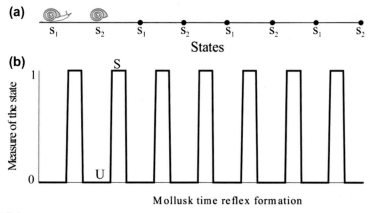

FIGURE 7.1

Mollusk time reflex formation: (a) states changed consistently in its organism as stable $[S_1 - S(1)]$ and unstable $[S_2 - U(0)]$, (b) symbolic diagram of its functional time.

7.2 **PRISONER TIME FORMATION IN THE CELL**

This example describes prisoner time formation in the cell—a small room in which a prisoner is locked up. He is trapped in the cell with two beds—one next to the left wall (LW), while another one next to the right wall (RW). In the well-isolated cell, the prison guard has the control of prisoner position, through the light signal that is visualized on the oscilloscope. Once a day prisoner has to follow strictly defined routine in accordance to "Prison Rules." They are: (1) start from the floor to lower bed next to the RW; (2) get on the upper bed floor; (3) get off from the upper bed to the lower bed; (4) get off from the lower bed to the floor; (5) move to the LW; (6) start from the floor to the lower bed next to LW and then repeat the same procedure with the bed next to this wall. Prisoner cannot go back until he does not finish the cycle as depicted in Fig. 7.2(a). Prison guard sees the lighting point on the oscilloscope marking out the shape as in Fig. 7.2(b).

In this example the functional system is a prisoner (Fig. 7.2(a)), which repeats a daily routine for a certain time. Here, we have symbolically $R = \{r_1, r_2\}$ where r_1 = prisoner is the prisoner as the one element of the set R and r_2 = repetition of the remembered routine the phenomenon as the second one. The prisoner is passing through states $S = \{s_1, \ldots, s_{40}, s_{41}, \ldots, s_{60}, s_{61}, \ldots, s_{100}, \ldots\}$ where $s_1, s_2, \ldots, s_{40} = $ on

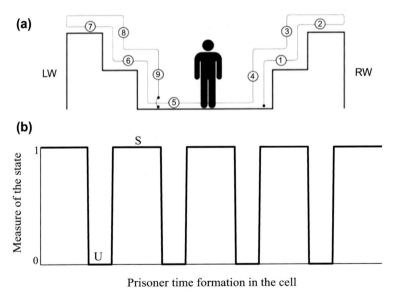

Prisoner time formation in the cell

FIGURE 7.2

Prisoner time formation in the cell: (a) states of the prisoner routine, (b) symbolic diagram of his functional time. The *numbers* indicate the pathway of the routine with states from which is passing through, while LW and RW are the left wall and the right wall, respectively. Letters U and S indicate unstable and stable states, respectively.

cell floor and s_{41}, s_{42},..., s_{60} = on bed, representing stable and unstable states, respectively, and so forth. The measure of the state in this example is $M = \{\ 1, ..., 1, ..., 1\ ,0, ..., 0, ..., 0,\ 1, ..., 1, ..., 1,\ ...\}$ where 0 and 1 correspond

$$\underbrace{}_{1-40}\ \underbrace{}_{41-60}\ \underbrace{}_{61-100}$$

to unstable (on bed) and stable (on cell floor) states, respectively. Thus, the guard can receive signals $I = \{i_1, i_2, ...\}$, on oscilloscope, like bars depicted in Fig. 7.2(b).

Unlike the previous example in this one, prisoner creates his own time in the cell by consciously knowing the rules. Again, this example is not a proof of the prisoner's astronomic time counting system since there is no such time in nature. Prisoner is moving in a "stairways" rhythm not due to the availability of counting systems of some universal time, but by following available rules.

7.3 FUNCTIONAL TIME FORMATION IN PROCESS OF BIOCHEMICAL SUBSTANCE EXCHANGE IN RING OF CELLS

Here we illustrate the formation of the functional time in the process of biochemical substance exchange between cells modeled by the system of coupled difference equations (Mihailović et al., 2011). This model comprises the following parameters: (1) c_i that represents coupling of two factors: concentration of molecules in intracellular environment and intensity of response they can provoke; (2) affinity p_i to uptake molecules, where this term is used as a measure of the degree of the cell capability to uptake the biochemical substance molecules, with condition $\sum_i p_i = 1$ where p_i is the affinity of the single cell ($p \in [0,1]$); and (3) parameter r that includes collective influence of environment factors which can interfere with the process of communication. Here, we consider a ring of coupled cells. Each cell is coupled to its neighbor through the mapping given in Mihailović et al. (2011) as it was similarly done in Suguna and Sinha (2005). In these approaches, cell moves locally in its environment without making long pathways, while the cell movement is considered in the π-cell coordinate system defined as

$$\pi = \frac{1 - \zeta}{1 - \zeta_0} \tag{7.1}$$

where ζ is the dimensionless radius of the cell within which it interacts with another cell, defined as $\zeta = R/R_{max}$ while R is the radius and R_{max} and R_{min} are its maximal and minimal values, respectively. Finally, $\zeta_0 = R_{min}/R_{max}$. The values of π-cell coordinate lie in the range 0 ($R = R_{max}$) and 1 ($R = R_{min}$). According to Mihailović and Balaž (2012a) the system of coupled difference equations for N cells exchanging the biochemical substance can be written in the form of matrix equation,

$$\mathbf{A} = (\mathbf{B} + \mathbf{C}) \cdot \mathbf{D} \tag{7.2}$$

where

$$
A = \begin{bmatrix} x_{1,n+1} \\ x_{2,n+1} \\ \cdot \\ x_{k-1,n+1} \\ x_{k,n+1} \\ \cdot \\ x_{N-1,n+1} \\ x_{N,n+1} \end{bmatrix}, \quad
D = \begin{bmatrix} x_{1,n} \\ x_{2,n} \\ \cdot \\ x_{k-1,n} \\ x_{k,n} \\ \cdot \\ x_{N-1,n} \\ x_{N,n} \end{bmatrix}, \quad
C = \begin{bmatrix}
0 & c_1 x_{2,n}^{p_1-1} & 0 & 0 & \cdot & 0 & 0 & 0 \\
0 & 0 & c_2 x_{3,n}^{p_2-1} & 0 & \cdot & 0 & 0 & 0 \\
\cdot & \cdot & \cdot & \cdot & \cdot & \cdot & \cdot & \cdot \\
0 & 0 & 0 & 0\ 0 & c_k x_{k+1,n}^{p_k-1} & \cdot & 0 \\
\cdot & \cdot & \cdot & \cdot & \cdot & \cdot & \cdot & \cdot \\
0 & 0 & 0 & 0 & \cdot & 0 & 0 & c_{N-1}x_{N,n}^{p_{N-1}-1} \\
c_N x_{1,n}^{p_N-1} & 0 & 0 & 0 & \cdot & 0 & 0 & 0
\end{bmatrix},
$$

$$
B = \begin{bmatrix}
(1-c_1)\,r(1-x_{1,n}) & 0 & 0\ 0 & \cdot & 0 & 0 & 0 \\
0 & (1-c_2)\,r(1-x_{2,n}) & 0\ 0 & \cdot & 0 & 0 & 0 \\
\cdot & \cdot & \cdot\ \cdot & \cdot & \cdot & \cdot & \cdot \\
0 & 0 & 0\ 0\ (1-c_k)\,r(1-x_{k,n})\ 0 & \cdot & 0 \\
\cdot & \cdot & \cdot\ \cdot & \cdot & \cdot & \cdot & \cdot \\
0 & 0 & 0\ 0 & \cdot & 0 & (1-c_{N-1})\,r(1-x_{N-1,n}) & 0 \\
0 & 0 & 0\ 0 & \cdot & 0 & 0 & (1-c_N)\,r(1-x_{N,n})
\end{bmatrix}
$$

$$(7.3)$$

with condition $\sum c_i = c$ with $0 \leq c \leq 1$ and r is the logistic parameter ($0 \leq r \leq 4$), while x_i represents concentration of molecules in cells. Solution of the system Eq. (7.3) gives the concentrations in all cells in time and space in π-cell coordinate system.

To demonstrate that each part of a functional system has its own intrinsic space time (Mihailović and Balaž, 2012a,b), i.e., functional time, we consider a functional system represented by a model consisting of three cells coupled in a ring, that exchange the biochemical substance, which is schematically shown in Fig. 7.3. Here, we have symbolically $R = \{r_1, r_2\}$ where $r_1 =$ ring of coupled cells— the ring of coupled cells (as a part of a tissue), as the one element of the set R and—$r_2 =$ process of exchange substance exchange between cells on diffusion-like manner. The system of the ring of coupled cells can pass through states $S = \{s_1, s_2,...\}$ where the process of substance exchange states can be either synchronized or unsynchronized, corresponding to stable and unstable states, respectively, and so forth. The measure of the state in this example is $M = \{m_1, m_2,...\}$ where m can take values 0 or 1 that correspond to unstable and stable states, respectively. Let us consider the signaling of this system. In Fig. 7.3 cells compose a multicellular system that is a complex one consisting of three components, i.e. (cell 1 vs. cell 2), (cell 2 vs. cell 3), and (cell 3 vs. cell 1). They exchange biochemical substances sending (1) single (Λ_{12}, Λ_{23}, and Λ_{31}), (2) double [(Λ_{12}, Λ_{23}), (Λ_{23}, Λ_{31}), and (Λ_{31}, Λ_{12})] and triple (Λ_{12}, Λ_{23}, Λ_{31}) signals. Signals that come from the system indicate that system remains functional until any signal reaches the location of an observer. In the model this condition is satisfied when there is synchronization in biochemical substance exchange between any two or three cells sending single, double, or triple signal (Table 7.1). Let it be noted here that in functional systems there exist components such that the system is terminated when they lose their functionality. Those components we will call cardinal components of the system. If the functional system does not send to an observer any signal that does not come from

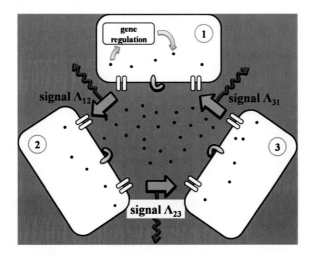

FIGURE 7.3

Schematic representation of a simple model of the ring of the three coupled cells exchanging the biochemical substance with the corresponding signaling.

Reprinted with permission from Mihailović, D.T., Balaž, I., 2012a. Forming the functional time in process of biochemical substance exchange between cells in a multicellular system. Mod. Phys. Lett. B 26, 1250175.

the cardinal component (noncardinal), it means that system is either (1) in function sending other signals but without signal from that component or (2) that component waits to be synchronized with other ones in order to send the signal (Mihailović and Balaž, 2012a). In a general case when a multicellular system has N cells, then the number of signals sent to an observer is $i = n(n - 1) + 1$ emitted by $j = C_2^N$ components. Thus $I = \{(i_1),\dots, (i_N), (i_1, i_2),\dots, (i_{N-1}, i_N), \dots, (i_1,\dots, i_N),\}$ signals will be send by the system. In our case when $N = 3$ then the number of signals is seven which can be sent through three groups of signals (Table 7.1).

We suppose that the strength of the single signal emitted from the considered system toward the cell is given by the functional form $\Lambda = \Lambda(\lambda, \pi)$, which is proportional to $\lambda_{ij}/\Delta\xi_{ij}$ (i and j are the neighboring cells), where (1) λ_{ij} is the largest Lyapunov exponent obtained for exchange between two cells (Mihailović and Balaž, 2012a) and (2) $\Delta\xi_{ij}$ expresses dependence of the signal strength on the distances

Table 7.1 Signals Sent by the Three Cells System in Fig. 7.3

Single	Double	Triple
Λ_{12}	$(\Lambda_{12}, \Lambda_{23})$	$(\Lambda_{12}, \Lambda_{23}, \Lambda_{31})$
Λ_{23}	$(\Lambda_{23}, \Lambda_{31})$	
Λ_{31}	$(\Lambda_{12}, \Lambda_{31})$	

between cells in the π-cell coordinate system. For three cell system strength signals having a logarithmic form we have

$$\Lambda_{12} = \left| \lambda_{12} \ln \frac{1}{|\pi_2(1 - \zeta_{02}) - \pi_1(1 - \zeta_{01})|} \right| \tag{7.4a}$$

$$\Lambda_{23} = \left| \lambda_{23} \ln \frac{1}{|\pi_3(1 - \zeta_{03}) - \pi_2(1 - \zeta_{02})|} \right| \tag{7.4b}$$

$$\Lambda_{31} = \left| \lambda_{31} \ln \frac{1}{|\pi_1(1 - \zeta_{01}) - \pi_3(1 - \zeta_{03})|} \right|. \tag{7.4c}$$

Lyapunov exponent, in Eq. (7.4a–c), is calculated for the biochemical substance exchange between each two cells in a multicellular system (Mihailović et al., 2011). For any $\lambda \geq 0$, in these equations, the biochemical substance exchange between cells is considered to be unsynchronized (Mihailović and Balaž, 2012a) and corresponding signal strength from that component is equal to zero.

According to Definition 3, the status of the functional system is also measurable through the computations. We consider the functional time of biochemical substance exchange in three cells system through the synchronization signal (0 and 1, representing either nonsynchronized state or synchronized one) that represents binary information about that exchange. The stability or synchronization in the functional system can be established, for example, over the Lyapunov exponent. Synchronization is a well-known collective phenomenon in various multicomponent functional systems (Pikovsky et al., 2001). The exchange of information (coupling) among the subsystems can be either global or local on multicell system (Mihailović and Balaž, 2012a; Ghosh et al., 2010). Here, the system is considered to be synchronized globally only when the largest Lyapunov exponent of the driven system is negative (Guireya et al., 2007). We study the stability of the fixed point by linearizing $n \geq 2$ component coupled system and obtain $\mathbf{Z}_{n+1} = \zeta_n \mathbf{Z}_n$ where ζ_n is the Jacobian of this system evaluated in $(0,0,0,\ldots,0)$ and $\mathbf{Z}_n = (x_{1,n}, x_{2,n}, \ldots, x_{N,n})$. By iterating we obtain

$$Z_{n+1} = \left(\prod_{s=0}^{n} \zeta_s \right) Z_0 \tag{7.5}$$

and thus we get the Lyapunov exponent

$$\lambda = \lim_{n \to \infty} \left(\ln \left\| \prod_{s=0}^{n} \xi_s \right\| \middle/ n \right). \tag{7.6}$$

Lyapunov exponent is calculated for the biochemical substance exchange between all cells in a multicell system (three cells in our case). For any $\lambda \geq 0$, in these equations, the biochemical substance exchange between cells is considered to be unsynchronized and the corresponding signal has the measure of the state zero ($m = 0$).

Having in mind coding with two numbers 0 and 1, representing either nonsynchronized state or synchronized one, we can establish a barcode of functional time in the process of biochemical substance exchange in a multicellular system. We call this code a *functional time barcode* which is a representation of system states through chronological time, which shows data about the process of biochemical substance exchange to which it is attached. It represents states by varying width of parallel lines and spaces between them, indicating whether system is synchronized or not, and may be referred to as linear.

As mentioned above, we deal with the system of three interacting cells that exchange the biochemical substance described by the mapping given by Eq. (7.2). So we have three cells in the system and one process (process of biochemical substance exchange), sending $3 \cdot 2 + 1$ signals (Table 7.1). This system has no cardinal components. Accordingly it cannot be terminated. The signal strengths are calculated by Eq. (7.4a−c) for each two interacting cells, while their π-cell coordinates were randomly chosen in the interval (0, 1). Since the functional time is related to the system state (Mihailović and Balaž, 2012a,b), we define the state in the following way: (1) the state is described with the set of parameters (c, p, r), which are randomly chosen; (2) in any state the system can be either synchronized or unsynchronized; (3) the system is synchronized if the cross-sample entropy of each system component is below the chosen threshold, i.e., when Lyapunov exponent is negative or close to zero (Mihailović and Balaž, 2012a); (4) if the system is in the state when it is synchronized then it sends the signal; and (5) transition from one state to another one is pursued by changes of state parameters and entropy threshold. In Fig. 7.4, functional time barcode is depicted for biochemical substance exchange for three cells, after 10,000 iterations. For each system component, (1) the largest Lyapunov exponent is calculated and (2) it checked whether the cross-sample entropy is below the entropy threshold following the procedure defined in Mihailović and Balaž (2012a). Then the corresponding value, either 0 (synchronized) or 1 (nonsynchronized), is associated to the state. This barcode shows a "history" of

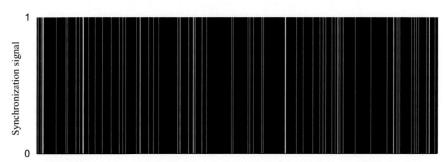

Functional time of biochemical substance exchange in the system of three cells (10,000 iterations)

FIGURE 7.4

Functional time barcode of biochemical substance exchange in three cells system.

Reprinted with permission from Mihailović, D.T., Balaž, I., 2012a. Forming the functional time in process of biochemical substance exchange between cells in a multicellular system. Mod. Phys. Lett. B 26, 1250175.

functional time of the considered process as a part of hierarchically established system of multicellular system consisting of three cells. Through the barcode, the proposed model of biochemical substance exchange between cells and corresponding mathematical tools give information of the evolution of the system during a given range of the functional time. It can be a useful tool in our attempts to know how the biological complex system will evolve, depending on either parameter values or initial conditions, if those attempts are directed rather toward its relative chance than predictability (Arshinov and Fuchs, 2003).

REFERENCES

Arshinov, V., Fuchs, C., 2003. Preface. In: Arshinov, V., Fuchs, C. (Eds.), Causality, Emergence, Self-Organisation. NIAPriroda, Moscow, Russia, pp. 1–18.

Ghosh, S., Rangarajan, G., Sinha, S., 2010. Stability of synchronization in a multi-cellular system. Europhys. Lett. 92, 40012.

Guireya, E.J., Beesb, M.A., Martina, A.P., Srokosza, M.A., Fashama, M.J.R., 2007. Emergent features due to grid-cell biology: synchronisation in biophysical models. Bull. Math. Biol. 69, 1401.

Jen, E., 2003. Stable or robust? What's the difference? Complexity 8, 12–18.

Lolaev, T.P., 1995. The philosophical and naturally scientific basis of time irreversibility. Issue of Moscow University, Seria 7 Philosophy 2, 80–90.

Lolaev, T.P., 1996. About the "Gear" of Time Flow. Problems of Philosophy, Moscow.

Lolaev, T.P., 1998. Time as a function of a biological system. In: Space-Time Organization of Ontogenesis. Moscow State University, Moscow, pp. 30–35.

Luhmann, N., 2012. Introduction to Systems Theory. Polity Press, Cambridge, UK.

Mihailović, D.T., Balaž, I., 2012a. Forming the functional time in process of biochemical substance exchange between cells in a multicellular system. Mod. Phys. Lett. B 26, 1250175.

Mihailović, D.T., Balaž, I., 2012b. An essay about the functional time of environmental interfaces regarded as complex biophysical systems. In: Proceedings of the 6th International Congress on Environmental Modelling and Software (iEMSs), 1–5 July 2012, Leipzig, Germany Meeting, Leipzig, Germany.

Mihailović, D.T., Balaž, I., Budinčević, M., Mihailović, A., 2011. Stability of intercellular exchange of biochemical substances affected by variability of environmental parameters. Mod. Phys. Lett. B 25, 2407.

Pikovsky, A., Rosenblum, M., Kurths, J., 2001. Synchronization: A Universal Concept in Nonlinear Sciences. Cambridge University Press, Cambridge.

Suguna, C., Sinha, S., 2005. Dynamics of coupled-cell systems. Physica A 346, 154–164.

Whitehead, A.N., 1978. Process and Reality, second ed. Free Press, New York.

Use of different coupled maps in the environmental interfaces modelling

III

Coupled logistic maps in the environmental interfaces modelling

8

8.1 COUPLING OF TWO LOGISTIC MAPS

There is a number of interesting environmental interface problems which can be described by the dynamics of coupled maps. Some of them are convection in conducting fluids (Hogg and Huberman, 1984), calculating the surface temperature in climate models (Mihailović et al., 2014), substance exchange between cells (Mihailović et al., 2011a), or effects of spatial heterogeneity on population dynamics (Kot, 1989; Lloyd, 1995). A coupled map is an ensemble of elements of a given discrete-time dynamics ("map") that interact ("couple") with other elements from a suitably chosen set. The dynamics of each element is given by a map. As a consequence, the coupled map is a discrete-time multidimensional dynamical system, in which usually all elements have identical map dynamics. However, coupled maps can also contain heterogeneous elements. In the aforementioned fields, it is of a great importance to understand the global dynamics of coupled systems as a function of both nonlinearity and coupling strength. We consider several types of coupling.

Let us consider two uncoupled maps. The system like this could describe two cells by maps

$$x_{n+1} = f(x_n), \quad y_{n+1} = f(y_n), \tag{8.1}$$

where

$$f(x) = rx(1 - x) \tag{8.2}$$

is the logistic equation and r the logistic parameter. Another example, could be two biological populations (let us say insects as in Fig. 1.1d), in which the number of individuals evolves from year to year according to Eq. (8.1) that is graphically depicted in Fig. 8.1.

One way to introduce the coupling is to consider the simplest spatially extended biologically realistic model using two coupled logistic maps (Hastings, 1993; Gyllenberg et al., 1993; Lloyd, 1995). In terms of dimensionless variables, this has the form

$$x_{n+1} = f(x_n) + c_1(f(y_n) - f(x_n)) \tag{8.3a}$$

$$y_{n+1} = f(y_n) + c_1(f(x_n) - f(y_n)). \tag{8.3b}$$

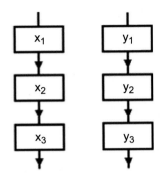

FIGURE 8.1

Two uncoupled logistic maps.

This model supposes that the environment consists of two patches between which the entities diffuse (Fig. 8.2a). Such coupling tends to equalize the instantaneous states of the entities (*diffusive* coupling). Let us assume that there is a density-dependent phase followed by a dispersal phase. The density-dependent phases are modeled by the logistic map, and the dispersal phase by a simple exchange of a fixed proportion of the populations. The parameter c_1 is a measure of the diffusion of individuals between the two patches, with $0 \leq c_1 \leq 1$. In Eqs. (8.3a) and (8.3b), it is assumed that the environment is homogeneous; hence the parameter r is the same for both patches. This model is designed similarly in spirit to that of Hassell et al. (1991) whose host—parasitoid model consists of a pair of variables at each site of at least 900 lattice sites. However, r does not need to be constant. Following Mihailović et al. (2012, 2011b), the parameter r is used to be different in Eqs. (8.3a) and (8.3b) that serves coupled maps representing energy exchange processes between two heterogeneous environmental interfaces regarded as biophysical complex systems. So, the dynamics of this simpler two-dimensional, two-parameter

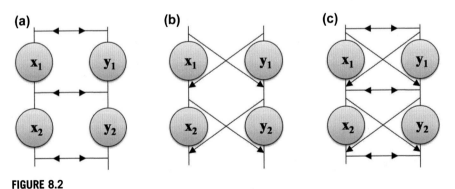

FIGURE 8.2

Schematic diagram of the diffusive (a), linear (b), and (c) combined coupling in environmental interfaces modelling.

system will be much easier to understand and the insights gained by studying it should shed light on the more complex systems.

There are mathematically simpler ways to couple two logistic maps. For example, we could have *linear* coupling (Fig. 8.2b)

$$x_{n+1} = f(x_n) + c_2(y_n - x_n), \tag{8.4a}$$

$$y_{n+1} = f(y_n) + c_2(x_n - y_n). \tag{8.4b}$$

Another popular form of coupling is a bilinear coupling, with the linear terms in (8.4) replaced by $c_2 x_n y_n$ terms (Lloyd, 1995). These forms of the coupled logistic maps have been studied previously using both numerical (Kaneko, 1983; Ferretti and Rahman, 1988; Satoh and Aihara, 1990) and analytic techniques (Sakaguchi and Tomita, 1990).

Finally, it is possible that both types of coupling are present, and then it will be the *combined* coupling (Fig. 8.2c), that can be written in the form

$$x_{n+1} = f(x_n) + c_1(f(y_n) - f(x_n)) + c_2(y_n - x_n), \tag{8.5a}$$

$$y_{n+1} = f(y_n) + c_1(f(x_n) - f(y_n)) + c_2(x_n - y_n). \tag{8.5b}$$

It appears that there is no necessity to invent some other types of coupling, in the same sense that this equation serves as a universal model of weakly coupled systems (Ivanova and Kuznetsova, 2002; Mihailović et al., 2012) that can be broadly used in environmental interfaces modelling.

8.2 AN EXAMPLE OF DIFFUSIVE COUPLING: INTERACTION OF TWO ENVIRONMENTAL INTERFACES ON THE EARTH'S SURFACE

As an example of diffusive coupling, here we consider the Earth's surface as an environmental interface. For this interface, visible radiation provides almost all of the received energy. Some of the radiant energy is reflected back to the space. The interface also radiates some of the energy received from the Sun. The quantity of the radiant energy remaining on the environmental interface is the net radiation, which drives physical processes important to our further considerations. Since all of the energy transfer processes occur in the finite time interval, the energy balance equation at any environmental interface can be written in terms of finite differences of ground and air temperatures and then, under some conditions, further transformed into the logistic equation (Mihailović et al., 2001; Mihailović, 2010).

We start with a simplified case of one bare soil—atmosphere interface (see Figure 1(f)). Our basic equation is the energy balance equation. Since all of the energy transfer processes occur in the finite time interval we shall immediately write this equation in terms of finite differences, i.e., in the form of difference equation

$$\mathbf{D}T_i = F_n, \tag{8.6}$$

where $\mathbf{D}$ is the finite difference operator defined as $\mathbf{D}T_i = (T_{i,n+1} - T_{i,n})/\Delta t$, T_i is the environmental interface temperature, n is the time level, Δt is the time step, $F_n = (R_n - H_n - E_n - S_n)/c_i$ is defined at the nth time level, R is the net radiation, H and E are the sensible and latent heat, respectively, transferred by convection and S is the heat transferred by conduction into deeper layers of underlying matter while c_i is the environmental interface soil heat capacity per unit area. In Eq. (8.6) the sensible heat flux is calculated as $C_H(T_i - T_a)$, where C_H is the sensible heat transfer coefficient and $T_a(t)$ is the gas temperature given as the upper boundary condition. The heat transferred into underlying matter is calculated as $C_D(T_i - T_d)$ where C_D is the heat conduction coefficient while $T_d(t)$ is the temperature of deeper layer of the underlying matter that is given as the lower boundary condition. Following Bhumralkar (1975) the net radiation term can be represented as $C_R(T_i - T_a)$ where C_R is the radiation transfer coefficient. According to Mihailović et al. (2001) for small differences of T_a and T_d, the expression for the latent heat flux can be written in the form $C_L f(T_a)$ $[b(T_i - T_a) + b^2(T_i - T_a)^2/2]$, where C_L is the latent heat transfer coefficient, $f(T_a)$ is the gas vapor pressure at saturation, and b is a constant characteristic for a particular gas. Calculation of time-dependent coefficients C_R, C_L, and C_D can be found in Monteith and Unswort (1990). After collecting the terms in Eq. (8.6) we get

$$\mathbf{D}T_i = A_1(T_{i,n} - T_{a,n}) - A_2(T_{i,n} - T_{a,n})^2 - A_3(T_{i,n} - T_{d,n}), \tag{8.7}$$

Where $A_1 = [C_R - C_H - bC_L\ f(T_a)]/c_i$, $A_2 = C_Lb^2f(T_a)/(2c_i)$, and $A_3 = C_D/c_i$ are coefficients also depending on Δt. With $\Delta t_p = 1/(A_1 - C_D/c_i)$, we indicate the scaling time range of energy exchange at the environmental interface including coefficients that express all kinds of energy reaching the environmental interface. For any chosen time interval, for solving Eq. (8.7), there always exists $\Delta t_{p,l} = \min[\Delta t_p(c_i, C_R, C_H, C_L)]$ when energy at the environmental interface which is exchanged in the fastest way by radiation, convection, and conduction. If we define dimensionless time $\tau = \Delta t_p/\Delta t_{p,l}$ and if we use for lower boundary condition $T_{d,n} = T_{a,n} - (c_i/C_D)\mathbf{D}T_a$ then Eq. (8.7) after some transformations take the form of a logistic equation, i.e.,

$$x_{n+1} = rx_n(1 - x_n), \tag{8.8}$$

where the symbols introduced have the following meaning: x is the dimensionless temperature (Mihailović, 2010), while the logistic parameter $r = 1 + \tau$ takes values from the interval $1 < r < 4$.

Under the aforementioned conditions Eq. (8.8) represents vertical energy exchange over the uniform environmental interface. However, in the nature usually we encounter mixture of two or more environmental interface, for example grid-box surface covered by different land covers. They will interact horizontally exchanging the energy between them. We consider the case for two interacting environmental interfaces represented by two logistic maps having the form

$$x_{n+1} = (1 - c)r_1x_n(1 - x_n) + cr_2y_n(1 - y_n) \tag{8.9a}$$

$$y_{n+1} = (1 - c)r_2y_n(1 - y_n) + cr_1x_n(1 - x_n), \tag{8.9b}$$

where x_n and y_n are dimensionless temperatures of surfaces covered, for example by the vegetation fractional covers c and $1 - c$, respectively.

We calculate the Lyapunov exponent λ to examine the behavior of the coupled maps given by Eqs. (8.9a) and (8.9b) in dependence on the coupling parameter c (Fig. 8.3). Each point in this graph is obtained by iterating many times from the initial condition to eliminate transient behavior and then averaging over another 50,000 iterations starting from the initial conditions $x_0 = 0.2$ and $y_0 = 0.25$ with $500c$ values. This simple analysis, where we consider only Lyapunov exponent, shows a very interesting feature of two coupled logistic maps representing interaction of two environmental interfaces through energy exchange between them. Thus, when the coupling parameter c is smaller and the logistic parameter r_2 is greater (closer to r_1), then the coupled maps are in the chaotic regime. Physically it means that strong vertical exchange over both environmental interfaces does not allow horizontal exchange (line $r_2 = 0.5r_1$ and line $r_2 = 0.75r_1$ in Fig. 8.3). Moreover, when the logistic parameter r_2 has the same value as r_1, the Lyapunov exponent is always positive expressing the fact that the considered dynamic system is in the chaotic regime for any value of c.

8.3 THE LINEAR COUPLING

For most of the biological populations, linear coupling is not the applicable model because it involves mixing of generations. However, in human populations for

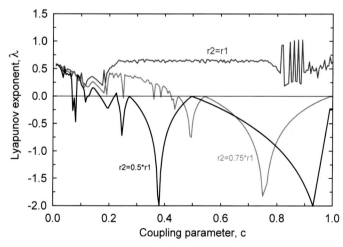

FIGURE 8.3

Lyapunov exponent of the coupled maps representing the interaction between two environmental interfaces, as a function of the coupling parameter c (ranging from 0 to 1) for different values of logistic parameters. The logistic equation with the logistic parameter $r_1 = 3.99$ is coupled with the logistic equation having the following logistic parameters: $r_2 = 0.5r_1$ and $r_2 = 0.75r_1$.

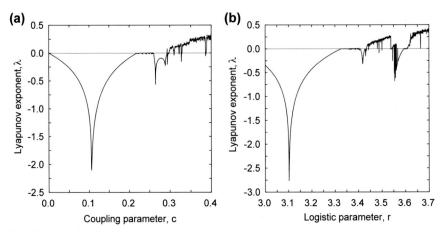

FIGURE 8.4

Lyapunov exponent for the coupled maps as a function of: (a) c ranging from 0 to 0.4 and with $r = 3.0$ and (b) r ranging from 3.0 to 3.7 and with $c = 0.06$. Each point was obtained by iterating many times from the initial condition to eliminate transient behavior and then averaging over another 50,000 iterations. Initial conditions $x_0 = 0.2, y_0 = 0.4$, with $5000c$ values for Fig. 8.4a and $200r$ values for Fig. 8.4b.

example, it can be used in representing dynamics of populations with permanent mixed flow of its members, especially in cases when inflow and outflow are of similar scale. The behavior of the Lyapunov exponent for two different initial conditions is shown in Fig. 8.4 for: (1) c in the range $0-0.4$ and $r = 3.0$ (Fig. 8.4a) and

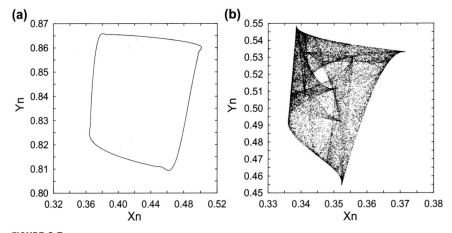

FIGURE 8.5

Plot of the iterates of the map (x_n, y_n) for initial point $x_0 = 0.2$, $y_0 = 0.4$; (a) quasiperiodic motion for $r = 3.378$, $c = 0.058$; (b) chaotic motion for $r = 3.613$, $c = 0.058$.

(2) r in the range 3.0−3.7 and $c = 0.06$ (Fig. 8.4b). From these figures is seen that for linear coupling chaotic behavior of the coupled maps occur for the higher values of c and r.

Plot of the iterates of the map (x_n, y_n) is depicted in Fig. 8.5. As it is seen from this figure, the coupled map displays quasiperiodic motion (Fig. 8.5a) and the chaotic one (Fig. 8.5b).

REFERENCES

Bhumralkar, C.M., 1975. Numerical experiments on the computation of ground surface temperature in an atmospheric general circulation model. J. Appl. Meteorol. 14, 1246−1258.

Ferretti, A., Rahman, N.K., 1988. A study of coupled logistic map and its applications in chemical physics. Chem. Phys. 119, 275−288.

Gyllenberg, M., Soderbacka, G., Ericsson, S., 1993. Does migration stabilize local population dynamics? Analysis of a discrete metapopulation model. Math. Biosci. 118, 25−49.

Hassell, M.P., Comins, H.N., May, R.M., 1991. Spatial structure and chaos in insect population dynamics. Nature 353, 255−258.

Hastings, A., 1993. Complex interactions between dispersal and dynamics: lessons from coupled logistic equations. Ecology 63, 1362−1372.

Hogg, T., Huberman, B.A., 1984. Generic behavior of coupled oscillators. Phys. Rev. A 29, 275−281.

Ivanova, A.S., Kuznetsov, S.P., 2002. Scaling at the onset of chaos in a network of logistic maps with two types of global coupling. Nonlinear Phenom. Complex Syst. 5, 151−154.

Kaneko, K., 1983. Transition from torus to chaos accompanied by frequency locking with symmetry breaking. Prog. Theor. Phys. 69, 1427−1442.

Kot, M., 1989. Diffusion-driven period-doubling bifurcations. BioSystems 22, 279−287.

Lloyd, A.L., 1995. The coupled logistic map: a simple model for effects of spatial heterogeneity on population dynamics. J. Theor. Biol. 173, 217−230.

Mihailović, D.T., 2010. Climate modelling beyond the complexity: challenges in model building. In: Alexandrov, V., Gajdusek, M.F., Knight, C.G., Yotova, A. (Eds.), Global Environmental Change: Challenges to Science and Society in Southeastern Europe. Selected Papers Presented in the International Conference Held 19−21 May 2008, Sofia (Bulgaria). Springer, Dordrecht, Heidelberg, London, New York.

Mihailović, D.T., Budincevic, M., Balaž, I., Mihailović, A., 2011a. Stability of intercellular exchange of biochemical substances affected by variability of environmental parameters. Mod. Phys. Lett. B 25, 2407−2417.

Mihailović, D.T., Budinčević, M., Kapor, D., Balaž, I., Perišić, D., 2011b. A numerical study of coupled maps representing energy exchange processes between two environmental interfaces regarded as biophysical complex systems. Nat. Sci. 1, 75−84.

Mihailović, D.T., Budinčević, M., Perišić, D., Balaž, I., 2012. Maps serving the combined coupling for use in environmental models and their behaviour in the presence of dynamical noise. Chaos Solitons Fractals 45, 156−165.

Mihailović, D.T., Kapor, D.V., Lalić, B., Arsenić, I., 2001. The chaotic time fluctuations of ground surface temperature resulting from energy balance equation for the soil-surface system. In: Abstracts of the 26th General Assembly of European Geophysical Society, 20−25 March, Nice, France.

Mihailović, D.T., Mimić, G., Arsenić, I., 2014. Climate models beyond the complexity: occurrence of the chaos and complexity in climate models. In: Ames, D.P., Quinn, N.W.T., Rizzoli, A.E. (Eds.), Proceedings of the 7th International Congress on Environmental Modelling and Software, June 15–19, San Diego, California, USA.

Monteith, J.L., Unsworth, M., 1990. Principles of Environmental Physics. Edward Arnold, London.

Sakaguchi, H., Tomita, K., 1990. Bifurcations of the coupled logistic map. Prog. Theor. Phys. 78, 305–315.

Satoh, K., Aihara, T., 1990. Numerical study on a coupled-logistic map as a simple model for a predator-prey system. J. Phys. Soc. Jpn. 59, 1184–1189.

Logistic difference equation on extended domain

9.1 LOGISTIC EQUATION ON EXTENDED DOMAIN: MATHEMATICAL BACKGROUND

With increased model complexity we need to introduce enough complexity to realistically model a process, but not so much that we cannot handle it. Various measures of complexity were developed to compare time series and distinguish regular (e.g., periodic), chaotic, and random behavior. The main types of complexity measures are Lyapunov exponent and entropies, among others. They are all defined for typical orbits of presumably ergodic dynamical systems, and there are profound relations between these quantities (Arshinov and Fuchs, 2003). For example, let us consider a dynamical system

$$X_{n+1} = S(X_n) \tag{9.1}$$

and make transformation T:T(X)=Y, where X and Y are vectors. If the Jacobi matrix is regular, either locally or globally, then for a transformed system

$$Y_{n+1} = G(Y_n) \tag{9.2}$$

information about the dynamics of this system can be obtained from the dynamics of the system (9.1) and vice versa. We deal with the difference equation

$$x_{n+1} = \rho x_n(1 - x_n); \quad \rho < 0, \tag{9.3}$$

whose dynamics can be completely described by the dynamics of the standard logistic difference equation

$$x_{n+1} = \mu x_n(1 - x_n); \quad 0 < \mu. \tag{9.4}$$

Namely, making successive transformations T_1 (symmetry), T_2 (homotety), and T_3 (translation) in Eq. (9.3), where $T_1(x) = -x$, $T_2(x) = (1 - 2/\rho)x$, and $T_3(x) = x + 1 - 1/\rho$, we get Eq. (9.4). Jacobian for all transformations is globally different from zero while μ and ρ are related by the equation $\mu = 2 - \rho$. Finally, for Eq. (9.3) we have the following properties: (1) $x = 0$ is the attractive fixed point for $-1 < \rho < 0$; (2) bifurcations start for $\rho < -1$ (Fig. 9.1a); (3) function $f(x) = \rho x(1 - x)$ maps interval $[1/\rho, 1 - 1/\rho]$ on itself for $-2 \leq \rho < 0$; (4) occurrence of

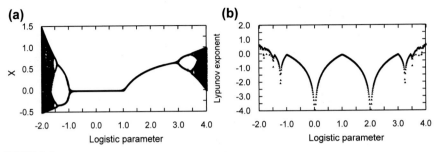

FIGURE 9.1

Bifurcation diagram (a) and Lyapunov exponent (b) of the difference Eqs. (9.3) and (9.4) as a function of the parameter of difference equation ranging in interval [−2, 4].

Reprinted with permission from Mihailović, D.T., Budinčević, M., Perišić, D., Balaž, I., 2012. Maps serving the combined coupling for use in environmental models and their behaviour in the presence of dynamical noise. Chaos Solitons Fractals 45, 156–165..

the chaotic behavior for $-2 \leq \rho < \rho_\infty$ where $\rho_\infty = 2 - \mu_\infty$ [$\mu_\infty \approx 3.56994$], and finally (5) orbits tend to infinity for $\rho < -2$. In Fig. 9.1a is depicted the bifurcation diagram of Eq. (9.3) on the whole domain [−2, 4].

We now analyze the occurrence of the chaos in solution of Eq. (9.3). Since a quantitative measure for identification of the chaos is the Lyapunov exponent λ, we will calculate its spectrum for Eq. (9.3) as a function of the parameter ρ ranging from −2 to 4. Their values are seen in Fig. 9.1b. This figure depicts two features of the Lyapunov exponent spectrum of Eq. (9.3): (1) its symmetry due the point logistic parameter having value 1 with the exact characteristics of the logistic equation spectrum going left and right toward to values −2 and 4, respectively, and (2) it is positive in the intervals $\rho \in [-2, 2 - \mu_\infty]$ and $\rho \in [\mu_\infty, 4]$ indicating chaotic fluctuations of x. However, inside $\rho \in [-2.0, 2.0 - \mu_\infty]$ and $[\mu_\infty, 4]$ intervals, there are a lot of opened periodical "windows" where $\lambda < 0$. Note, that if the logistic parameter belongs to the interval [−1, 1], then 0 is the attractive point, while if it belongs to the interval [1, 3] then the attractive point is $1 - 1/\mu$.

To measure the complexity and uncertainties of quantity time series described by Eq. (9.3), we use the sample entropy (SampEn) and the permutation entropy (PermEn). Sample Entropy, as a measure quantifying regularity and complexity, is believed to be an effective analyzing method of diverse settings that include both deterministic chaotic and stochastic processes. It can be very useful in the analysis of physiological, sound, climate, and environmental interface signals that involve relatively small amount of data (Kennel et al., 1992; Richman and Moorman, 2000; Lake et al., 2002). SampEn(m, r, N) is the negative natural log of the conditional probability that two sequences similar within a tolerance r for m points remain similar at the next point, where N is the total number of points and self matches are not included, i.e., SampEn(m, r, N) $= -\ln$ (A^m/B^m) where $A^m(r) = \sum_{i=1}^{N-m} A_i^m(r)/(N - m)$ and $B^m(r) = \sum_{i=1}^{N-m} B_i^m(r)/(N - m)$. A low value of SampEn is interpreted as one showing increased regularity or order in the data series. The threshold factor or filter r is an important parameter. In principle, with an infinite amount of data, it should approach zero. With finite amounts of data, or with measurement noise, r value typically varies between 10% and 20% of the time series standard deviation (Pincus, 1991).

Permutation Entropy (PermEn) of order $n \geq 2$ is defined as PermEn $= \sum p(\pi) \ln p(\pi)$ where the sum runs over all $n!$ permutations π of order n. This is the information contained in comparing n consecutive values of the time series. Consider a time series $\{x_t\}_{t=1,\ldots T}$. We consider all $n!$ permutations π of order n which are considered here as possible order types of n different numbers. For each π we determine the relative frequency $p(\pi) = \#\{t \mid 0 \leq t \leq T - n, (x_{t+1}, \ldots, x_{t+n})$ has type $\pi\}/(T - n + 1)$. This estimates the frequency of π as good as possible for a finite series of values. To determine $p(\pi)$ exactly, we have to assume an infinite time series $\{x_1, x_2, \ldots\}$ and take the limit for $T \to \infty$ in the above formula. This limit exists with probability 1 when the underlying stochastic process fulfills a very weak stationary condition: for $k \leq n$, the probability for $x_t < x_{t+k}$ should not depend on t. Permutation entropy as a natural complexity measure for time series behaves similar as Lyapunov exponents and is particularly useful in the presence of dynamical or observational noise (Brandt and Pompe, 2002).

Fig. 9.2 depicts SampEn of a single time series obtained from Eq. (9.3) as a function of the parameter ρ ranging from -2 to -1.4 (Fig. 9.2a) and from 3.4 to 4

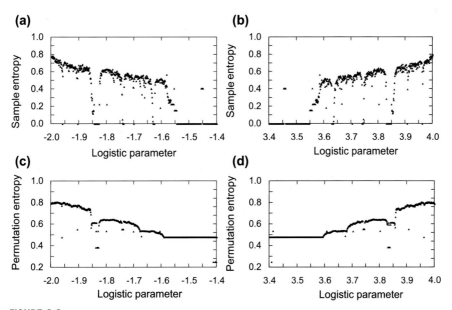

FIGURE 9.2

Sample (a and b) and permutation entropy (c and d) of the difference Eqs. (9.3) and (9.4) as a function of the parameter ρ ranged in intervals [2.0, 1.4] and [3.4, 4.0].

Reprinted with permission from Mihailović D.T., Budinčević, M., Kapor, D., Balaž, I., Perišić, D., 2011. A numerical study of coupled maps representing energy exchange processes between two environmental interfaces regarded as biophysical complex systems. Nat. Sci. 1, 75–84..

(Fig. 9.2b). Those two figures show output for this equation over a range of growth values, for sample length $m = 2$. It is clearly seen that there are some regions of stability around -1.83 and 3.83, respectively. We also computed permutation entropy. The test case used was, again, Eq. (9.3). Fig. 9.2c and d plot the computed PermEn versus the growth rate of parameter ρ, which is periodic for some regions and chaotic for others and some regions of stability around -1.83 and 3.83, respectively. Let us note that PermEn is very similar to the positive Lyapunov exponent (Figs. 9.2a vs 9.2c and 9.2b vs 9.2d).

9.2 LOGISTIC EQUATION ON EXTENDED DOMAIN IN COUPLED MAPS SERVING THE COMBINED COUPLING: A DYNAMICAL ANALYSIS

We consider the logistic equation on extended domain in coupled maps serving the combined coupling (8.5), i.e.,

$$x_{n+1} = (1 - c_1)\mu x_n(1 - x_n) + c_1\mu y_n(1 - y_n) + c_2(y_n - x_n) \tag{9.5a}$$

$$y_{n+1} = (1 - c_1)\mu y_n(1 - y_n) + c_1\mu x_n(1 - x_n) + c_2(x_n - y_n). \tag{9.5b}$$

We use Lyapunov exponent and cross-sample entropy (Cross-SampEn), included in the archive of dynamical analysis, to analyze the system (9.5). We calculate the Lyapunov exponent λ to see the behavior of the coupled maps given by Eqs. (9.5a) and (9.5b) depending on different values of the coupling parameters c_1 and c_2. Fig. 9.3 depicts Lyapunov exponent for the coupled maps as a function of these parameters ranging from 0 to 0.9, with the increment of 0.001, and the parameter μ ranging from 1.95 to -1.4 and from 3.4 to 3.9 with the increment of 0.01. Each point was obtained by iterating 1000 times from the initial condition to eliminate transient behavior and then averaging over another 600 iterations starting from initial condition $x_0 = 0.20$ and $y_0 = 0.25$. This simple analysis, where we consider Lyapunov exponent, shows a very interesting feature of these two coupled maps. From this figure is seen that there exist two distinguished regions with positive as well as negative values of λ where for c_2 below 0.5 the Lyapunov exponent of the coupled maps is always negative.

Cross-SampEn measure of asynchrony is a recently introduced technique for comparing two different time series to assess their degree of asynchrony or dissimilarity (Pincus and Singer, 1995; Pincus et al., 1996). Let $u = [u(1), u(2),...u(N)]$ and $v = [v(1),v(2),...v(N)]$ fix input parameters m and r. Vector sequences: $x(i) = [u(i),\quad u(i + 1),...u(i + m - 1)]\quad$ and $\quad y(j) = [v(j), v(j + 1),...v(j + m - 1)]$ and N is the number of data points of time series, $i, j = N - m + 1$. For each $i \le N - m$ set $B_i^m(r)(v\|u) = $ (number of $j \le N - m$ such that $d[x_m(i), y_m(j)] \le r)/(N - m)$, where j ranges from 1 to $N - m$. And then $B^m(r)(v\|u) = \sum_{i=1}^{N-m} B_i^m(r)(v\|u)/(N - m)$ which is the average value of $B_i^m(v\|u)$. Similarly we define A^m and A_i^m as $A_i^m(r)(v\|u) = $ (number of

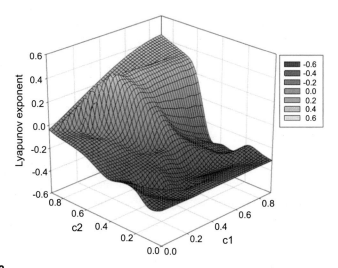

FIGURE 9.3

Lyapunov exponent for the combined coupling given by Eqs. (9.5a) and (9.5b) for values of c_1 and c_2 ranged between 0 and 0.9 and μ taken in intervals $[-1.95, -1.4]$ and $[3.4, 3.9]$.

Reprinted with permission from Mihailović, D.T., Budinčević, M., Perišić, D., Balaž, I., 2012. Maps serving the combined coupling for use in environmental models and their behaviour in the presence of dynamical noise.
Chaos Solitons Fractals 45, 156–165..

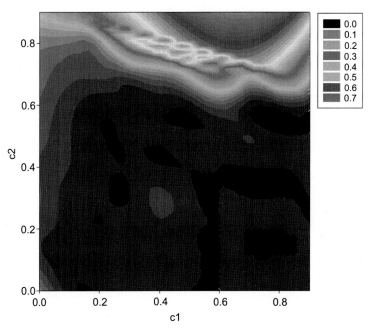

FIGURE 9.4

Cross-SampEn for the combined coupling given by Eq. (9.5a) and (9.5b) for values of parameters c_1 and c_2 ranged between 0 and 0.9 and μ taken in intervals $[-1.95, -1.4]$ and $[3.4, 3.9]$.

Reprinted with permission from Mihailović, D.T., Budinčević, M., Perišić, D., Balaž, I., 2012. Maps serving the combined coupling for use in environmental models and their behaviour in the presence of dynamical noise.
Chaos Solitons Fractals 45, 156–165..

$j \leq N - m$ such that $d[x_m(i), y_m(j)] \leq r)/(N - m)$ and $A^m(r)(v\|u) = \sum_{i=1}^{N-m} A_i^m(r)(v\|u)/(N - m)$ which is the average value of $A_i^m(v\|u)$. Finally, we have

$$\text{Cross-SampEn} = -\ln\{A^m(r)(v\|u)/B^m(r)(v\|u)\} \tag{9.6}$$

We applied Cross-SampEn with $m = 5$ and $r = 0.05$ for x and y time series. Fig. 9.4 depicts that (c_1, c_2) phase space is covered with values of Cross-SampEn equal or very close to zero, corresponding to the region in Fig. 9.3, where λ is negative. It points out on a high synchronization between the coupled maps in that region. In the rest of the (c_1, c_2) phase space, the entropy is greater than zero corresponding to positive values of λ.

For analysis of influence of the dynamical noise in system (9.5a) and (9.5b) we chose system with the same values of coupling parameters $c_1 = c_2 = 0.5$. The Lyapunov exponent and Cross-SampEn for this choice of the parameters as a function of logistic parameter μ are depicted in Fig. 9.5.

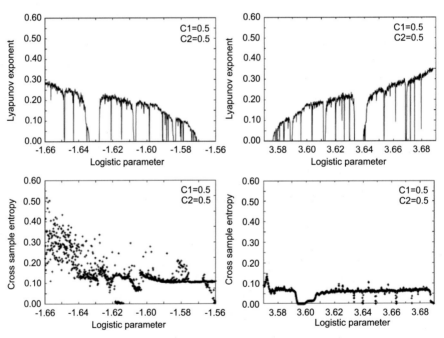

FIGURE 9.5

Lyapunov exponent and Cross-SampEn for the combined coupling given by Eq. (9.5a) and (9.5b) for $c_1 = c_2 = 0.5$ and logistic parameter μ taken in intervals $[-1.66, -1.56]$ and $[3.57, 3.69]$.

Reprinted with permission from Mihailović, D.T., Budinčević, M., Perišić, D., Balaž, I., 2012. Maps serving the combined coupling for use in environmental models and their behaviour in the presence of dynamical noise. Chaos Solitons Fractals 45, 156–165.

REFERENCES

Arshinov, V., Fuchs, C., 2003. Preface. In: Arshinov, V., Fuchs, C. (Eds.), Causality, Emergence, Self-organisation. NIAPriroda, Moscow, Russia, pp. 1–18.

Bandt, C., Pompe, B., 2002. Permutation entropy: a natural complexity measure for time series. Phys. Rev. Lett. 88, 174102.

Kennel, M.B., Brown, R., Abarbanel, H.D.I., 1992. Determining embedding dimension for phase-space reconstruction using a geometrical construction. Phys. Rev. A 45, 3403–3411.

Lake, D.E., Richman, J.S., Griffin, M.P., Moorman, J.R., 2002. Sample entropy analysis of neonatal heart rate variability. Am. J. Physiol. Regul. Integr. Comp. Physiol. 283, R789–R797.

Mihailović, D.T., Budinčević, M., Kapor, D., Balaž, I., Perišić, D., 2011. A numerical study of coupled maps representing energy exchange processes between two environmental interfaces regarded as biophysical complex systems. Nat. Sci. 1, 75–84.

Mihailović, D.T., Budinčević, M., Perišić, D., Balaž, I., 2012. Maps serving the combined coupling for use in environmental models and their behaviour in the presence of dynamical noise. Chaos Solitons Fractals 45, 156–165.

Pincus, S.M., 1991. Approximate entropy as a measure of system complexity. Proc. Natl. Acad. Sci. U.S.A. 88, 2297–2301.

Pincus, S., Singer, B.H., 1995. Randomness and degrees of irregularity. Proc. Natl. Acad. Sci. U.S.A. 93, 2083–2088.

Pincus, S.M., Mulligan, T., Iranmanesh, A., Gheorghiu, S., Godschalk, M., Veldhuis, J.D., 1996. Older males secrete luteinizing hormone and testosterone more irregularly, and jointly more asynchronously, than younger males. Proc. Natl. Acad. Sci. U.S.A. 93, 14100–14105.

Richman, J.S., Moorman, J.R., 2000. Physiological time-series analysis using approximate entropy and sample entropy. Am. J. Physiol. Heart Circ. Physiol. 278, H2039–H2049.

Generalized logistic equation with affinity: its use in modelling heterogeneous environmental interfaces

10.1 GENERALIZED LOGISTIC MAP WITH AFFINITY: MATHEMATICAL BACKGROUND

As mentioned previously, the exchange of biochemical substance is defined in a diffusion-like manner. The dynamics of intracellular behavior is expressed as a logistic map $f(x) = rx(1 - x)$, where x is the concentration of a given substance in a cell, while r is a logistic parameter, $0 < r \leq 4$ (Devaney, 2003; Gunji and Kamiura, 2004; Mihailović et al., 2013). However, instead of this map we use another form, which includes a parameter p that represents the cell affinity. By introducing this parameter we formalize an intrinsic property of the cell that includes (1) affinity of genetic regulators toward arriving signals which determine intensity of cellular response and (2) affinity for uptake of signaling molecules as it is shown in Fig. 7.3 according to Mihailović et al. (2011). Namely, from the logistic equation follows that the level of intracell dynamics is the most intensive for the concentration $x_{max,p=1} = 0.5$ that comes from $df(x)/dx = r(1 - 2x) \equiv 0$ and $p = 1$. If we wish to generalize this condition one possible way of introducing the parameter p is to postulate the condition $df(x, p)/dx = rpx^{p-1}(1 - 2x^p)$. It means that the level of concentration $x_{max,p}$ when the intracellular dynamics is the most intensive depends on the cell affinity p, i.e., $x_{max,p} = 1/2^{1/p}$. Calculating the integral $f(x, p) = \int rpx^{p-1}(1 - 2x^p)dx$ we get $f(x) = rx^p(1 - x^p)$. We will call this map: the map with the *cell affinity*, where $0 < x \leq 1$ and $0 < p \leq 1$. Fig. 10.1 depicts that the intensity of the intracellular dynamics starts to grow when the cell affinity p is around 0.1. Note, that this map can be employed in modelling other environmental interface systems, in which parameter p formalizes some of their intrinsic properties (hereafter referred to as *affinity* parameter).

To avoid double approximation (see Section 3.2) the cell dynamics is expressed here as a difference equation, i.e.,

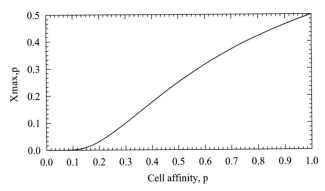

FIGURE 10.1

Dependence of concentration $x_{max,p}$ on the cell affinity p (Mihailović et al., 2013).

$$f\left(x_{i,n}\right) = rx_{i,n}^{p}\left(1 - x_{i,n}^{p}\right). \tag{10.1}$$

The dynamics of this map (Eq. 10.1) is governed by two parameters, p and r, which express cell affinity and influence of environment, respectively. We analyze this map using Lyapunov exponent, λ, and sample entropy (SampEn). Fig. 10.2 shows the variations of λ and SampEn against cell affinity p, with: (1) r randomly chosen in the interval (3,4), (2) initial condition $x_0 = 0.25$, and (3) $m = 5$ and $r_s = 0.05$. For each x, 10^4 iterations of the map (Eq. 10.1) are applied, and the first 10^3 steps are abandoned. Fig. 10.2(a) indicates that for $p < 0.2$, λ takes negative values while for higher values of p there is a frequent occurrence of regions with instability ($\lambda > 0$). As it is seen from Fig. 10.2(a) the SampEn entropy with a biased statistics is closer to 1 for larger values of p. From this figure it is also seen that SampEn (Fig. 10.2(b)) follows the Lyapunov coefficient (Fig. 10.2(a)), i.e., takes values nearly zero when $\lambda < 0$. Above analysis indicates that for the lower levels of affinity, the generalized logistic map with the cell affinity (Eq. 10.2) can better simulate the biochemical substance exchange in cell than it is possible by the ordinary logistic equation. This is particularly pronounced for $p < 0.2$.

10.2 UNCERTAINTIES IN MODELLING THE TURBULENT ENERGY EXCHANGE OVER THE HETEROGENEOUS ENVIRONMENTAL INTERFACES — SCHMIDT'S PARADOX

The Earth's climate system, joining the physical and chemical components of the atmosphere, ocean, land surface, and cryosphere is the target of global climate models

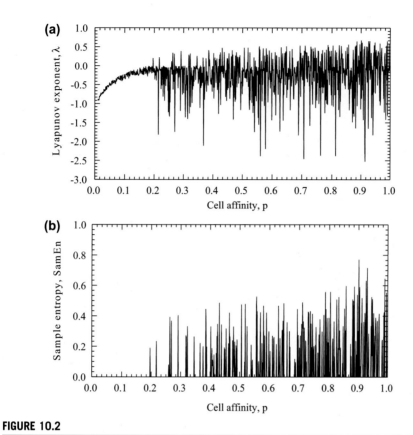

FIGURE 10.2

The dependence on the cell affinity p of (a) the Lyapunov exponent, λ, and (b) the sample entropy (SampEn) of intracell dynamics, simulated by Eq. (10.1) (Mihailović et al., 2013).

whose objective is to correctly simulate the spatial variation of climate in some average sense. The main current issues in modelling the global climate system can be summarized as follows. (1) Chaos. This is a deterministic chaos whose sources are nonlinearities in the Navier–Stokes equations and their sensitivity to initial conditions. However, in addition, in climate models, coupling of a nonlinear model over one environmental interface to a nonlinear model over another environmental interface (for example, land and ocean), gives rise to something much more complex than the deterministic chaos of the weather model, leading to bifurcation, instability, and chaos (Annan and Connolley, 2005). (2) Confidence in climate models. This issue is how well the climate model reproduces reality, that is, whether the model works and is it fit for its intended purpose (Curry, 2011). (3) Climate model imperfection. The meaning of this issue can be addressed to the fact that our understanding of, and ability to simulate, the Earth's climate is rather limited. The climate model imperfection is divided into two types: uncertainty and inadequacy. The term model uncertainty

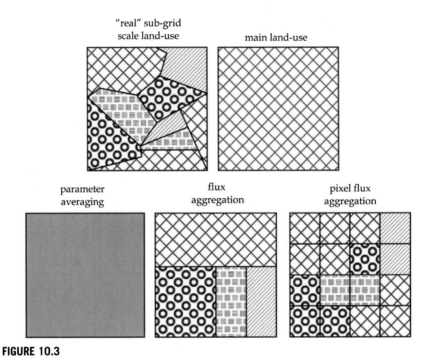

FIGURE 10.3

Schematic diagram illustrating how the subgrid scale surface patch-use classes are treated within the different parameterization schemes for subgrid scale surface fluxes in climate and other models of different scales (Mihailović et al., 2015).

means that we cannot reliably choose parameters, which will give the most informative results; the term inadequacy means that before we run any simulation of the future, we know in advance that models are not realistic representations of many key aspects of the real system (Stainforth et al., 2007). (4) Subgrid scale parameterization. Except uncertainty in model parameters and initial conditions, the model uncertainty is associated with subgrid scale parameterizations (e.g., boundary layer turbulence, cloud microphysics) generating systematic errors in meteorological fields which are obtained by the downscaling procedure (Mihailović et al., 2015).

The approaches for calculating the turbulent transfer of momentum, heat, and moisture from a grid-box composed of heterogeneous surfaces to the atmosphere can be classified as follows (Fig. 10.3). (1) Main land-use. In this approach surface fluxes are calculated based on the characteristics of the largest fraction of the grid-box. This method is computationally very economical but in pronouncedly heterogeneous areas it should not be applied. (2) Parameter averaging, where grid-box mean radiation, aerodynamic, physiological, morphological, and soil parameters are averaged for all patches in the grid-box. Note that this approach, which is the favorable one concerning time-consuming, can produce even worse results than the main land-use approach, especially for very coarse resolutions. (3) Flux aggregation is the method where the fluxes are averaged over the grid-box. (4) Pixel flux aggregation. This mosaic flux aggregation method is optimal for future pixel-based patch-use data sets, although it is computationally relatively expensive (Bohnenstengel, 2012). (5) Combined method is a combination of (2) and (3) methods (Hess and McAvaney, 1998).

To take into account the effects of heterogeneity of the underlying surface using the method of *parameter averaging* we shortly describe (1) an expression for a general equation for the wind speed profile in a roughness sublayer under neutral conditions and (2) method for calculating the aggregated values of the aerodynamic characteristics that is elaborated in Mihailović et al. (1999). As suggested by Mihailović et al. (1999), who introduced an expression for the mixing length over a grid-box consisting of vegetated and nonvegetated surfaces, the aggregated mixing length l_m^a at level z above a grid-box consisting of the heterogeneous surface defined above, might be represented by some combination of their single mixing lengths. If, as a hypothesis, we assume a linear combination weighted by fractional cover, according to mixing length theory we can define l_m^a as

$$l_m^a = k \left[\sum_{i=1}^{K} \sigma_i \alpha_i (z - d_i) + \sum_{i=1}^{L} \delta_i z + \sum_{i=1}^{M} \nu_i z \right], \tag{10.2}$$

where k is the von Karman's constant taken to be 0.41; σ_i, δ_i, and ν_i are partial fractional covers for vegetation, bare/water surface, and urban part, respectively, while d_i is zero displacement height for the ith vegetative part in the grid-box (Mihailović et al., 2002). The nonuniformity of the vegetative part is expressed with the surface vegetation fractional cover σ_i representing the i type of vegetation cover that fills the grid-box. Their sum takes values from 0 (when only solid surface or water are present) to 1 (when the ground surface is totally covered by plants). The nonuniformity of bare soil and water portion (bare soil, sea, river, lake, water catchments) of the grid-box will be denoted by symbols δ_i while ν_i represents the urban fractional cover; the total sum of all these fractional covers must be equal to 1. Parameter α_i is the dimensionless constant that depends on morphological and aerodynamic characteristics of the vegetative cover whose values vary according to the type of vegetative cover (Mihailović et al., 1999).

Starting from the expression for the momentum transfer coefficient K_m for nonuniform surface in the grid-box, i.e., $K_m = l_m^a u_*^a$, where u_*^a is the friction velocity, Mihailović et al. (2002) derived a wind profile $u(z)$ in the roughness sublayer above the heterogeneously built grid-box under neutral conditions, which can be written in the form

$$u(z) = \frac{u_*^a}{k\Lambda} \ln \frac{z - D}{Z_0} \tag{10.3}$$

where, Z_0 and D are roughness length and displacement height above nonhomogeneously covered grid-box, and Λ is a parameter describing the departure of a real wind profile in roughness sublayer from the classical logarithmic relationship. If we consider a nonuniform underlying surface, particularly over the urban area, whose nonuniformity is expressed with the surface fractional covers σ_i (for vegetation), δ_i (for bare soil or water surface) and ν_i (for urbanized surface), and K, L, and M are total numbers of homogeneous patches with vegetation, bare soil, and urban land, respectively, then aggregated values for Z_0, D, and Λ can be calculated from the following expressions:

$$Z_0 = \frac{1}{\Lambda}\left(\sum_{i=1}^{K}\frac{\sigma_i\alpha_i^2}{\sigma_i(\alpha_i-1)+1}z_{0v,i} + \sum_{i=1}^{L}\delta_i z_{0u,i} + \sum_{i=1}^{M}\nu_i z_{0w,i}\right) \tag{10.4}$$

$$D = \frac{1}{\Lambda}\sum_{i=1}^{K}\sigma_i\alpha_i d_i \tag{10.5}$$

$$\Lambda = \sum_{i=1}^{K}\sigma_i\alpha_i + \sum_{i=1}^{L}\delta_i + \sum_{i=1}^{M}\nu_i, \tag{10.6}$$

where $z_{0v,i}$, $z_{0u,i}$, and $z_{0w,i}$ are roughness lengths for vegetation, solid, and water and urban part, respectively.

When using either method (3) or (5) then we encounter the occurrence of Schmidt's paradox (Lettau, 1979; Mihailović and Kapor, 2012). It describes situation when small regions of evident surface heterogeneity, with intense upward-directed turbulent-sensible heat fluxes can take over the grid-area averaged value of these fluxes, while the mean gradient of potential temperature still indicates an overall stable stratification between the surface and the lowest climate model level. Further, this situation causes arising of counter-gradient heat transfer. Thus, the subgrid scale surface flux parameterization has to capture this phenomenon to derive a representative grid-box averaged flux and further yield the correct mean temperature gradient by allowing a transport of heat in the direction opposite to the mean gradient. To avoid this situation many attempts have been made (Lamb and Durran, 1978; Hess and McAvaney, 1998). However, regardless of which approach is applied, the physics of the countergradient relationship must still be accounted for.

In this section we describe a combined method that combines the parameter averaging (2) and the flux aggregation (3) approaches in calculating the surface temperature of the grid-box (Mihailović et al., 2005). In the further text we use angular brackets to indicate an average of certain physical quantity A over the grid-box, i.e.,

$$\langle A \rangle = \sum_{i=1}^{NP}\xi_i A_i \tag{10.7}$$

where NP is the number of patches within a grid-box and ξ_i is the fractional cover for the ith surface type. In parameter aggregation approach the mean sensible heat flux $\langle H_0 \rangle$ and latent heat flux $\langle \lambda E_0 \rangle$, calculated over the grid-box, where λ is the latent heat of vaporization, are found by assuming, for example, the aerodynamic resistance representation, i.e.,

$$\langle H_0 \rangle = \rho c_p \frac{\langle T_0 \rangle - T_a}{\langle r_a \rangle} \tag{10.8}$$

and

$$\lambda \langle E_0 \rangle = \frac{\rho c_p}{\gamma} \frac{\langle e_0 \rangle - e_a}{\langle r_a \rangle}, \tag{10.9}$$

where ρ is the air density, c_p is the specific heat of air at constant pressure, γ the psychrometric constant, r_a the resistance between canopy air or ground surface and the atmospheric lowest model level, T_a the air temperature, and e_a the water vapor pressure. The subscript a indicates the atmospheric lowest model level and the subscript 0 indicates the surface or environment inside the canopy. The $\langle r_a \rangle$ is defined as

$$\langle r_a \rangle = \langle r_s \rangle \delta \mu + \frac{1}{k \langle u_* \rangle} \ln \frac{z_a - \langle d \rangle (1 - \delta)}{z_b - \langle d \rangle (1 - \delta)} \tag{10.10}$$

where $\langle r_s \rangle$ is the bare soil surface resistance, δ ($\delta = 1$ for the bare soil, water and urban fraction; $\delta = 0$ for vegetative surface) and μ ($\mu = 1$ for the bare soil fraction; $\mu = 0$ for vegetative surface, water and urban fraction) the parameters, u_* the friction velocity, z_a is the height of the lowest atmospheric model level, z_b a height taking values z_0 and H (canopy height) for the barren/urban/water and vegetative part, respectively. We parameterize $\langle r_s \rangle$ following Sun (1982) and Mihailović and Kallos (1997). If the surface fluxes aggregation approach is applied then the mean surface fluxes are given by

$$\langle H_0 \rangle = \rho c_p \sum_{i=1}^{NP} \xi_i \frac{T_{m,i} - T_a}{r_{a,i}} \tag{10.11}$$

$$\lambda \langle E_0 \rangle = \frac{\rho c_p}{\gamma} \sum_{i=1}^{NP} \xi_i \frac{e_{m,i} - e_a}{r_{a,i}}, \tag{10.12}$$

where the subscript m refers to the single patch in the grid-box (vegetation, bare soil, water and urbanized area) whose temperature is calculated by the surface scheme. However, according to Hess and McAvaney (1998), it seems that averaging temperatures over different patches in the grid-box, rather than the sensible heat flux, can be the source of problems. There is an alternative method for calculation of temperature and water vapor pressure diagnostically from Eqs. (10.8) and (10.9), when the grid-averaged fluxes are known from Eqs. (10.11) and (10.12). Since we have three unknowns, it is necessary to introduce the associated parameter and flux aggregation equations for momentum

$$\langle u_*^2 \rangle = \left[\frac{k\Lambda}{\ln \frac{z_a - \langle D \rangle}{\langle Z_0 \rangle}} \right]^2 \langle F(\langle Ri_b \rangle, u_a, \langle T_0 \rangle, T_a) \rangle u_a^2 \tag{10.13}$$

$$\langle u_*^2 \rangle = \sum_{i=1}^{NP} \xi_i \left[\frac{k\Lambda_i}{\ln \frac{z_a - D_i}{Z_{0,i}}} \right]^2 F_i\left[Ri_{b,i}, u_a, T_{m,i}, T_a\right] u_a^2 \tag{10.14}$$

where Z_0, D, and Λ are given by Eqs. (10.4)–(10.6), F represents the nonneutral modification, Ri_b the bulk Richardson number, while u_a and T_a are the wind speed and air temperature, respectively, at the lowest model level. Now, the mean averaged momentum flux is calculated from Eq. (10.14). If this value is substituted into Eq. (10.13), the resulting equation can be solved for $\langle F \rangle$. The aggregation parameter version of the aerodynamic resistance r_a can be now determined (since $\langle F \rangle$, $\langle Z_0 \rangle$, $\langle D \rangle$, and $\langle H \rangle$ are all known). Thus,

$$\langle r_a \rangle = r_s \delta \mu + \frac{\frac{k\Lambda}{\ln \frac{z_a - \langle D \rangle}{\langle Z_0 \rangle}}}{\left\{ \sum_{i=1}^{NP} \xi_i \left[\frac{k\Lambda_i}{\ln \frac{z_a - D_i}{Z_{0,i}}} \right]^2 F_i\left[Ri_{b,i}, u_a, T_{m,i}, T_a\right] \right\}^{1/2}} \ln \frac{z_a - \langle D \rangle (1 - \delta)}{z_b - \langle D \rangle (1 - \delta)}. \tag{10.15}$$

Hence, the grid-averaged surface values of temperature and water vapor pressure can be found from Eqs. (10.8) and (10.9), i.e.,

$$\langle T_0 \rangle = \frac{\langle r_a \rangle \langle H_0 \rangle}{\rho c_p} + T_a \tag{10.16}$$

$$\langle e_0 \rangle = \frac{\langle r_a \rangle \gamma \lambda \langle E_0 \rangle}{\rho c_p} + e_a. \tag{10.17}$$

Fig. 10.4 depicts comparison differences in surface temperatures obtained by two different land surface schemes incorporated in the 1-D model (Mihailović et al., 2005). The first one was a simulation with the MM5 land-surface parameterization (Dudhia, 1993) while the second simulation was with the LAPS land surface scheme (Mihailović, 1996) according to Eq. (10.16). In this equation the average total heat flux over the grid-box $\langle H_0 \rangle$ is calculated using a simple linear average of partial sensible heat fluxes (Fig. 10.4(a)), while the aggregated values of the corresponding aerodynamic characteristics over the grid-box are calculated by Eqs. (10.4)–(10.6). The simulated values of the surface sensible heat fluxes and surface temperature are presented in Fig. 10.4(a) and (b), respectively. Fig. 10.4 depicts the temporal variation of the surface temperature obtained by the LAPS and MM5 schemes of 1-D model that are compared with the observations. The observations are derived from the temperature profile measured over the Baxter site. From this figure is seen that no significant differences occur between compared cover types during the period between midnight and 0800 UTC. The minimum of the surface temperature, for both parameterizations, occurs approximately at the same time, i.e., around 0500 UTC. Also, is seen that there is no huge difference in their values, although the minimum obtained by the LAPS scheme is slightly lower than minimum calculated by MM5 scheme. However, during the 0800–1400 UTC time interval the surface temperature, for both parameterizations, overestimates the observed

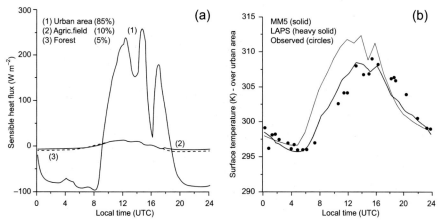

FIGURE 10.4

(a) The diurnal variation of sensible heat fluxes, obtained by the LAPS land surface scheme (Mihailović, 1996) and (b) the surface temperatures over the grid-box (85% urban part, 15% forest, and 5% agricultural field) representing the Baxter site, Philadelphia, PA (USA), for July 17, 1999. The calculations of surface temperatures are performed using the MM5 surface parameterization (main land use, Dudhia, 1993) and LAPS land surface scheme (combined method, Mihailović et al., 2005) in 1-D model.

Reprinted with permission from (Mihailović, D.T., Rao, S.T., Alapaty, K., Ku, J.Y., Arsenic, I., Lalic, B., 2005. A study on the effects of subgrid-scale representation of land use on the boundary layer evolution using a 1-D model. Environ. Model. Softw. 20, 705–714.).

values although the LAPS scheme gives values that are much closer to them. After 1600 UTC, both schemes underestimate the surface temperature in comparison with the observations.

10.3 USE OF THE GENERALIZED LOGISTIC EQUATION WITH AFFINITY IN MODELLING THE TURBULENT ENERGY EXCHANGE OVER THE HETEROGENEOUS ENVIRONMENTAL INTERFACES

The main difficulties arising from the inclusion of subgrid scale heterogeneities into surface flux parameterizations come from the nonlinear dependence of the surface fluxes on the surface layer characteristic. It means: (1) that turbulent flux spatially averaged over the grid-box can significantly differ from the flux determined from the "mean" surface characteristics and then applying the flux function to the artificial but homogeneous surface characteristic value and (2) that unique effect of several different fluxes on the main flow is not the result of simple averaging of spatial fluxes which represent the individual effects in the grid-box (Giorgi and Avissar, 1997; Bohnenstengel, 2012; Mihailović et al., 2015). These two deficiencies are known

FIGURE 10.5

The grid-box over the Prospect Park, NY, USA, illustrating the heterogeneous grid-box used in environmental model simulations. It consists from the following environmental interface surfaces (patches): (1) urban part (surrounding buildings), (2) vegetative part (mixture of trees and grass), and (3) water surface (lake).

Taken from the Google Maps (Mihailović, D.T., Kostić, V., Mimić, G., Cvetković, L., 2015. Stability analysis of turbulent heat exchange over the heterogeneous environmental interface in climate models. Appl. Math. Comput. 265, 79–90.).

as the aggregation effect. The inaccuracies can arise due to heterogeneity induced subgrid scale circulations, which are not explicitly resolved on the grid scale (dynamical effect). This effect needs to be accounted for by parameterizations for horizontal resolutions used for global and regional climate or weather forecast models.

In this chapter we demonstrate a use of the generalized logistic equation with affinity in analysis of the flux subgrid parameterization and its stability, which will be done in Section 22.2. For that purpose, here we add one more effect to these already mentioned. We call it patch size effect [environmental interface (EI) size effect] or shortly *EI size effect*, which comes from the size of a single EI in the grid-box (Fig. 10.5). Although this effect is implicitly included into two other ones we will consider it explicitly.

In nature, there are three mechanisms of energy transfer: radiation, conduction, and convection. For our purposes we will need: incoming and outgoing shortwave and longwave radiation, convection of heat and moisture into the atmosphere, and conduction of heat into deeper soil layers of ground. The dynamics of energy exchange, by the above-described mechanisms, under the energy balance equation can be expressed as a logistic map $f_r(x) = rx(1 - x)$, where x is the dimensionless temperature and r is a logistic parameter, $r \in (0, 4.0]$, (see Mihailović et al., 2012; Mimić et al., 2013). Here, r is the parameter representing a diversity of the aggregated energy flux intensities over the grid-box and, thus, it represents aggregation effect.

As a measure of diffusion of the energy exchange between jth and ith EI is the coupling parameter c_{ij} ($c_{i,j} \in [0.0, 1.0]$) which drives the dynamics between different EI in the grid-box (dynamical effect). Here, a natural requirement of such a model is that $\sum_{j \in N} c_{i,j} = 1$, for all $i \in N : \{1, 2, ..., n\}$ where n is the total number of EI patches in the grid-box. Therefore, the coupling parameter in ith EI can be written as

$$c_{i,i} = 1 - \sum_{j \neq i} c_{i,j}. \tag{10.18}$$

To complete the model we have to consider the EI size effect, i.e., contribution of the EI size to the energy exchange given by the logistic equation $f_r(x) = rx(1 - x)$. For that purpose we use the generalized form of this map

$$f_{r,p}(x) = rx^p(1 - x^p), \tag{10.19}$$

which includes a parameter p that represents the total turbulent energy exchange between a single EI and the surrounding environment. Since we take $p \in [0.0, 1.0]$, we have $f_{r,p}: [0.0, 1.0] \rightarrow [0.0, 1.0]$.

Given a $p \in [0.0, 1.0]$, a measure of this exchange is an integral of Eq. (10.19) taken over all dimensionless energies p

$$\frac{I(p)}{r} = \int_0^1 \frac{\Phi_{r,p}(x)}{r} = \frac{p}{(p+1)(2p+1)}. \tag{10.20}$$

Note that from this point the notation $f_{r,p}$ (Eq. 10.19) will be replaced by $\Phi_{r,p}$. From Fig. 10.6 it is seen that the energy exchange rate grows up with increase of the parameter p, i.e., when the size of a single EI becomes larger. The energy of exchange dynamics uncoupled EI is expressed here as a difference equation (de Vaart, 1973; Kreinovich and Kunin, 2003), i.e.,

$$x^{(k+1)} = \Phi_{r,p}\left(x^{(k)}\right), \quad \text{for } k \in \mathbf{N}. \tag{10.21}$$

Therefore, we are interested in the evolution of $n \in \mathbf{N}, n \geq 2$ (generally), coupled EIs, in other words, in the following discrete dynamical system

$$\mathbf{x}^{(k+1)} = C\Phi_{r,p}\left(\mathbf{x}^{(k)}\right), \tag{10.22}$$

where $\mathbf{x}^{(k)} = [x_1^{(k)} x_2^{(k)} ... x_n^{(k)}]^T \in [0, 1]^n$ is a state vector at time step $k \in \mathbf{N}$, $C = [c_{i,j}] \in \mathbf{R}^{n,n}$ is diffusive coupling matrix, and is a global logistic map given by

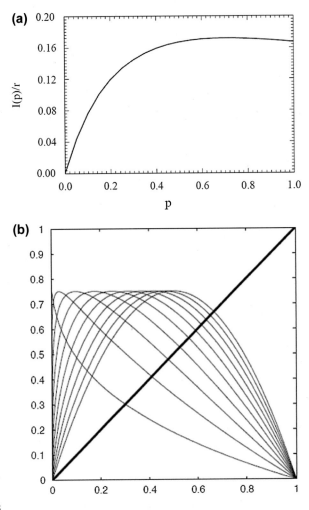

FIGURE 10.6

The dependence on the parameter p of (a) total energy exchange between a single EI and the surrounding environment, $I(p)/r$, and (b) the plot of the function $y = \phi_{r,p}$ for $r = 3$ and $p = 0.1, 0.2,..., 0.9, 1$ with the bolded line $y = x$.

$\Phi(\mathbf{x}) = [\phi_{r1,p1}\phi_{r2,p2}... \phi_{rn,pn}]^T$. The parameters of the discrete dynamical system Eq. (10.22) are: (1) $r_i \in [0.1,1.0]$—logistic parameter for ith EI, (2) $p_i \in [0.1,1.0]$—affinity parameter for ith EI, and $c_{i,j} \in [0.1,1.0]$—measure of diffusion of energy from jth EI to ith EI.

REFERENCES

Annan, J., Connolley, W., 2005. Chaos and Climate. http://www.realclimate.org/index.php/archives/2005/11/chaos-and-climate/.

Bohnenstengel, S.I., 2012. Can a Simple Locality Index Be Used to Improve Mesoscale Model Forecasts? (Ph.D. thesis) Department Geowissenschaften, Universitat Hamburg.

Curry, J., 2011. Reasoning about climate uncertainty. Clim. Chang. 108, 723−732.

Devaney, R.L., 2003. An Introduction to Chaotic Dynamical Systems, second ed. Westview Press, Boulder.

Dudhia, J., 1993. A nonhydrostatic version of the Penn State-NCAR mesoscale model: validation tests and simulation of an Atlantic cyclone and cold front. Mon. Weather Rev. 121, 1493−1513.

Giorgi, F., Avissar, R., 1997. Representation of heterogeneity effects in earth system modelling: experience from land surface modeling. Rev. Geophys. 35 (4), 413−438.

Gunji, Y.-P., Kamiura, M., 2004. Observational heterarchy enhancing active coupling. Physica D 198, 74−105.

Hess, G.D., McAvaney, B.J., 1998. Realisability constraints for land-surface schemes. Glob. Planet. Chang. 19, 241−245.

Kreinovich, V., Kunin, I., 2003. Kolmogorov complexity and chaotic phenomena. Int. J. Eng. Sci. 41, 483−493.

Lamb, R.G., Durran, D.R., 1978. Eddy diffusivities derived from a numerical model of the convective boundary layer. Il Nuovo Cimento 1C, 1−17.

Lettau, H.H., 1979. Wind and temperature profile prediction for diabatic surface layer including strong inversion cases. Bound. Lay. Meteorol. 17, 443−464.

Mihailović, D.T., 1996. Description of a land-air parameterization scheme (LAPS). Glob. Planet. Chang. 13, 207−215.

Mihailović, D.T., Balaž, I., Arsenić, A., 2013. Numerical study of synchronization in the process of biochemical substance exchange in a diffusively coupled ring of cells. Cent. Eur. J. Phys. 11, 440−447.

Mihailović, D.T., Budinčević, M., Balaž, I., Mihailović, A., 2011. Stability of intercellulular exchange of biochemical substances affected by variability of environmental parameters. Mod. Phys. Lett. B 25, 2407−2417.

Mihailović, D.T., Budinčević, M., Perišić, D., Balaž, I., 2012. Maps serving the combined coupling for use in environmental models and their behaviour in the presence of dynamical noise. Chaos Solitons Fractals 45, 156−165.

Mihailović, D.T., Kallos, G., 1997. A sensitivity study of a coupled soil-vegetation boundary layer scheme for use in atmospheric modelling. Bound. Lay. Meteorol. 82, 283−315.

Mihailović, D.T., Kapor, D., 2012. Modelling of flux exchanges between heterogeneous surface and atmosphere. In: Gualtieri, C., Mihailovic, D.T. (Eds.), Fluid Mechanics of Environmental Interfaces, second ed. CRC Press/Balkema, pp. 79−105.

Mihailović, D.T., Kostić, V., Mimić, G., Cvetković, L., 2015. Stability analysis of turbulent heat exchange over the heterogeneous environmental interface in climate models. Appl. Math. Comput. 265, 79−90.

Mihailović, D.T., Lalic, B., Rajkovic, B., Arsenic, I., 1999. A roughness sublayer wind profile above non-uniform surface. Bound. Lay. Meteorol. 93, 425−451.

Mihailović, D.T., Rao, S.T., Alapaty, K., Ku, J.Y., Arsenic, I., Lalic, B., 2005. A study on the effects of subgrid-scale representation of land use on the boundary layer evolution using a 1-D model. Environ. Model. Softw. 20, 705–714.

Mihailović, D.T., Rao, S.T., Hogefre, C., Clark, R., 2002. An approach for the aggregation of aerodynamic parameters in calculating the turbulent fluxes over heterogeneous surfaces in atmospheric models. Environ. Fluid Mech. 2, 315–337.

Mimić, G., Mihailović, D.T., Budinčević, M., 2013. Chaos in computing the environmental interface temperature: nonlinear dynamic and complexity analysis of solutions. Mod. Phys. Lett. B 27, 1350190.

Stainforth, D.A., Downing, T.E., Washington, R., Lopez, A., New, M., 2007. Issues in the interpretation of climate model ensembles to inform decisions. Philos. Trans. R. Soc. A 365, 2163–2177.

Sun, S.F., 1982. Moisture and Heat Transport in a Soil Layer Forced by Atmospheric Conditions (M.S. thesis). Department of Civil Engineering, University of Connecticut, 72 pp.

van der Vaart, H.R., 1973. A comparative investigation of certain difference equations and related differential equations. Bull. Math. Biol. 35, 195–211.

Maps serving the different coupling in the environmental interfaces modelling in the presence of noise

11.1 BEHAVIOR OF A LOGISTIC MAP DRIVEN BY FLUCTUATIONS

In this chapter we investigate the behavior of the coupled maps in the presence of the fluctuations or other noise. Note that in statistical mechanics, the term fluctuations is often a synonym for thermal fluctuations, which are random deviations of a system from its average state, that occur in a system at equilibrium. It has been shown in the case of uncoupled nonlinear oscillators that the introduction of external or parametric fluctuations has a pronounced effect on the dynamics of such systems (Hogg and Huberman, 1984). First, we consider behavior of a logistic map driven by fluctuations. A detailed overview of literature about logistic map driven by periodic signal, quasiperiodic signal, or noise is given in Zheng-Ling et al. (2009).

Because there is always noise in the real world, we consider effects of fluctuations on a simple nonlinear dynamical system, i.e., the logistic equation

$$x_{n+1} = rx_n(1 - x_n), \tag{11.1}$$

which we model by adding random noise $\Delta \xi_n$. The effect of additive noise $\Delta \xi_n$ we model by adding uniformly distributed random numbers to the map of Eq. (11.1), i.e.

$$x_{n+1} = rx_n(1 - x_n) + \Delta \xi_n. \tag{11.2}$$

Here $\Delta \xi_n = \mathbf{D}\delta_n$ measures the noise intensity while δ_n are random numbers uniformly distributed in the interval [0,1] and $\mathbf{D}$ is the amplitude of the noise.

We deal with the nondivergent interval and chaos excited by fluctuations of system (11.2). In that sense, the Lyapunov exponent λ in the presence of noise according to the necessary and sufficient condition of convergence of a function series, the system (11.2), will be nondivergent if

$$\lim_{n \to \infty} \frac{x_{n+1}}{x_n} < 1. \tag{11.3}$$

The condition (11.3) says that the difference boundary condition is

$$\lim_{n \to \infty} \left| \frac{rx_n - rx_n^2 + \Delta \xi_n}{x_n} \right| = 1. \tag{11.4}$$

Its solution is

$$x_n = \frac{\pm r \mp \sqrt{(r-1)^2 + 4r\Delta \xi_n}}{2r}. \tag{11.5}$$

If x_n is a real number, then

$$r^2 - 2r + 1 + 4r\Delta \xi_n \geq 0 \tag{11.6}$$

is necessary and sufficient.

The plots of the Lyapunov exponent of system (11.2) are shown in Fig. 11.1. We calculate x_n 10,000 times from the initial value $x_0 = 0.6$ for each step of change of r,

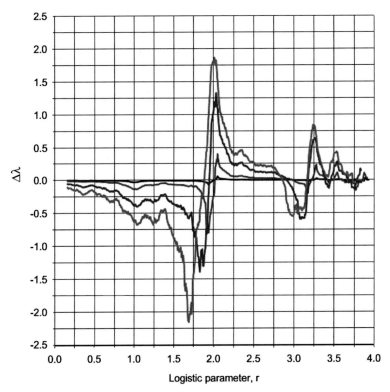

FIGURE 11.1

Comparison of Lyapunov exponents between systems (11.1) and (11.2) with $\Delta \lambda$ being the difference of λ between system (11.2) and system (11.1): $D = 0.001$(black), $D = 0.01$(red), $D = 0.05$ (blue), and $D = 0.1$.

which was 0.005. The latter 3000 points are used to draw the figures of the Lyapunov exponent.

From Fig. 11.1 is seen that if the logistic parameter r is in the intervals (1,75; 2,75) and (3.3,4) the additive noise significantly affects stability of the system (11.2) making it more unstable.

11.2 BEHAVIOR OF THE COUPLED MAPS SERVING THE COMBINED COUPLING IN THE PRESENCE OF DYNAMICAL NOISE

As it said, the dynamical noise can dramatically change the dynamics of the coupled maps. We use the term dynamical noise for a situation where the output of a dynamical system corrupted with noise is used as an input during the next iteration. Consequential analyses of real systems in environment in terms of chaos theory should take into account the effect of dynamical noise on the system's dynamics. In fact, as Ruelle (1994, p. 27) set it, real systems can in general be described as deterministic systems with some added noise. This descriptive approach is sufficiently indistinctive that it appears to cover everything. In economics, for example, such a description is familiar and the noise is called "shocks." A first remark concerning the above picture is that the separation between noise and the deterministic part of the evolution is indeterminate, because one can always interpret "noise" as a deterministic time evolution in infinite dimension (Serletis et al., 2007a). Serletis et al. (2007b) argue that dynamical noise (noise that acts as a driving term in the equations of motion) can noticeably change the dynamics of nonlinear dynamical systems. In reality, dynamical noise can make the recognition of chaotic dynamics very difficult. Additionally the dynamical noise can shift bifurcation points and produce noise-induced transitions, making the determination of bifurcation boundaries very difficult (Serletis and Shahmoradi, 2006).

In this section we examine how dynamical noise can affect the structure of the bifurcation diagram of the coupled maps. Many authors, dealing with spatial heterogeneity in population dynamics, are considering different coupling forms for logistic maps, what is well elaborated by Savi (2007). This noise enters in two specific ways: it disturbs either the parameter r (parametric excitation) or the deterministic law by an additive "shock" (external excitation). Specifically, we deal with coupled maps given by Eqs. (8.5a) and (8.5b), where $c_1 = c_2 = 0.5$, now written in the form

$$x_{n+1} = (1 - c_1)rx_n(1 - x_n) + c_1 ry_n(1 - y_n) + c_2(y_n - x_n) + \Delta\xi, \qquad (11.7a)$$

$$y_{n+1} = (1 - c_1)ry_n(1 - y_n) + c_1 rx_n(1 - x_n) + c_2(x_n - y_n) + \Delta\eta. \qquad (11.7b)$$

We analyze first the randomness influence on the above-coupled maps by adding random noise. Here $\Delta\xi_n = \mathbf{D}\delta_n^{(1)}$ and $\Delta\eta_n = \mathbf{D}\delta_n^{(2)}$ measure the noise intensity while $\delta_n^{(1)}$ and $\delta_n^{(2)}$ are random number uniformly distributed in the interval [0,1] and $\mathbf{D}$ is the amplitude of the noise. To illustrate their dynamics we plot their bifurcation diagrams in the absence of noise (Fig. 11.2). We focus now on the influence of noise in $\Delta\xi_n$ and $\Delta\eta_n$. Bearing in mind that $\mathbf{D} = 0.1$ and $\mathbf{D} = 0.2$, we analyze results from bifurcation diagrams depicted in Fig. 11.3, which may be compared with Fig. 11.2. It is perceptible that noise destroys some periodic windows, changing some expected behavior, already when $\mathbf{D} = 0.1$ corresponding to low amplitude of the additive noise (Serletis et al., 2007b), as it is seen on the first four upper panels in Fig. 11.3. Further, for the doubled amplitude, i.e., $\mathbf{D} = 0.2$, the noise (the last four lower panels in Fig. 11.3), exceedingly destroyed the pictures of bifurcation diagrams in Fig. 11.3.

In the case of uncoupled nonlinear oscillators, it has been shown that the addition of parametric fluctuations has a pronounced effect on the dynamics of such systems (Hogg and Huberman, 1984; Liu and Ma, 2005; Thattai and van Oudenaarden, 2001).

It is therefore of interest to investigate the effect of noise on either environmental processes or events represented by the system of two maps serving the combined

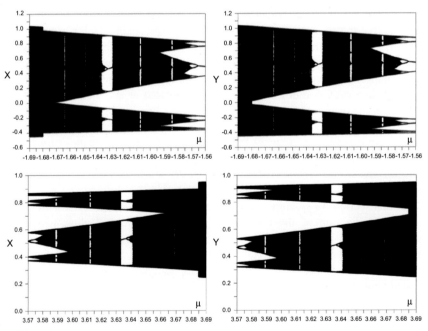

FIGURE 11.2

Bifurcation diagrams of the coupled maps given by Eqs. (11.7a) and (11.7b) for $c_1 = c_2 = 0.5$ in the absence of noise $\mathbf{D} = 0$.

Reprinted with permission from Mihailović, D.T., Budinčević, M., Perišić, D., Balaž, I., 2012. Maps serving the combined coupling for use in environmental models and their behaviour in the presence of dynamical noise. Chaos Solitons Fractals 45, 156–165.

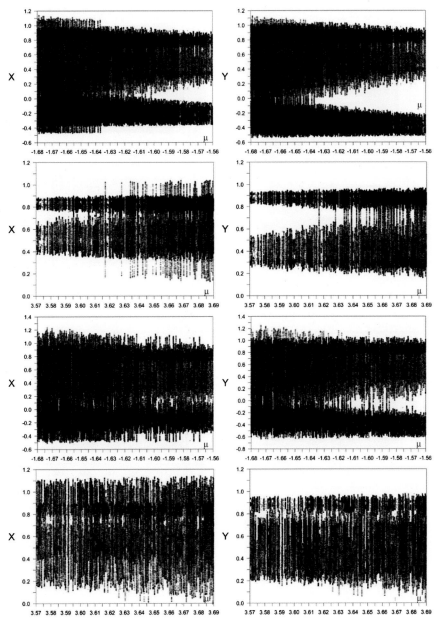

FIGURE 11.3

Bifurcation diagrams of the coupled maps given by Eqs. (11.7a) and (11.7b) for $c_1 = c_2 = 0.5$ 'shocked' by added noise. The four upper panels are for **D** = 0.1, while the last four ones are for **D** = 0.2.

Reprinted with permission from Mihailović, D.T., Budinčević, M., Perišić, D., Balaž, I., 2012. Maps serving the combined coupling for use in environmental models and their behaviour in the presence of dynamical noise.
Chaos Solitons Fractals 45, 156–165.

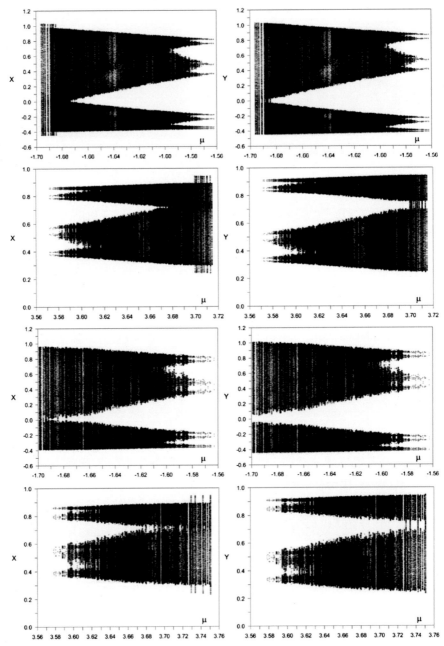

FIGURE 11.4

Bifurcation diagrams of the coupled maps given by Eqs. (11.8a) and (11.8b) for
$c_1 = c_2 = 0.5$ when forcing is done by the parametric noise. The first four panels are for
$\mathbf{D} = 0.01$, while the last four lower ones are for $\mathbf{D} = 0.025$.

Reprinted with permission from Mihailović, D.T., Budinčević, M., Perišić, D., Balaž, I., 2012. Maps serving the
combined coupling for use in environmental models and their behaviour in the presence of dynamical noise.
Chaos Solitons Fractals 45, 156—165.

coupling. In modelling environmental interfaces especially interesting cases are those where, due to either internal or external noise, parameters of the oscillators have small, random variations. These so-called parametric fluctuations can be simulated by modulating the values of the nonlinearity parameters by uniform random numbers in a small interval. Specifically, in our case, when coupling parameters c_1 and c_2 are fixed, it gives the following map:

$$x_{n+1} = (1 - c_1)r_n^{(1)}x_n(1 - x_n) + c_1 r_n^{(2)}y_n(1 - y_n) + c_2(y_n - x_n), \tag{11.8a}$$

$$y_{n+1} = (1 - c_1)r_n^{(2)}y_n(1 - y_n) + c_1 r_n^{(1)}x_n(1 - x_n) + c_2(x_n - y_n), \tag{11.8b}$$

where $r_n^{(1)} = r(1 + \Delta\xi_n)$ and $r_n^{(2)} = r(1 + \Delta\eta_n)$. Bifurcation diagrams in Fig. 11.4 depicts the change in their structure comparing to Fig. 11.2, when the parametric noise is introduced with amplitudes $\mathbf{D} = 0.01$ and $\mathbf{D} = 0.025$, corresponding to the low intensity of additive noise. It seems that the parametric forcing produces larger changes in the bifurcation diagrams than in the case of the added noise. Namely, looking at Fig. 11.3 (the first four upper panels when $\mathbf{D} = 0.1$) and Fig. 11.4 (the last four lower panels when $\mathbf{D} = 0.025$) we can see similar changes in bifurcation diagrams.

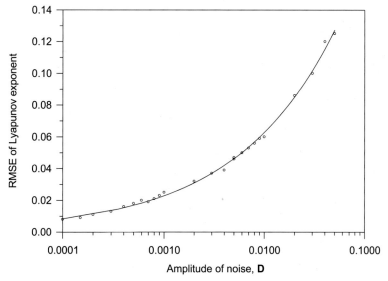

FIGURE 11.5

RMSE of the Lyapunov exponent for the coupled maps given by Eqs. (11.9a) and (11.9b) as a function of the amplitude **D** of the noise that is introduced by the parametric forcing. Values of c_1 and c_2 and r are the same as in Fig. 3.8.

Finally, we consider its behavior when the parametric noise is introduced in all parameters in Eqs. (8.5a) and (8.5b). Hence,

$$x_{n+1} = (1 - c_{1,n}) r_n^{(1)} x_n (1 - x_n) + c_{1,n} r_n^{(2)} y_n (1 - y_n) + c_{2,n} (y_n - x_n), \qquad (11.9a)$$

$$y_{n+1} = (1 - c_{1,n}) r_n^{(2)} y_n (1 - y_n) + c_{1,n} r_n^{(1)} x_n (1 - x_n) + c_{1,n} (x_n - y_n), \qquad (11.9b)$$

where $c_{1,n} = c_1 (1 + \Delta \alpha_n)$ and $c_{2,n} = c_2 (1 + \Delta \beta_n)$, $\Delta \alpha_n = \mathbf{D} \delta_n^{(3)}$ and $\Delta \beta_n = \mathbf{D} \delta_n^{(4)}$ measure the noise intensity while $\delta_n^{(3)}$ and $\delta_n^{(4)}$ are random numbers uniformly distributed in the interval $[-1,1]$. Now, we set center of attention on the changes of the Lyapunov exponent depending on the amplitude of the noise introduced (Liu and Ma, 2005). We calculated the $RMSE = \sum_{i=1}^{N} \{[\lambda^c(c_{1,n}, c_{2,n}, r_n^{(1)}, r_n^{(2)}) - \lambda^0(c_1, c_{2,n}, r)]/N\}^{1/2}$ of the Lyapunov exponent, for the coupled maps Eqs. (11.9a) and (11.9b), where λ^c and λ^0 are values calculated in the presence and absence of the noise, respectively. In calculations of the parametric noise, the amplitude $\mathbf{D}$ is ranged from 0.0001 to 0.05 while the other parameters are used as in Section 9.2. The results of calculations are shown in Fig. 11.5. This figure clearly shows that the increase of $RMSE$ is growing up with the amplitude, like a power function. Similar result, but for $RMSE$ of the Cross-SampEn, for the maps representing biochemical substances exchange between cells, was obtained by Mihailović and Balaž (2011).

REFERENCES

Hogg, T., Huberman, B.A., 1984. Generic behavior of coupled oscillators. Phys. Rev. A 29, 275–281.

Liu, Z., Ma, W., 2005. Noise induced destruction of zero Lyapunov exponent in coupled chaotic systems. Phys. Lett. A 343, 300–305.

Mihailovic, D.T., Balaz, I., 2011. A model representing biochemical substances exchange between cells. Part II: effect of fluctuations of environmental parameters to behavior of the model. J. Appl. Funct. Anal. 6, 77–84.

Mihailović, D.T., Budinčević, M., Perišić, D., Balaž, I., 2012. Maps serving the combined coupling for use in environmental models and their behaviour in the presence of dynamical noise. Chaos Solitons Fractals 45, 156–165.

Ruelle, D., 1994. Where can one hope to profitably apply the ideas of chaos? Phys. Today 47, 24–30.

Savi, M.A., 2007. Effects of randomness on chaos and order of coupled maps. Phys. Lett. A 364, 389–395.

Serletis, A., Shahmoradi, A., 2006. Comment on "Singularity bifurcations" by Yijun He and William A. Barnett. J. Macroecon. 28, 23–26.

Serletis, A., Shahmoradi, A., Serletis, D., 2007a. Effect of noise on the bifurcation behavior of nonlinear dynamical systems. Chaos Solitons Fractals 33, 914–921.

Serletis, A., Shahmoradi, A., Serletis, D., 2007b. Effect of noise on estimation of Lyapunov exponents from a time series. Chaos Solitons Fractals 32, 883–887.

Thattai, M., van Oudenaarden, A., 2001. Intrinsic noise in gene regulatory networks. Proc. Natl. Acad. Sci. U.S.A. 98, 8614–8619.

Zheng-Ling, Y., Yang, G., Yong-Tao, G., Jun, Z., 2009. Behavior of a logistic map driven by white noise. Chin. Phys. Lett. 26, 060506.

Heterarchy and exchange processes between environmental interfaces

IV

Heterarchy as a concept in environmental interfaces modelling

12

12.1 HIERARCHY AND HETERARCHY

Hierarchy, as a necessary precondition in forming organized, functional systems, came under the focus of natural sciences relatively recently. Since it refers to a very broad spectrum of phenomena it is hard to make one, uniform, definition which is at the same time applicable to social, ecological, living, or any other organized systems. Even within mentioned groups, there is a large number of different kinds of systems, with different organization and different hierarchical schemes (for an overview see Pattee, 1973; Allen and Starr, 1982; Salthe, 1989; Ahl and Allen, 1996). More particularly, in the domain of modelling living systems, the problem of creating dynamical hierarchies have been postulated as one of open and very challenging tasks for the future development of artificial life modelling (Bedau et al., 2000). For a general case, it has been stated that the ordering of hierarchical levels is ruled by several criteria: (1) being the context of, (2) offering constraint to, (3) behaving more slowly or at a lower frequency than, (4) being populated by entities with greater integrity and higher bond strength than, and (5) containing and being made of lower levels (Ahl and Allen, 1996). In some relatively stable and fixed system, we can make straightforward characterization by enumerating important properties of elements and classifying them according to their role in the hierarchy. For example, if we consider metabolism from the perspective of its constituent elements (i.e., enzymes), we can accurately characterize it by giving amino acid constitution of each enzyme, its spatial structure, and its mechanism of action. In short, we can use fixed set of relations to represent the system. However, if we want to consider metabolism as a dynamical, self-organized system, where processes are not in the form of predefined procedures but are generated during the process itself and decisions about the next step are always local and context dependent, characterization by the set of fixed properties is questionable. In that case, local interpretation of a given situation is the key factor in the process of establishing hierarchy (e.g., existence of some enzyme at a particular place determines in which way substrate will be transformed and which metabolic pathway will take primacy in a given moment, at the given place). Also, at the cellular level, most obvious hierarchical order is established between the so-called information storage domain and metabolic domain,

Developments in Environmental Modelling, Volume 29, ISSN 0167-8892, http://dx.doi.org/10.1016/B978-0-444-63918-9.00012-0

which is usually represented in the form of the following chain: DNA → RNA → Proteins. Such ordering fulfills almost all of above-mentioned criteria for ordering of hierarchical levels. However, closer examination reveals several facts which disturb such straightforward scheme. First, some classes of proteins and some classes of RNAs are able to influence the pattern of expression of genes in DNA. Also, all three domains themselves contain internal hierarchical relations so the chain is transformed into a web of feed-forward and feedback loops. From these simple examples is obvious that establishing one satisfactory hierarchical division of functional compartments in dynamical complex systems where hierarchies can interact with each other and reinterpret their current role could be a very challenging task. Therefore, to cope more efficiently with the problem of modelling hierarchies, we believe that the very idea of "proper" hierarchies should be relativized by acknowledging that hierarchical systems themselves are often composed of embedded internal observers, who are able to reinterpret hierarchical levels in accordance to the current context (Salthe, 1989). This notion lead us to the concept of heterarchy that may be defined as the relation of elements to one another when they are unranked or when they possess the potential for being ranked in a number of different ways (Crumley, 1995).

The concept of the heterarchy may be illustratively presented by either analyzing interrelationships of small groups or the society as a whole. In that way the hierarchy−heterarchy relation offers a new approach to the study of agency, conflict, and cooperation (Crumley, 1995). As an example we consider a heterarchy of relationship between environmental modelling group (GROUP) and environmental protection institute (INSTITUTE) to which the GROUP members are employed (Fig. 12.1a). The GROUP has a leader (LEADER). Her/his action, therefore, affects both the GROUP and the INSTITUTE, simultaneously. Let us suppose that, pressed by the INSTITUTE, the LEADER accepted to finish the design of the model before the already fixed deadline. Reason for accelerating the work on completion of the model was faster spending of the financial support than it was expected. Although the LEADER's acceptance of such condition is good for the INSTITUTE, it is bad for her/his GROUP. If someone listens to this topic she/he might think that it satisfies the condition of heterarchy, i.e., simultaneous interaction among levels (Gunji and Kamiura, 2003). On the other hand, someone who listens to this topic has to discern that such a simultaneous interaction outcomes just from a hidden specific operation such that bad (or good, respectively) for the GROUP is mapped to good (bad, respectively) for the INSTITUTE. To recognize "simultaneousness" in interaction, one has to grasp both independency of two levels (GROUP and INSTITUTE) and simultaneous interaction. Because of independency of those levels, one must have in mind all possible operations between them but with the focus on the method of choice of *one* operation. If we define a set of values for the GROUP and the INSTITUTE as $S = \{(0, bad),(1, good)\}$ (Fig. 12.1a) then all possible operations from the GROUP to the INSTITUTE we express through corresponding interpretations (Int_0, Int_1, Int_2, and Int_3), which are depicted on the body of the upper arrow in 12.1b.

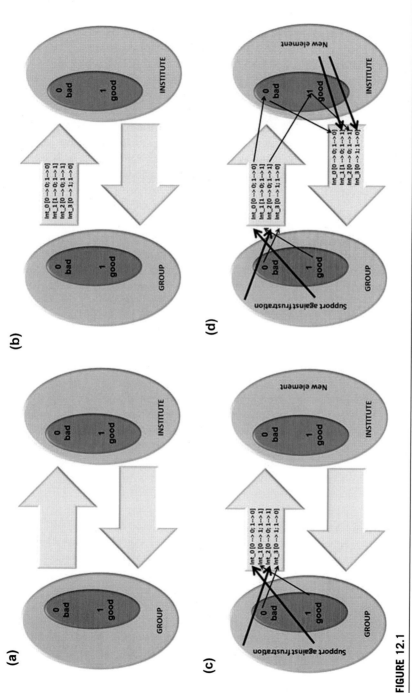

FIGURE 12.1

Images showing hetarchy consisting of GROUP and INSTITUTE. (a) Set of values for the GROUP and the INSTITUTE as S={(0, *bad*), (1, *good*)}; (b) Operation from the GROUP to the INSTITUTE expressed through corresponding interpretations (Int_0, Int_1, Int_2 and Int_3); (c) appearance of new value "support against frustration" within S; (d) new state of the GROUP "support against frustration".

At one moment, the LEADER gathers the people from the group to announce that she/he intends to accept the INSTITUTE's demand to shorten the deadline for finishing the model, while the GROUP members working on the model design are angry and frustrated after they had heard about her/his intention. The LEADER, hesitating to accept INSTITUTE's suggestion, is thinking that acceptance is bad for the GROUP but is good for the INSTITUTE. The thinking (i.e., choice—Int_3) is going on in a finite time. Therefore, such a process itself can have the value in S, in the GROUP. Now, we have the following situation. The GROUP members begin to feel that their LEADER is under the huge pressure to accept INSTITUTE's demand, and they think that their own attitudes gives her/him too much feeling of guiltiness. Therefore, they decide to support their LEADER. As a final result, they gave him green light—with a support against their frustration. The prolonged hesitation of the LEADER drives the process of choosing an interpretation in the GROUP and that activates appearance of new value, "support against frustration" within S. As a final result, the value in the GROUP changes from $S = \{0,1\}$ to $S = \{0,1,2(\textit{support against frustration})\}$. Now we will demonstrate what choice of an interpretation, having a sense in a particular level (GROUP), changes the structure of the level.

An observer has to describe a LEADER's decision and corresponding action—to finish design of the model before the already fixed deadline—as a simultaneous process of choosing one interpretation. Evidently, it makes sense if a chosen interpretation has a value of S. This situation can be described in the following way.

The definition of heterarchy given by Crumley (1995) can be concisely written as simultaneous interaction among some levels. However, now we have *simultaneous choice* between two dynamics, i.e., *intralevel* dynamics and *interlevel* dynamics. In our example, the intralevel dynamics is just a choice of a value of S (0 or 1 representing a value of a particular level) and the interlevel dynamics is a choice of an interpretation (Int_0—Int_3). The simultaneous choice is defined by two properties: (1) a map-property and (2) simultaneous making value (Gunji and Kamiura, 2003). The map-property is defined as: for all elements of S, there exists an interpretation (one-to-one). For example, for 0 in a GROUP, the LEADER chooses Int_3, and for 1 she/he chooses Int_1 (indicated by thin arrows in Fig. 12.1c). It makes a map. On the contrary, if for 0 she/he chooses both Int_3 and Int_1, a map is set to be one-to-many and the map-property fails. The property (2) is defined as follows. Each possible chosen interpretation has to have a value in a level (GROUP). The map-property looks natural but it needs all possible correspondences between an element of S and all interpretations. Even if somebody observes only one correspondence between 0 and Int_1, an observer has to decide the correspondence for 1 because of the map-property. The simultaneous making value is defined so as to expand such a standpoint.

Let us imagine that a map is defined in the following way: $0 \rightarrow$ Int_3, $1 \rightarrow$ Int_1. Simultaneous choice means that each interpretation has a value of S in this choice. For this choice, somebody can recognize that Int_1 has a value 1 and Int_3 has a value 0, once each interpretation is chosen. On the other hand, the property of simultaneously making value demands that it should be performed for all interpretations. While Int_0 and Int_2 (indicated by thick arrows in Fig. 12.1c) are not chosen, they also have to have values of S. Suppose that Int_0 has a value 0. If so, then the map-property failed because a value 0 is mapped both to Int_0 and Int_3. As a result, the map-property and simultaneous making value constitutes a tradeoff relationship (a situation that involves losing one quality or aspect of something for gaining another quality or aspect).

To formalize this story algebraically, let us suppose that each level is defined by a set, S, while a set of the interlevel operations are defined by $Hom\,(S,S)$ that is a set of functions from S to S. The map-property of simultaneous choice is defined by $f{:}S \rightarrow Hom\,(S,S)$ that is a map. Additionally, the property of making value is defined by f that covers all elements of $Hom\,(S,S)$. As a result, simultaneous choice requires that a map f is surjective (i.e., every element y in Y has a corresponding element x in X such that $\forall y \in Y, \exists x \in X, f(x) = y$). Such a requirement is hopeless, in principle since the number of elements of $Hom\,(S,S)$ is N^N where N is the number elements of S. Thus, the map cannot cover all elements of $Hom\,(S,S)$.

Now we have a situation that simultaneous choice is collapsed while heterarchy proceeds as a real system. In this state of affairs, somebody has to focus on the concept of heterarchy as a real system against the collapse of observer's frame. In our example the collapse can explain the appearance of emergent state of "support against frustration," instead of perpetual change of frames. Change of competing interpretations in the GROUP is expressed as an assumption of a surjective map from S to $Hom\,(S,S)$ (Fig. 12.1d). If somebody wants to make a system that avoids collapse and keeps simultaneous choice, she/he has to find *new source* that is mapped to possible elements of $Hom\,(S,S)$ out of S. In Fig. 12.1d, a map called choice from S to $Hom\,(S,S)$ is indicated by a thin arrow, and emergent arrows required by simultaneous choice are indicated by thick arrows. To avoid one-to-many mapping, a new source of an arrow is constructed out of $\{good, bad\}$. In our example it is represented as a new state of the GROUP, "support against frustration." The collapse-assumption, named simultaneous choice makes re-organization of the system possible (Gunji and Kamiura, 2003). This example shows how the engine of heterarchy works.

12.2 OBSERVATIONAL HETERARCHY AND FORMALIZATION OF HETERARCHICAL LEVELS

The engine of heterarchy works due to the fact that components of the system possess the potential for being ranked in a number of different ways. For example, if in a system consisting of two different subsystems (intrasubsystem operations and intersubsystems operations) both intraoperations and interoperations are allowed, then the system is called heterarchy. In a hierarchical system, one layer depends on the other layer in a strictly defined order, while a heterarchical system can switch the dependence relations of each layer. Seen through the optics of modelling the complex systems, heterarchy reveals that in some systems it is impossible to determine to which subsystems an element appertains (Gunji and Kamiura, 2004).

At this place, we will make one comment on the history of the notion of heterarchy. According to Gunji and Kamiura (2004) two Santa Fe scientists dealing with the complex system problems, in about same days have launched two notions regarding to the heterarchy: (1) that the most important problem in complex systems is to describe the agent who can adjust the way of measurement by result of measurement (Kauffman, 2002) and (2) its significance with respect to the difference between stability and robustness (Jen, 2003).

The idea for (1) results from Maxwell's demon, and in 1980s it was described as the notion of internal measurement (Matsuno, 1989) and endo-physics (Rössler, 1989). Namely, in Maxwell's thought experiment, the demon creates a temperature difference simply from information about the gas molecule temperatures and without transferring any energy directly to them and thus violating the second law of thermodynamics (a partition with a small trapdoor is placed in the box, and the trapdoor is guarded by the demon who, without expending energy, selects which molecules go through to the other side). Note that long after the paper on Maxwell's demon, Szilard (1929) showed that the thought experiment does not actually violate the laws of physics because the demon must exert some energy in determining whether molecules were hot or cold. In (2) the robustness and stability are used in the following context: "Robustness is an approach to feature persistence in systems that compels us to focus on perturbations, and assemblages of perturbations, to the system different from those considered in the design of the system, or from those encountered in its prior history. To address feature persistence under these sorts of perturbations, we are naturally led to study the coupling of dynamics with organizational architecture; implicit rather than explicit assumptions about the environment" (Jen, 2003). Finally, having in mind previous text and example from the Section 12.1, in this book we use the following definition of heterarchy: "If a system consists of two different subsystems, intra subsystem operations and inter-subsystems operations, and if the mixture between intra- and inter-operations is permitted, the system is called heterarchy" (Gunji and Kamiura, 2003, 2004).

Communication between two environmental interfaces (for example, forest and grove) satisfies the definition of heterarchy. Given two environmental forest interfaces of different ecological types, each interface is expressed as a map by which birds are classified in a term of their attributes (flying, hunting). These maps represent intrasubsystem operations (blue arrow in Fig. 12.2). There is a communication (exchange) between two maps and that represents intersubsystem operation (violet arrow in the same figure). In this communication, i.e., migration of birds leads to the mixing between interlevel operation and intralevel operation. That is why the system is heterarchy.

The next question is how often heterarchy arises in the actual experience. Although hierarchical structures are universally found, most of them are not heterarchies. Their hierarchical levels are clearly separated, and interlayer and intralayer reinterpretation does not occur. Heterarchies, on the other hand, have more dynamic

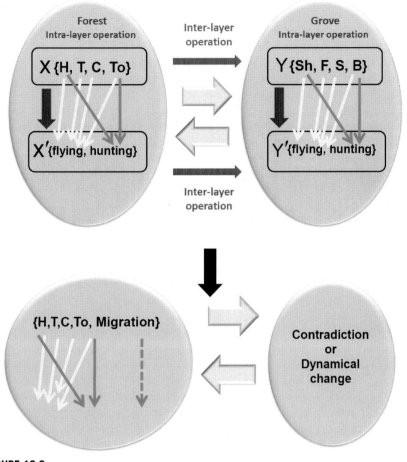

FIGURE 12.2

Example of heterarchy illustrating interenvironmental interfaces (forest and grove)
communication with the mixture of interlayer operation and intralayer operation. It consists of
intralayer operation such as a mapping a bird to forest or grove, interlayer operation such as
the migration, and the mixture such that the migration needs a new property. The mixture
demands to infinite regression of redefinition of a system and makes known impossibility to
describe a heterarchy. List of forest birds: hawk (H), turtledove (T), cuckoo (C), and tawny owl
(To). List of grove birds: sparrow hawk (Sh), finch (F), skylark (S), and buzzard (B).

structure which emerges from the existence of internal observers. In other words, the
internal measurement is introduced by an internal object that can perform reinterpretation, and such heterarchy has been defined as *observational heterarchy* (Gunji and
Kamiura, 2004). Although such structures are not as universal as ordinary hierarchies, they are ubiquitous in nature.

In our consideration, first let us define two perspectives, intent-perspective and extent-perspective. Given a system (phenomenon, concept), the extent-perspective consists of all objects belonging to the concept, while the intent-perspective is the collection of all attributes shared by the objects (Gunji and Kamiura, 2004). For example, intent of odd number (concept) is expressed as $2n + 1$, and extent of it is expressed as (1, 3, 5 …), or another one. Thus, at the forest level (or similarly at the grove level) hunting, flying birds describe a subconcept of the concept of the flying birds (Fig. 12.2). The extent of this subconcept consists only of the hawk, and the intent consists of the three attributes hunting, flying, and bird. It is often difficult to list all the objects belonging to a concept and usually impossible to list all its attributes; therefore, it is natural to work with a specific context in which the objects and attributes are fixed. A context is a triple (G,M,I) in which G and M are sets and $I \subseteq G \times M$. The elements of G and M are called objects and attributes, respectively. $(g,m) \in I$ means "the object g has attribute m."

In Section 2.4 we have elaborated in detail the meaning of the formal concept. Therefore, here we will shortly repeat some definitions to define two perspectives, intent-perspective and extent-perspective using Fig. 12.2. A concept is considered to be determined by its extent and intent: the Extent consists of all objects belonging to the concept, while the Intent is the collection of all attributes shared by the objects. It is often difficult to list all the objects belonging to a concept and usually impossible to list all its attributes; therefore, it is natural to work with a specific context in which the objects and attributes are fixed. A triplet (G,M,I) we call a context in which G and M are sets and $I \subseteq G \times M$. The elements of G and M are called objects and attributes, respectively. $(g,m) \in I$ means "the object g has attribute m." For $X \subseteq G$ and $Y \subseteq M$, we define $X' = \{m \in M \mid (\forall g \in X)\ (g,m) \in I\}$ and $Y' = \{g \in G \mid (\forall m \in Y)\ (g,m) \in I\}$ Therefore, X' is the set of attributes common to all objects in X, and Y' is the set of objects possessing the attributes of Y. The concept of the context (G,M,I) is a pair (X,Y), in which $X \subseteq G$, $Y \subseteq M$, $X' = Y$ and $Y' = X$. The Extent of the concept (X,Y) is X, while the Intent is Y. A subset $X \subseteq G$ is the extent of some concept if and only if $X'' = (X')' = X$, in which case the unique concept of which X is an extent is (X,X'). The corresponding statement applies to these subsets B of M, which are the intents of some concept.

For a general system, however, two perspectives are inconsistent with each other. As we said, in general, Intent-perspective is assumed to be equivalent to Extent-perspective, as well as a concept in a set theory or formal concept in a concept lattice. However, in the modelling procedure it is usual practice that the equivalence between Intent-perspective and Extent-perspectives results just from an approximation and/or and hypothesis (Gunji, Kamiura, 2004). In summary, for the observational heterarchy we can say that it is a two-level entity which includes interlevel operations. It also encompasses simultaneous communication among levels through simultaneous choice that is stated as surjective map from a set of one level to a set of interlevel operations. This choice is a source of the collapse of the logical background thus the heterarchy is regarded as a system which inherits logical collapse. For the sake of the logical collapse, heterarchy gives leap to reorganization of the structure. Thus, the engine of heterarchy provides the dynamics of the system. Nevertheless, the heterarchy results from the interaction on the relation object—observer with two essential levels, i.e., intent-perspective and extent-perspective.

REFERENCES

Ahl, V., Allen, T.F.H., 1996. Hierarchy Theory. Columbia University Press, p. 208.

Allen, T.F.H., Starr, T.B., 1982. Hierarchy: Perspectives for Ecological Complexity. University Chicago Press, p. 328.

Bedau, M., McCaskill, J., Packard, P., Rasmussen, S., Green, D., Ikegami, T., Kaneko, K., Ray, T., 2000. Open problems in artificial life. Artif. Life 6 (4), 363—376.

Crumley, C.L., 1995. Heterarchy and the analysis of complex societies. Archeol. Pap. Am. Anthropol. Assoc. 6, 1—5.

Gunji, Y.-P., Kamiura, M., 2003. An observational heterarchy as phenomenal computing. In: Proceeding of CRPIT '03 Selected Papers From Conference on Computers and Philosophy, vol. 37, pp. 39—44.

Gunji, Y.-P., Kamiura, M., 2004. Observational heterarchy enhancing active coupling. Phys. D 198, 74—105.

Jen, E., 2003. Stable or robust? what's the difference? Complexity 8, 12—18.

Kauffman, S.A., 2002. Investigations. Oxford University Press.

Matsuno, K., 1989. Protobiology: Physical Basis of Biology. CRC Press, Boca Raton, MI.

Pattee, H.H. (Ed.), 1973. Hierarchy Theory: The Challenge or Complex Systems. Braziller, New York, p. 156.

Rössler, O.E., 1989. Explicit observers. In: Plath, P.J. (Ed.), Optimal Structures in Heterogeneous Reaction Systems. Springer-Verlag, New York, pp. 123—138.

Salthe, S., 1989. Evolving Hierarchical Systems: Their Structure and Representation. Crossroad Publishing Company, New York, p. 343.

Szilard, L., 1929. Über die Entropieverminderung in einem thermodynamischen System bei Eingriffen intelligenter Wesen. Z. für Phys. (in German). ISSN: 0044-3328 840—856. http://dx.doi.org/10.1007/BF01341281 (Available Online in English at: Aurellen.org).

Heterarchy and biochemical substance exchange in a diffusively coupled ring of cells

13

13.1 OBSERVATIONAL HETERARCHY AND BIOCHEMICAL SUBSTANCE EXCHANGE BETWEEN TWO CELLS

Understanding how local intracellular biochemical exchange processes and global features, like environment and system size, influence the robustness, adaptability and evolution of the collective behavior of multicell systems is one of the most challenging topics in the biology of complex systems today (Levin, 2006; Pikovsky et al., 2001; Arenas et al., 2008; Chen et al., 2003; Ghosh et al., 2010; Mihailović et al., 2013). Information coupling and the exchange of biophysical substances among the components of multicell systems are both driven by a range of intrinsic and extrinsic factors. Several authors have made significant contributions to the understanding of multicell system dynamics through studies of the stability of the synchronized state, which is required for robust functioning of the multicell system in the face of noise and perturbation (Pikovsky et al., 2001; Arenas et al., 2008; Chen et al., 2003; Ghosh et al., 2010; Rajesh et al., 2007; Rajesh and Sinha, 2008). However, these authors considered cells as completely uniform particles, without internal structure and without the ability to change their behavior. It is well known that in actuality, in natural conditions, bacterial cells spend most of the time in the stationary phase which is (in contrast to the exponential phase) characterized by a decrease in growth rate, slowdown of all metabolic processes, and increase in resistance to several stress conditions (Jones, 1985; de Groot and Littauer, 1989; Kolter at al., 1993; Spector and Kenyon, 2012).

In considering these problems, we have to include observational heterarchy, a challenging topic when dealing with complex systems. For the topic of this chapter, let us briefly summarize the points of the Section 12.2. In essence, observational heterarchy discloses that it is impossible to explicitly determine to which subsystems an element belongs (Gunji and Kamiura, 2003, 2004; Mihailović et al., 2013). It is based on the notion of an agent carrying the adjustment of measurement (Kauffman, 2002; Jen, 2003). Therefore, the dynamics of the complex system are articulated in terms of two kinds of dynamics, Intent and Extent dynamics, and the communication between them, where Intent corresponds to an attribute of a given phenomenon and Extent corresponds to a collection of objects satisfying that phenomenon (Gunji and Kamiura, 2004). Gunji and Kamiura (2003) have introduced the concept of observational heterarchy, pointing up that the process of measurement and description cannot be disjointed from what is observed and measured. In other words, the epistemology cannot be separated from ontology, resulting in a dynamical description and a dynamical ontology as noticed by Bickhard and Terveen (1996).

Observational heterarchy consists of two sets of intralayer maps, called Intent-perspective and Extent-perspective, and interlayer operations satisfying the following conditions: (1) the interlayer operations inherit the mixture of intralayer and interlayer operations and (2) there is a procedure by which the interlayer operation can be regarded as an adjoint functor. If the interlayer operation satisfies the conditions (1) and (2), it is called a prefunctor. According to Gunji and Kamiura (2004), preserving the above composition occurs as follows: A prefunctor, $\langle F \rangle$: Int $\rightarrow$ Ext is mapping a set X to a set $\langle F \rangle X$, and a map Φ to a map $f^*\Phi f$, where $f^*f(x) = x$ for all x in $f(X)$ with $f(X)$: $\langle F \rangle X \rightarrow X$. In this sense we call f^* the pseudo-inverse of f. Because applying a prefunctor to a map is expressed as composition of maps, it satisfies the conditions (1) and (2). The approximation is defined by the assumption that f is a one-to-one map. If one accepts that the approximation $f^* = f^{-1}$ holds, then a prefunctor can become a functor. Given two maps Φ, Ψ: $X \rightarrow X$,

$$\langle F \rangle(\Phi)\langle F \rangle(\Psi) = (f^*\Phi f)(f^*\Psi f) = f^*\Phi(ff^*)\Psi f = f^*\Phi(ff^{-1})\Psi f = f^*\Phi\Psi f\langle F \rangle(\Phi f).$$

(13.1)

It implies that F preserves the composition of maps, Φ and Ψ. However, there is inconsistency between Intent and Extent (Gunji and Kamiura, 2004), illustrated, for example, in adaptive mutation in the *Lactose operon* (Cairns and Foster, 1991; Shapiro, 1992, 1995). Thus, in the phenomenon of the protein population, the Intent, given by an ordinary differential equation, ignores its differences, while Extent, consisting of individual proteins, focuses on differences. Their equivalence comes from the approximation alone, and otherwise cannot happen.

Since these and many other processes in a cell are defined as diffusion-like manner (Devaney, 2003; Gunji and (Kamiura, 2004), looking from the Intent-perspective and Extent-perspective in a cell we address the synchronization of the *passive* coupling for two cells given by Eqs. (13.1) and (10.1), i.e., the generalized logistic equation by the affinity. The time development of the intracellular dynamics for two cells, is expressed as

$$x_{i,n+1} = (1 - c)\Phi(x_{i,n}) + f(\Phi(x_{i,n})),$$

(13.2)

where n is the time iteration, $i,j = 1,2$, $x_{i,n} \in [0,1]$, c is the coupling parameter (concentration of the substrate), f is the map representing the flow of the material from cell to cell, $f(x)$ is defined by a map that can be approximated by a linear map, and Φ is one of the maps in the pair (Φ,Ψ) whose composition is preserved by a prefunctor

$\langle F \rangle$ (Gunji and Kamiura, 2004). If $f(x) = cx$, the interaction is expressed as a linear coupling between two cells. Here, we apply the framework of an observational heterarchy to the two cell system. If Intent and Extent are denoted by Φ and Ψ, respectively, the time development of the concentration is expressed as $x_{i,n+1} = (1 - c)\Phi(x_{i,n}) + \Psi(x_{i,n})$. For $\Psi(X) = f(\Phi(X))$, this expression is reduced to Eq. (13.2).

Synchronization is a well-known collective phenomenon in various multicomponent biological systems (Pikovsky et al., 2001; Mihailović et al., 2013). The exchange of information (coupling) among the components can be either global or local. This is also considered on the cell level, for example, in mechanisms of (1) cell cycle synchronization (Guireya et al., 2007) or (2) intercellular biochemical substance exchange with intracellular dynamics described by a logistic equation (Balaž and Mihailović, 2010a,b, 2011; Mihailović and Balaž, 2011a; Mihailović et al., 2011; Mihailović and Balaž, 2012; Mihailović et al., 2014).

In connection with the synchronization, an interesting question is whether a coupled map system called *active* coupling can achieve synchronization surrounded by perturbations, or whether perturbation enhances robust synchronization of many cells. Rosen (1985, 1991) and Varela (1979) pointed out that perturbation influences not only state but also function, because the disjointing between state and function results just from the framework of a set theory (Gunji and Kamiura, 2004). Therefore, the question mentioned is replaced by: how can the influence of perturbation be formalized in a term of function? According to Gunji and Kamiura (2003) the answer to this question is yielded by observational heterarchy.

First, we address the synchronization of the simplest *passive* coupling which follows Eqs. (13.2) and (3.16) and $f(x) = cx$. Following analysis by Fujisaka and Yamada (1983) we obtain bifurcation diagram with respect to the deviation of synchronized state over the coupling in the range $0 < c < 0.5$ (Fig. 13.1). Fig. 13.1a shows the diagram of bifurcation for $x_0 - x_1$ against the coupling c, where logistic parameter is 4. For each c, 10^4 iterations of the map are applied for a random initial state, and the first 10^3 steps are abandoned. It is seen that synchronization is stable in the range that $0.25 < c < 0.5$. Fig. 13.1b shows enlarged bifurcation diagram in the region, $0 < c < 0.25$. There are some windows in a chaotic region.

Further, we will consider the influence of perturbation in passive coupling. It is implemented by nonmonotonous (and/or nonlinear) flow (i.e., coupling parameter), $f : [0,1] \rightarrow [0,1]$, simply expressed as (Gunji and Kamiura, 2004)

$$f(x_{i,n}) = \begin{cases} cx_{i,n} & 0 < x_{i,n} < a \\ \dfrac{(ac - bd)(x_{i,n} - a)}{a - b} + ac & a < x_{i,n} < b \\ \dfrac{(bd - c)(x_{i,n} - 1)}{b - 1} + c & b < x_{i,n} < 1 \end{cases} . \tag{13.3}$$

It discloses the simple fluctuated flow, whereas it is nonmonotonous. The first and second crooked point is laid on the line $f(x) = cx$ and $f(x) = dx$, respectively.

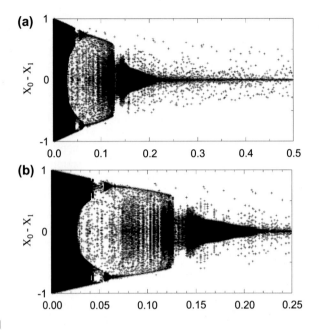

FIGURE 13.1

(a) Diagram of bifurcation with respect to the deviation of synchronized state over the coupling parameter in the range $0 < c < 0.5$, as for the passive coupling. Horizontal axis represents coupling parameter. (b) Enlarged diagram of (a), focusing on the range, $0 < c < 0.25$. The parameters have the following values: $r = 4$ and $p = 0.5$.

Eq. (13.3) can be approximated to $f(x) = cx$. In this sense, we can utilize c as the coupling parameter and call it apparent coupling parameter. Due to the fluctuated coupling parameter, the coupled map system is replaced by

$$x_{0,n+1} = (1 - c)\Phi(x_{0,n}) + f(\Phi(x_{1,n})) \tag{13.4a}$$

$$x_{1,n+1} = (1 - c)\Phi(x_{1,n}) + f(\Phi(x_{0,n})). \tag{13.4b}$$

Fig. 13.2 shows the diagram of bifurcation solution for $x_0 - x_1$ against the apparent coupling parameter, c. For each c, 10^4 iterations of the map, given by Eqs. (13.4a) and (13.4b), is applied for a random initial state, and the first 10^3 steps are abandoned. Fig. 13.2 is calculated for $a = 0.733$, $b = 0.8$, and $d = 0.75c$. Due to the nonmonotonous function of f, efficient coupling parameter is widely distributed beyond the apparent coupling parameter. That is a reason why stability of synchronization is lost, even if the apparent coupling parameter is in the stable region of synchronization, $0.25 < c < 0.5$. Therefore, within the framework of passive coupling, stormy perturbations disturb synchronization.

Here, we deal with synchronization of two passively coupled cells (Eqs. (13.2) and (10.1)) as it is shown in Fig. 7.3 (using any two in the ring of coupled cells),

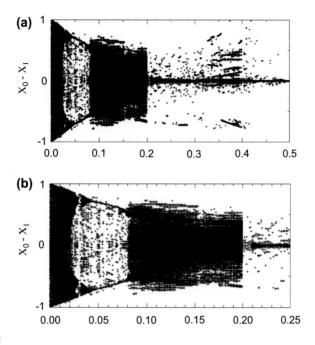

FIGURE 13.2

(a) Diagram of bifurcation for $x_0 - x_1$ against the apparent coupling parameter, c, as for the fluctuated passive coupling defined by Eqs. (13.4a) and (13.4b). Horizontal axis represents coupling parameter. (b) Enlarged diagram of (a), focusing on the range, $0 < c < 0.25$. The parameters have the following values: $r = 4$ and $p = 0.5$.

which will be considered as synchronized only when the largest Lyapunov exponent is negative (Zhou and Lai, 1998; Guireya et al., 2007). We calculate this exponent using Eqs. (7.5) and (7.6).

Fig. 13.3 depicts the normalized frequency of synchronization $F_p(\lambda < 0)$ for a system of two passively coupled cells (Eqs. (10.1) and (13.2)), as a function of cell affinity p, averaged over all values of the coupling parameter c and logistic parameter r. The value of the normalized frequency of synchronization F_p is calculated as

$$F_p = \frac{\sum N_n(\lambda < 0)}{\sum N_n(\lambda < 0) + \sum N_p(\lambda > 0)}, \tag{13.5}$$

where $\sum N_n(\lambda < 0)$ and $\sum N_p(\lambda > 0)$ are the numbers of negative and positive values of the Lyapunov exponent λ, respectively. These numbers were calculated for fixed values of p, with c and r changing in the intervals (0,1) and (1,4), respectively, with a step of 0.05. From this figure it is seen that for $p > 0.2$, F_p starts to decline, indicating a decrease of number of states which are synchronized.

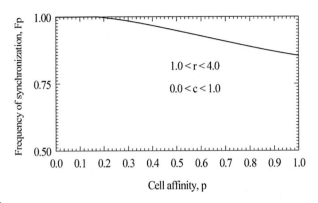

FIGURE 13.3

Normalized frequency of synchronization, $F_p(\lambda < 0)$ for system of two cells passively coupled as a function of cell affinity p. An averaging was done over all values of coupling parameter c and logistic parameter r (Mihailović et al., 2013).

13.2 SIMULATIONS OF ACTIVE COUPLING IN A MULTICELL SYSTEM

As we mentioned in Section 13.1, in nature, microscopic biochemical substrates are perpetually influenced by stormy perturbations, and these perturbations affect not only the state but also the function of cells. Therefore, we address the behavior of active coupling and estimate whether the coupled map system described above can achieve synchronization in a multicell system under the influence of perturbations (Mihailović et al., 2013). The active coupling dynamics of the two-cell system used in the simulations are defined by the following equations:

$$x_{i,n+1} = (1-c)\Phi_n(x_{i,n}) + \Psi_n(x_{i,n}) \tag{13.6a}$$

$$\Phi_{n+1} = f\Psi_n ff^* \tag{13.6b}$$

$$\Psi_{n+1} = f^*\Phi_{n+1} \tag{13.6c}$$

$$\Phi^0(x_{i,n}) = rx_{i,n}^p\left(1 - x_{i,n}^p\right). \tag{13.6d}$$

We note that the dynamical system defined by Eqs. (3.16) and (13.2) is called the passive coupling and that is the usual coupled map system. The active coupling can be approximated to passive coupling, where the approximation is defined by adjunction or the equivalence between Intent and Extent. Compared with passive coupling, the behavior of active coupling is much more complex (Gunji and Kamiura, 2004; Mihailović et al., 2013). In Eqs. (13.6a)–(13.6d), because of a pseudo-inverse map, f^*, all calculations are defined to be approximations. In simulations, the Intent map was a discontinuous map, expressed by $\Psi_{n+1} = f^*\Phi_{n+1}$. In order to see how perturbation enhances robust behavior in the framework of observational heterarchy in a multicell system represented by a ring of coupled cells (Fig. 7.3), we consider the following model.

In our approach, a cell moves locally in its environment without making long pathways. According to Mihailović and Balaž (2012), the system of coupled difference equations for a set of N cells exchanging biochemical substance, can be written in the form of a matrix equation

$$\mathbf{XN1} = (\mathbf{A} + \mathbf{B}) \cdot \mathbf{XN}. \tag{13.7}$$

The elements in the matrices in Eqs. (13.6a)−(13.6d) are

$$XN1_{i,n+1} = x_{i,n+1}, \quad XN_{i,n} = x_{i,n},$$
$$A_{i,k} = (1-c)\Phi(x_{i,n})\delta_{i,k} \tag{13.8}$$

and

$$B_{i,k} = \begin{cases} \Psi_n(x_{k,n}) & k = i+1, i < N \\ 0 & k \neq i+1, i < N \\ \Psi_n(x_{k,n}) & k = 1, i = N \\ 0 & k \neq 1, i = N \end{cases} \tag{13.9}$$

where $i = 1, 2, \ldots, N$ and is the Kronecker delta.

We perform simulations with active coupling in the ring of $N = 100$ cells, defined by Eqs. (13.6a)−(13.6d), with and without the perturbation given in Fig. 13.4. The results of the simulations are shown in Fig. 13.5. In this figure the

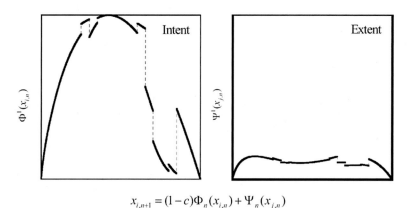

$$x_{i,n+1} = (1-c)\Phi_n(x_{i,n}) + \Psi_n(x_{j,n})$$

FIGURE 13.4

A pair of Intent and Extent maps in fluctuated active coupling expressed as Eqs. (13.6a)−(13.6d). The left diagram represents the Intent map, $\Phi^1(x_{i,n})$ with $i = 0, 1$, and the right diagram represents the Extent map, $\Psi^1(x_{i,n})$ with $i = 0, 1$. The Intent map is replaced by a discontinuous map f^* (Mihailović et al., 2013).

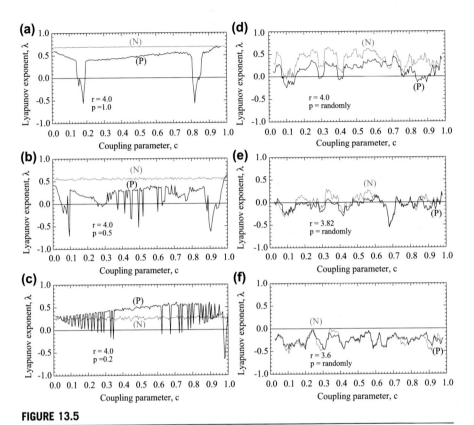

FIGURE 13.5

Diagram of Lyapunov exponent, λ, against coupling parameter c for the fluctuated active coupling defined by Eqs. (13.6a)–(13.6d)—P (*black line*) compared to passive coupling—N (*gray line*) for different values of cell affinity p and logistic parameter r. In (a–c) p takes the fixed values (1, 0.5, 0, 2), while $r = 4$. In (d–f) p is randomly chosen, while r takes values 4, 3.82, and 3.6, respectively. Simulations are performed with the ring of $N = 100$ cells (Mihailović et al., 2013).

Lyapunov exponent λ is plotted against the coupling parameter c for active coupling with perturbation (black line) compared to the passive coupling (gray line), for different values of the cell affinity p and logistic parameter r. We calculated the Lyapunov exponent using Eqs. (7.5) and (7.6).

In calculating λ, for each c from 0.0 to 1.0 with step 0.005, we apply 10^4 iterations for an initial state, and then the first 10^3 steps are discarded. To see how the active coupling modifies the synchronization property of the model, we perform two kinds of simulations. Firstly, we use $r = 4.0$ and a fixed value of the cell affinity p (Fig. 13.5a–c); secondly, we use a randomly chosen c and a logistic parameter r with values of 4, 3.82, and 3.6, respectively (Fig. 13.5d–f). Fig. 13.5a–c shows that in the chaotic regime ($r = 4.0$), regardless of the value of p, the Lyapunov exponent

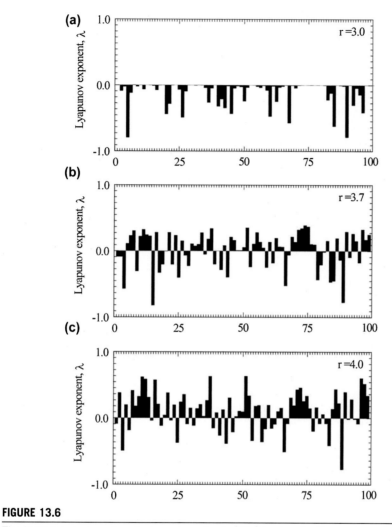

FIGURE 13.6

The Lyapunov exponent against number of cells N in the ring, for three values of r: 3.0 (a), 3.7 (b), 4.0 (c); c takes values in the interval (0,1) while p is randomly chosen for each c (Mihailović et al., 2013).

is always positive ($\lambda > 0$) and therefore the process of biochemical substance exchange in a multicell system is always unsynchronized. However, the stormy perturbation disturbs this state (Fig. 13.5a–c). Although the logistic parameter is settled at $r = 4$ for chaotic behavior, the coupling parameter c tunes the interaction and leads to synchronization in some intervals, particularly for $p = 1$ and $p = 0.5$. This behavior is most pronounced in Fig. 13.5d–f where p is randomly chosen; here the process of biochemical substance exchange in a multicell system exhibits a strong tendency toward synchronization, even though the logistic parameter r is in the chaotic region ($r = 4$, 3.82, and 3.6).

The dynamics of the coupled multicell system (Eqs. (13.6a)−(13.6d) are governed by four main parameters: the number of cells N (ring size), the coupling parameter c, the logistic parameter r, and the cell affinity p. Here we present the collective dynamics of the multicell system by varying the number of cells from $N = 1$ to 100 for (1) c taking values in the interval (0,1) and (2) p randomly chosen for each c. We perform simulations for three values of r: 3.0, 3.7, and 4.0. We calculate Lyapunov exponent as in previous experiments. Fig. 13.6a−c depicts the Lyapunov exponent against number of cells N in the ring for three values of r. From this figure it is seen that, regardless of the number of cells, the process of biochemical substance exchange in a multicell system is much more synchronized for lower values of the logistic parameter r. A similar simulation with the dynamics of the coupled multicell system was done in Ghosh et al. (2010), but for just two parameters (N and c).

REFERENCES

Arenas, A., Diaz-Guilera, A., Kurths, J., Moreno, Y., Zhou, C., 2008. Synchronization in complex networks. Phys. Rep. 469, 93−153.

Balaz, I., Mihailovic, D.T., 2010a. Modeling the intercellular exchange of signaling molecules depending on intra- and inter-cellular environmental parameters. Arch. Biol. Sci. 62 (4), 947−956. Belgrade.

Balaz, I., Mihailovic, D.T., 2010b. A short essay about modeling local interactions and functional robustness in living systems. In: Mihailovic, D.T., Lalic, B. (Eds.), Advances in Environmental Modeling and Measurements. Nova Science Publishers, Inc, New York, pp. 77−88.

Balaz, I., Mihailović, D.T., 2011. A model representing biochemical substances exchange between cells. Part I: model formalization. J. Appl. Funct. Anal. 6 (1), 70−76.

Bickhard, M.H., Terveen, L., 1996. Foundational Issues in Artificial Intelligence and Cognitive Science. Elsevier, New York.

Cairns, J., Foster, P.L., 1991. Adaptive reversion of a frameshift mutation in *Escherichia coli*. Genetics 128, 695−701.

Chen, Y., Rangarajan, G., Ding, M., 2003. General stability analysis of synchronized dynamics in coupled systems. Phys. Rev. E 67, 026209.

Devaney, R.L., 2003. An Introduction to Chaotic Dynamical Systems, second ed. Westview Press, Boulder.

Fujisaka, H., Yamada, T., 1983. Stability theory of synchronized motion in coupled-oscillator systems. Prog. Theor. Phys. 69, 32−47.

de Groot, H., Littauer, A., 1989. Hypoxia, reactive oxygen and cell injury. Free Radic. Biol. Med. 6, 541−551.

Ghosh, C., Rangarajan, G., Sinha, S., 2010. Stability of synchronization in a multi-cellular system. Europhys. Lett. 92, 40012.

Guireya, E.J., Beesb, M.A., Martina, A.P., Srokosza, M.A., Fashama, M.J.R., 2007. Emergent features due to grid-cell biology: synchronisation in biophysical models. Bull. Math. Biol. 69, 1401.

Gunji, Y.-P., Kamiura, M., 2003. Observational heterarchy as phenomenal computing. In: Selected Papers from the Computers and Philosophy Conference 2003, Sydney, Australia (Australian Computer Science Communications, Sydney 2004) 39.

Gunji, Y.-P., Kamiura, M., 2004. Observational heterarchy enhancing active coupling. Phys. D 198, 74−105.

Jen, E., 2003. Stable or robust? What's the difference? Complexity 8, 12−18.

Jones, D.P., 1985. The role of oxygen concentration in oxidative stress: hypoxic and hyperoxic models. In: Sies, H. (Ed.), Oxidative Stress. Academic Press, London, pp. 151−195.

Kauffman, S.A., 2002. Investigations. Cambridge University Press, Cambridge.

Kolter, R.D., Siegele, A., Tormo, A., 1993. The stationary phase of the bacterial life cycle. Annu. Rev. Microbiol. 47, 855−874.

Levin, S.A., 2006. Fundamental questions in biology. PLoS Biol. 4, e300.

Mihailović, D.T., Balaž, I., Arsenić, I., 2013. A numerical study of synchronization in the process of biochemical substance exchange in a diffusively coupled ring of cells. Cent. Eur. J. Phys. 11, 440−447.

Mihailović, D.T., Balaz, I., 2011. A model representing biochemical substances exchange between cells. Part II: effect of fluctuations of environmental parameters to behavior of the model. J. Appl. Funct. Anal. 6 (1), 77−84.

Mihailović, D.T., Budinčević, M., Balaž, I., Mihailović, A., 2011. Stability of intercellular exchange of biochemical substances affected by variability of environmental parameters. Mod. Phys. Lett. B 25, 2407−2417.

Mihailović, D.T., Balaž, I., 2012. Synchronization in biochemical substance exchange between two cells. Mod. Phys. Lett. B 26, 1150031−1.

Mihailović, D.T., Kostić, V., Balaz, I., Cvetković, L., 2014. Complexity and asymptotic stability in the process of biochemical substance exchange in a coupled ring of cells. Chaos Solitons Fractals 65, 30−43.

Pikovsky, A., Rosenblum, M., Kurths, J., 2001. Synchronization: A Universal Concept in Nonlinear Sciences. Cambridge University Press, Cambridge.

Rajesh, S., Sinha, S., 2008. Measuring collective behaviour of multicellular ensembles: role of space-time scales. J. Biosci. 33, 289−301.

Rajesh, S., Sinha, S., Sinha, S., 2007. Synchronization in coupled cells with activator-inhibitor pathways. Phys. Rev. E 75, 011906.

Rosen, R., 1985. Theoretical Biology and Complexity, Three Essays on the Natural Philosophy of Complex Systems. Academic Press, Orlando.

Rosen, R., 1991. Life Itself. Columbia University Press, New York.

Shapiro, J.A., 1995. Adaptive mutation: who's really in the garden? Science 268, 373−374.

Shapiro, J.A., 1992. Natural genetic engineering in evolution. Genetica 86, 99−111.

Spector, M.P., Kenyon, W.J., 2012. Resistance and survival strategies of *Salmonella enterica* to environmental stresses. Food Res. Int. 45, 455−481.

Varela, F.J., 1979. Principles of Biological Autonomy. North-Holland, Amsterdam.

Zhou, C., Lai, C.-H., 1998. Synchronization with positive Lyapunov exponents. Phys. Rev. E 58, 5188−5191.

Heterarchy and albedo of the heterogeneous environmental interfaces in environmental modelling

14

14.1 HETERARCHY AND AGGREGATION OF ALBEDO OVER HETEROGENEOUS ENVIRONMENTAL INTERFACES

The shortwave radiation albedo is an important boundary condition for environmental models comprising the atmospheric solar radiative transfer module. It is defined as the ratio of diffuse upward and global (i.e., direct plus diffuse) downward shortwave radiation at the Earth's surface. Among the tasks in the grid-based environmental models, the basic one is the determination of albedo over the heterogeneous environmental interface with height difference between the patches (Mihailovic and Balaž, 2007; Kapor et al., 2002, 2010a,b, 2012; Mihailović et al., 2004; Ćirišan et al., 2010; Kreuter et al., 2014). The structure of the grid-box on the environmental interface can remarkably vary both spatially and temporally, consisting of the patches covered with the surfaces of different origin. Here, we concentrate ourselves on environmental interface between the atmosphere and the land. Land part is composed of the patches of plant communities, bare soil, rocky ground, or all water surfaces and other natural ones, providing us with a very heterogeneous picture in the grid-box. It is essential to stress that our main interest is the situation when these patches differ in height, so that there appears a geometrical effect influencing the value of the shortwave radiation leaving the surface, directly affecting the value of the albedo seen by the instruments above the surface.

The most common approach for calculation of the albedo, in the case of patches of equal height, is to make a simple averaging to determine the albedo as the grid-box average albedo (in the further text referred as the SA). However, a physics-based analysis indicates that there is a significant deviation of the albedo above varying height heterogeneous surface from that calculated by simple averaging (McComiskey et al., 2006), seriously affecting the calculated values of quantities describing surface biophysical processes like land surface energy budgets, canopy photosynthesis and transpiration, urban surface physics and snow melt, among others (Hu et al., 1999; Jacobson, 1999; Kapor et al., 2002; Schwerdtfeger, 2002; Wendisch et al., 2004). It is, therefore, important to understand the general behavior and limitations of the approaches used for aggregating the albedo over a heterogeneous grid-box in current environmental models. The assumptions for aggregating the albedo over a very heterogeneous surface where various surfaces occur at different heights were theoretically considered in the paper by Mihailovic et al. (2004), and later by Kapor et al. (2010a,b) and Mihailović et al. (2012).

As mentioned above, the airborne measurements of the surface albedo indicated that the values obtained are lower than the ones parameterized in environmental models (Wendisch et al., 2004). Now, we consider the heterarchy in the context of surface albedo parameterization over heterogeneous environmental interfaces. Fig. 14.1 schematically depicts two hierarchical lines H_A and H_B in the

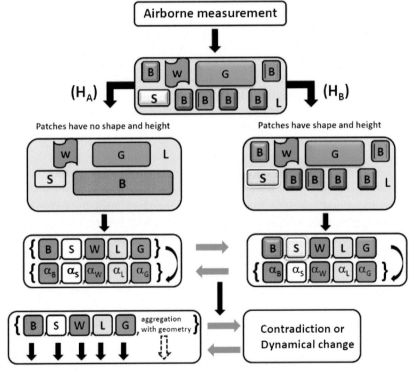

FIGURE 14.1

Example of heterarchy illustrating intersurface albedo parameterization methods over grid-box (two-dimensional and three-dimensional geometry) communication with the mixture of interlayer operation and intralayer operation. List of patch abbreviations: green area (G), buildings (B), water surface (W), area covered by snow (S), and bare soil and short grass area (L). At the bottom squares are symbols for surface albedos of patches.

parameterization of surface albedo over a heterogeneous grid-box consisting of patches [green area (G), buildings (B), water surface (W), area covered by snow (S), and bare soil and short grass area (L)]. Boxes on the hierarchical line H_A are depicted two-dimensionally, while on the H_B hierarchical line, they are drawn with a relief structure indicating the presence of the third dimension. The H_A and H_B lines correspond to methods of parameterization, which could be either the simple SA approach, or it could take into account the geometrical effect (in the further text referred as the PA), respectively. Both gird-boxes are represented as the sets of patches (objects)—upper squares and albedos (attributes)—bottom squares. Here we have intersurface albedo parameterization methods (two-dimensional and three-dimensional geometry) communication with the mixture of interlayer operation and intralayer operation. It consists of intralayer operation such as mapping a patch to two-dimensional or three-dimensional images, interlayer operation such as the aggregation, and the mixture such that the aggregation needs a new property.

Here we briefly summarize the main theoretical features of the method for aggregating the albedo over a very heterogeneous surface where various surfaces occur at different heights, suggested in Mihailovic et al. (2004). The basic idea of the approach relies on the fact that a part of the radiation reflected from the lower surface is absorbed by the lateral sides S_3 of the surface lying on a higher level (Fig. 14.2). The albedo is measured at a level above all the surfaces so the absorbed radiation is not being registered by the instruments, although it would appear in any simple averaging of the total albedo. The amount (ratio) of the reflected energy lost is obtained by taking into account the solid angle within which these lateral sides are seen from each point of the lower surface.

There are several basic assumptions for this calculation. First, it is assumed that the radiation reflected from a given surface is diffuse and homogeneous, neglecting the multiple scattering effects and the dependence of the albedo on the zenithal angle. This is based on the assumption that the reflecting surface is sufficiently rough so that the reflected radiation is isotropic and amount of energy reflected within any solid angle is proportional only to the solid angle itself. In this way we can neglect the orientation of the radiation falling to that surface.

To define the flux that is lost due to the absorption, we define a loss coefficient k_l, determined by the radiant energy flux ratio. Following Liou (2002), the basic expression to calculate the radiant energy flux (dE/dt) is

$$(dE/dt) = IdS \cos \theta d\Omega \qquad (14.1)$$

where I is the total intensity of radiation obtained from the monochromatic intensity by integrating it over the entire range of the spectrum, $dS = dxdy$ is the infinitesimal element of surface from which radiation is reflected, $\cos \theta$ gives the direction of the radiation stream and $d\Omega$ is the element of solid angle within which the infinitesimal amount of radiant flux emitted from the infinitesimal surface element $dxdy$ is confined to. The amount of flux emitted from the lower horizontal surface into the upper space is $(dE/dt)_h = IS_1\pi$, where S_1 is the area from which the radiation is reflected. The total energy coming from the lower horizontal surface toward the

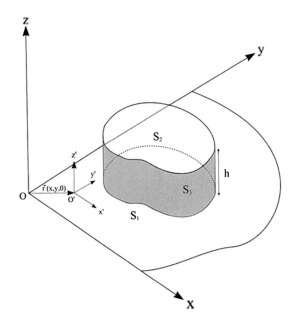

FIGURE 14.2

Schematic representation of an arbitrary grid-box geometry consisting of two surfaces differing for the height h.

Reprinted with permission from Mihailović, D.T., Kapor, D., Ćirišan, A., Firanj, A., 2012. Parametrization of the albedo over the heterogeneous surfaces for different geometries in a land surface scheme by the Monte Carlo ray-tracing method. Atmos. Res. 107, 51–68. Notation follows the text (Mihailovic et al., 2004; Kapor et al., 2010a).

lateral surface is $(dE/dt)_l$ and is derived using the following expression where the boundaries of the integration for a given point are determined over the local azimuthal (Φ_l, Φ_u) and zenithal (θ_l, θ_u) angles in terms of the x,y coordinates (Figs. 1 and 2 in Kapor et al. (2010a), or in more detail Figs. 3, 4, 6, and 7 in Mihailovic et al. (2004)). Thus, we have

$$(dE/dt)_l = I \iint_S dxdy \int_{\Phi_l(\overrightarrow{r})}^{\Phi_u(\overrightarrow{r})} d\Phi \int_{\theta_l(\overrightarrow{r},\varphi)}^{\theta_u(\overrightarrow{r},\varphi)} \cos\theta \sin\theta, \qquad (14.2)$$

where subscript l denotes the lower bound of integration, while subscript u its upper one. The ratio of the expression (14.2) and $(dE/dt)_h = IS_1\pi$ gives the loss coefficient k_l $(0 < k_l \leq 1)$

$$k_l = (dE/dt)/IS_1\pi. \qquad (14.3)$$

If we assume that the grid-box of the area S is divided into two parts having the areas S_1 and S_2 with corresponding albedos α_1 and α_2, respectively, the average albedo over the grid-box following the standard approach, commonly used in environmental models, is

$$\overline{\alpha}_c = \alpha_1 \sigma_1 + \alpha_2 \sigma_2 \tag{14.4}$$

where σ_i is the fractional cover, calculated as a ratio of patch's area S_i and the total grid-box area $S(\sigma_i = S_i/S, \, i = 1, 2)$. Taking into account the geometry expressed through the loss coefficient, the average albedo is then calculated as

$$\overline{\alpha}_n = (1 - k_l)\alpha_1 \sigma_1 + \alpha_2 \sigma_2. \tag{14.5}$$

The loss coefficient definition is conceptually analogous to the idea of the sky-view factor introduced by Oke (1987). Detailed overview of the recently proposed evaluations of k_l needed for calculating the average albedo, for different geometries is available, in Mihailovic et al. (2004) and Kapor et al. (2010a). It is important to notice that for simple geometries one can derive k_l completely analytically, and we shall outline one such calculation later.

However, with more complex geometries, a different approach is needed. In the Monte Carlo approach to different simulations in environment, individual quanta or particles are subjected to the same physical and chemical processes and events in the computer as in the physical world. It can be thought of as a direct simulation of environmental processes not requiring explicit equations and their solution as the basis for the simulation. The use of the Monte Carlo method in atmospheric optics is summarized in the pioneering work by Marchuk et al. (1980). It is broadly used in both theoretical and applied environmental sciences. So far, quite a few researchers have applied the Monte Carlo method into the evaluation of the surface albedo when the land surface is characterized by complex geometry in land surface models. However, Mayer et al. (2010) used similar logic in MYSTIC three-dimensional radiative transfer model.

In the case of very complicated grid-box geometry, when the analytical solution for the loss coefficient is not accessible, the Monte Carlo Ray Tracing (MCRT) method is reliable and most efficient method for calculating the loss coefficient. Note that the simple Monte Carlo method is not too efficient (Sanchez, 1998) compared to the MCRT method which reproduces analytical results up to a high precision, as shown by our studies. Let us first explain the general idea of the calculation procedure. The main idea of the MCRT method is to follow the path of the appropriately chosen ray of light, after it had undergone diffuse, single scattering from the lower surface S_1 of the grid-box. We are interested in the possibility that the ray may be absorbed by the vertical boundary, that is the lateral side of surface lying on a higher level, if it reaches it in accordance with our single scattering assumption. Averaging in this way, the observed behavior over a large number ($N = 10^6$) of the followed light paths, we can conclude about the value of loss coefficient k_l as the origin of radiative flux loss, within a given grid-box geometry. The details of a particular Monte Carlo procedure strongly depend on the geometry of the grid-box, so we will illustrate the particular procedure related to the simplest grid-box geometry shown in Fig. 14.3.

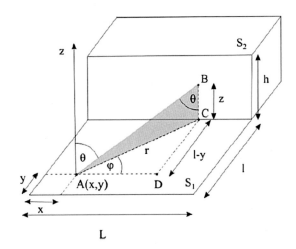

FIGURE 14.3

Schematic diagram of the procedure for application of the MCRT method for calculating the aggregated albedo.

Reprinted with permission from Mihailović, D.T., Kapor, D., Ćirišan, A., Firanj, A., 2012. Parametrization of the albedo over the heterogeneous surfaces for different geometries in a land surface scheme by the Monte Carlo ray-tracing method. Atmos. Res. 107, 51–68.

The point $A(x,y)$, which belongs to the lower surface and represents a point of the intercept of this surface and the incoming beam, is randomly sampled by generating two random numbers r_1 and r_2 uniformly distributed in the interval $(0,1)$. The area of the lower surface is $S_1 = L \times l$. So we write that $x = r_1 \times L$ and $y = r_2 \times l$. In agreement with our basic assumption—diffusive and single ray scattering, we choose a random direction (θ,φ) in the upper half-space [with $\theta \in (0,\pi/2)$; $\varphi \in (0,2\pi)$] to simulate the trace of scattered beam. Using the uniform random numbers r_3 and r_4 from the range $(0,1)$ gives us the way to choose this random direction as $\varphi = r_3 \times 2\pi$ and $\theta = r_4 \times \pi/2$. Further approach was based on the idea of line-plane intersection, where the reflected beam was treated as a straight line while the vertical area had the role of a plane. The intersection of the line and the plane can be derived using general expression of the analytical geometry (McCrea, 1960). Now, if the point $B(x,y,z)$ lies within the borders of vertical area, then diffusively scattered beam will be absorbed (in the single scattering approximation) and this case is positive for absorption. This procedure was repeated $N = 10^6$ times and the loss coefficient was estimated as

$$k_l = N_a(\text{number of cases which were positive for absorption})/ \\ N(\text{number of conducted numerical experiments}). \quad (14.6)$$

Cases treated in this chapter can be decomposed into sets of planes which might have different orientation with respect to the reference frame, which is taken into account in the equation of the plane, i.e., its normal.

14.2 INFLUENCE OF THE ALBEDO CALCULATION ON THE EFFECTIVE TEMPERATURE OF THE HETEROGENEOUS GRID-BOX CONSISTING OF DIFFERENT COVERS

Our first example is the geometry which allows the analytical solution. It is the case of a "rectangular prism" ("propagating building") geometry consisting of a rectangular prism with quadratic basis (side l) placed in the center of the grid-box (side L) (Fig. 14.4a and e), having different albedo and height from the surrounding area. Changing the relative dimensions of this prism influences significantly the value of the reflected energy lost, leading to the various aggregated albedo results.

The detailed procedure of loss coefficient calculation in case of this geometry is given in Appendix A of Mihailovic et al. (2012). The final analytical result for the loss coefficient k_l as a function of the dimensionless quantities is: reduced relative height $\widehat{h} = h/L$ and the reduced relative length $\widehat{l} = l/L$ (in the further text will be indicated as reduced height and length, respectively) is

$$k_l\left(\widehat{l}, \widehat{h}\right) = \left(4k_{l1}\left(\widehat{l}, \widehat{h}\right) + 4k_{l2}\left(\widehat{l}, \widehat{h}\right)\right) \big/ \left(1 - \widehat{l}^{\,2}\right)\pi\right), \tag{14.7}$$

where subscripts 1 and 2 indicate corresponding surfaces, l is the length of the "propagating building" edge, while L is the size of the grid-box. k_{l1} denotes the contribution coming from the points on the lower surface from which the radiation can "hit" only one side of the building, while k_{l2} describes the contribution of the points from which the radiation can reach two sides of the building.

The comparison of the results for the loss coefficient obtained from the analytical expression and by the MCRT method is given in (Fig. 14.5), which depicts the dependence of the loss coefficient on different values of the reduced length (0.0, 0.1, 0.25, 0.5, 0.707, 0.866, 1.0) and height of the "propagating building." It can be seen from this figure that the values of the loss coefficient obtained by the MCRT method are highly close to the ones calculated by the analytical expression. We have also used a particular form of the numerical integration for the fourfold integration and these results give lower values for the loss coefficient compared to the analytical ones, and also longer CPU time for the calculation, thus justifying the use of MCTR as a reliable method for our calculations.

The loss coefficient evaluated by the MCRT method is used to evaluate the ratio of the albedos calculated with geometrical effect included (Eq.(14.5)) and by the standard method (Eq. (14.4)). Then, for the further analysis, we have calculated the ratio Γ defined as

$$\Gamma = \overline{\alpha}_n / \overline{\alpha}_c. \tag{14.8}$$

Here, we plot a series of values of Γ for the "propagating building" made of concrete and surrounded by different underlying surfaces (water, forest, garden, concrete, agricultural land, and snow, that are commonly met in the environmental modelling) as a function of the reduced side length of "propagating building" for various heights. The results can be summarized in the set of plots in Fig. 14.6.

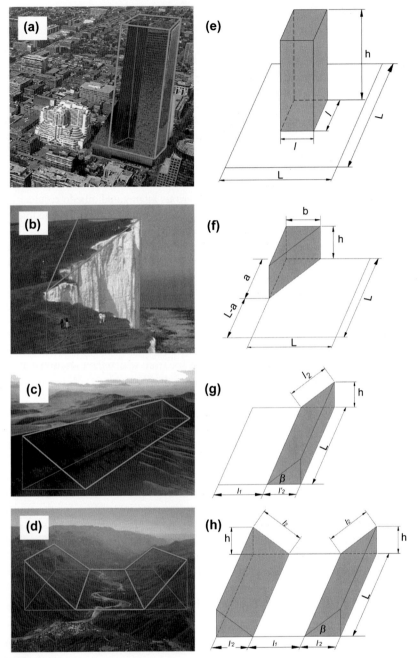

FIGURE 14.4

Schematic representation of natural and artificial solid surfaces (a–d) and their approximate geometries in parameterization of the albedo over the heterogeneous surfaces (e) "rectangular prism," (f) "trilateral prism," (g) "slope," and (h) "canyon" geometry.

Reprinted with permission from Mihailović, D.T., Kapor, D., Ćirišan, A., Firanj, A., 2012. Parametrization of the albedo over the heterogeneous surfaces for different geometries in a land surface scheme by the Monte Carlo ray-tracing method. Atmos. Res. 107, 51–68.

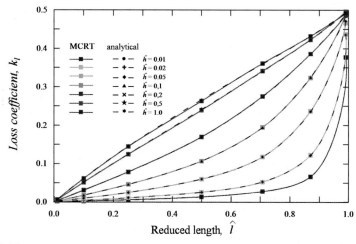

FIGURE 14.5

Dependence of the loss coefficient on the reduced length $\widehat{l}$ for different values of the reduced height $\widehat{h}$ of the "propagating building" geometry obtained by an analytical expression (A6) in Appendix A (Mihailovic et al., 2012) and by the MCRT method.

Reprinted with permission from Mihailović, D.T., Kapor, D., Ćirišan, A., Firanj, A., 2012. Parametrization of the albedo over the heterogeneous surfaces for different geometries in a land surface scheme by the Monte Carlo ray-tracing method. Atmos. Res. 107, 51–68.

We notice here a fascinating fact: the existence of a minimum of the ratio Γ, which lies at the value between 0.87 and 0.73 of the value of the standard albedo. Although it is rather difficult to examine this phenomenon analytically, let us try to reason out the general features. We are aware of the fact that for $\widehat{l} = 0$, there exists a single surface with albedo α_1, and an increase of the dimension of higher surface obviously leads to a decrease of the albedo. However, for $\widehat{l} = 1$, there again exists a single surface with albedo α_2 and just before this final limit we should have a smaller albedo. Obviously, in both limiting cases we have $\Gamma = 1$, while albedo is lower between them so there must exist a minimum. These are the first, qualitative results, and one has to examine further the behavior of this minimum, its variation with albedo and height.

Now, we continue by analyzing first the geometry with an elevated corner area having triangular base, the so-called trilateral prism geometry (Fig. 14.4b and f). In a square grid-box, in the top left corner, there is an elevated surface at the height h with respect to the rest of the grid-box. This area of triangle shape has the dimensions $a \times b$, where a is the fixed arm, exactly half of the length of the grid-box length, while the arm b takes different values depending on the angle of the triangle. This choice of dimensions was just for convenience, to simplify the analytical expression.

In our studies (Mihailovic et al., 2012) we also included the so-called slope (Fig. 14.4c and g) and canyon geometry (Fig. 14.4d and h) which are relevant for

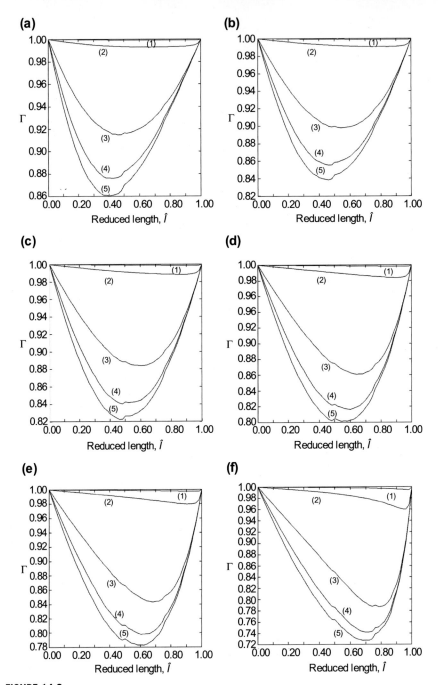

FIGURE 14.6

Dependence of Γ (the ratio of the albedo calculated by the PA method to the SA albedo calculation method) on the reduced length $\hat{l}$ for the grid-box consisting of the "propagating building." The reduced height $\hat{h}$ in all plots corresponds to the values: $\hat{h} = 0.001$ (1); 0.01 (2); 0.2 (3); 0.5 (4); 1 (5). The plots correspond to the following values of the lower surface albedo α_1: (a) 0.1 (water); (b) 0.15 (forest); (c) 0.20 (garden); (d) 0.3 (concrete); (e) 0.4 (agricultural land); (f) 0.95 (snow) which are taken from Oke (1987). All calculations of the albedo using the PA method were performed by the MCRT.

Reprinted with permission from Mihailović, D.T., Kapor, D., Ćirišan, A., Firanj, A., 2012. Parametrization of the albedo over the heterogeneous surfaces for different geometries in a land surface scheme by the Monte Carlo ray-tracing method. Atmos. Res. 107, 51–68.

practice. The analytical solution to these complex geometries was too complicated to be found, so a direct system simulation was done by the MCRT method.

In Fig. 14.7, the ratio of the loss coefficient calculated by the MCRT method is presented for different values of the reduced lengths $\widehat{b}$ and $\widehat{l}_1$ (0.0, 0.1, 0.25, 0.5, 0.75, 0.99) and height of the surface lying on a higher level. The reduced dimensionless quantities $\widehat{b} = b/L$ and $\widehat{h}$ were introduced for the simplicity of the loss coefficient calculation, where $L \times L$ is already the mentioned area of the grid-box.

It seems obvious that the changes in the aggregated albedo value with respect to the conventional approach may lead to the differences in the effective surface temperature calculated over the grid-box. To illustrate these differences, we have performed a set of experiments based on both approaches in calculating the albedo. In these experiments, we computed the effective surface temperature over the grid-box using the land surface scheme LAPS (Land Air Parameterization Scheme) designed to be run either as a standalone model or as the part of an atmospheric model (Mihailovic et al., 2010). The numerical tests over the grid-box with different geometries have been performed with the forcing meteorological data for July 17, 1999, in Philadelphia, PA. The grid-box used in these simulations, represents the Baxter site, with the prevailed synoptic conditions described in Zhang et al. (2001). The forcing data were used from the lowest level of MM5 model (Dudhia, 1993). The initial conditions for prognostic variables were the same as in Zhang et al. (2001), number of vertical layers was 32, with the lowest level set at 10 m, while the time step was 600 s. For the sake of the simplicity, the experiments were performed with two surfaces of different albedos with the use of the above-mentioned geometries. The idea of experiments was to establish changes in daily course of the effective surface temperature over the heterogenous grid-box caused by the difference in calculation of the albedo by the standard approach and the one taking into account the geometrical effect. The runs of the effective surface temperature were done using the time step of 600 s. We must repeat here that we assume isotropic diffuse reflection so that the reflected light is isotropic and does not depend on the zenithal angle of incoming light. Of course in a more detailed study, one should take care also of daily variation of zenithal angle when one follows the daily changes of temperature.

We start with an example of the "rectangular prism" geometry ("propagating building") with the dimensions of the grid-box 100 m × 100 m. The simulations were done for several patch areas, with different surface types. Accordingly, each of the subregions had different and corresponding albedo. We analyzed three situations. In all of them, the central solid area consisted of the concrete having the albedo value of 0.30 (Jacobson, 1999). The other patch, surrounding the "propagating building," of the grid-box, in these three simulations, was covered with the grass (G), the forest (F), and the concrete (C) with the values of the albedo 0.20, 0.15, and 0.30, respectively. The albedo was calculated taking the following values of the reduced lengths ($\widehat{l}$) 0.50, 0.707, and 0.866 providing a central square, that takes 25%, 50%, and 75% of the grid-box area. We wanted to analyze the daily effective surface temperature course differences obtained by both the methods in

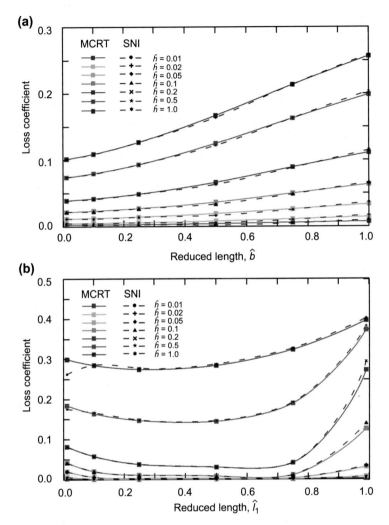

FIGURE 14.7

Dependence of the loss coefficient on the reduced lengths $\hat{b}$ and $\hat{l}_1$, for different values of the reduced height $\hat{h}$, obtained by the standard numerical integration (SNI) approach and by the MCRT method: (a) the "trilateral prism" geometry and (b) the "slope" geometry.

Reprinted with permission from Mihailović, D.T., Kapor, D., Ćirišan, A., Firanj, A., 2012. Parametrization of the albedo over the heterogeneous surfaces for different geometries in a land surface scheme by the Monte Carlo ray-tracing method. Atmos. Res. 107, 51–68.

a case of the grid-box covered with each type of the surfaces, i.e., G, F, and C (as the fraction of the grid-box in amounts of 25%, 50%, or 75%, each of them) surrounding the "propagating building" taking 75%, 50%, or 25% area of the grid-box, respectively. The altitude of a central patch was case sensible. In the G—C simulations, the building was 2 m high, while the grass was 0.5 m. For the case of F—C simulations, the forest height is taken to be 2.5 m, while the building was 10 m. The same height of a central area concrete building (10m) is used when both patches in the grid-box were consisting from the concrete (C—C simulations).

To define more precisely the difference between the daily effective surface temperature course obtained by two methods, the root mean square error (RMSE) is calculated. In Fig. 14.8 are depicted the averaged values of the RMSE for the G—C, F—C, and C—C for 24 h simulations. It can be seen from the figure that maximal increase in the effective surface temperature and decrease in albedo is for 50—50% area coverage when both patch areas are covered with the concrete. In F—C simulations we also obtained a significant RMSE, especially for case when the central area occupies 25% of grid-box. The lowest values of the RMSE were obtained for the G—C simulations.

We now deal with the "trilateral prism" geometry with the same dimensions of the grid-box as the previous one. We considered the prism having a triangular shape

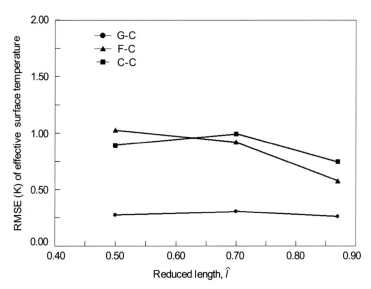

FIGURE 14.8

Dependence of RMSE of the effective surface temperature, obtained by using the PA (performed by the MCRT) and SA albedo calculation methods, on the reduced length $\hat{l} = 1$.

Reprinted with permission from Mihailović, D.T., Kapor, D., Ćirišan, A., Firanj, A., 2012. Parametrization of the albedo over the heterogeneous surfaces for different geometries in a land surface scheme by the Monte Carlo ray-tracing method. Atmos. Res. 107, 51–68.

with the area of $\widehat{a} \times \widehat{b}/2$ and the reduced height $\widehat{h}$. The reduced value $\widehat{a} = 0.5$ was fixed on the half size of the grid-box while the reduced length $\widehat{b} = b/L$ was taking values 0.25, 0.5, 0.75, and 1. The fractional covers of the elevated surface and the rest of the grid-box were $\sigma_2 = \widehat{a} \times \widehat{b}/2$ and $\sigma_1 = 1 - \widehat{a} \times \widehat{b}/2$, respectively. The simulations were done similar to the ones for rectangular prism, i.e., with the same surface types (G and F), while instead of C we used R surface type.

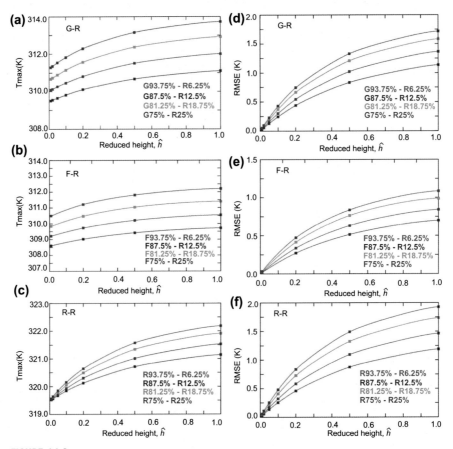

FIGURE 14.9

The maximum of the effective surface temperatures, obtained by the PA (performed by the MCRT) (left panels, a–c) and RMSE (right panel, d–f) on the reduced height $\widehat{h}$, for different values of the fractional covers, for the "trilateral prism" geometry. The RMSE is calculated on the basis of differences between the PA and SA albedo calculation methods for all points in the simulated daily courses.

Reprinted with permission from Mihailović, D.T., Kapor, D., Ćirišan, A., Firanj, A., 2012. Parametrization of the albedo over the heterogeneous surfaces for different geometries in a land surface scheme by the Monte Carlo ray-tracing method. Atmos. Res. 107, 51–68.

We also analyzed the daily effective surface temperature course differences obtained by both methods in the case of the grid-box covered with each type of the surfaces, i.e., G, F, and R, that fills the surface next to the "trilateral prism" which consists of the concrete and taking the following fractions of σ_1 and σ_2: 0.9375 (σ_1)—0.6250 (σ_2), 0.8750 (σ_1)—0.1250 (σ_2), 0.8125 (σ_1)—0.1875 (σ_2), and 0.7500 (σ_1)—0.2500 (σ_2). The simulations for which the LAPS model was run were G—R, F—R, and R—R with same values for the albedo as in the previous simulations. The altitude of the trilateral prism patch was case sensible with changes in the reduced heights $(\hat{h})$ as 0.01, 0.02, 0.05, 0.1, 0.2, 0.5, and 1.0. In the G—R simulations, the grass was 0.5 m tall, while in the F—R simulations the forest had height of 14 m. However, in these simulations the values of the reduced height $\hat{h}$ were only 0.2, 0.5, and 1.0, respectively.

Fig. 14.9 show the difference in the maximum of the surface effective temperature obtained by the MCRT method for different values of the reduced height and the fractional cover of each patch in a case of "trilateral prism" geometry. When the grid-box is covered with the grass and the rock or if whole grid-box is covered with the rock, the increase in the maximum temperature is from about 1.5—2.5K (Fig. 14.9a and c). In the F—R simulations (Fig. 14.9b), there is an increase of 1K when the grid-box is more covered with forest till 2K when the larger part of the grid-box is covered with the rock. Considering the RMSE (Fig. 14.9d—f), that is calculated on the basis of differences between two albedo calculation methods for all points in the simulated daily courses, the case when the whole grid-box area is covered by the rock gives the largest difference between the temperature obtained by two methods of calculating the albedo. As the area of the surface lying on a higher level is larger, it increases the RMSE value for the observed case.

REFERENCES

Ćirišan, A.M., Mihailović, D.T., Kapor, D.V., 2010. Implementation of a new approach to albedo calculation over heterogeneous interfaces for different geometries. In: Mihailović, D.T., Lalić, B. (Eds.), Advances in Environmental Modeling and Measurements. Nova Science Publishers Inc., New York, pp. 47—55.

Dudhia, J., 1993. A nonhydrostatic version of the Penn State-NCAR mesoscale model: validation tests and simulation of an Atlantic cyclone and cold front. Mon. Wea. Rev. 121, 1493—1513.

Hu, Z., Islam, S., Jiang, L., 1999. Approaches for aggregating heterogeneous surface parameters and fluxes for mesoscale and climate models. Bound. Layer Meteorol. 93, 313—336.

Jacobson, M.Z., 1999. Fundamentals of Atmospheric Modelling. Cambridge University Press, The Edinburgh Building, Cambridge, p. 645.

Kapor, D., Cirisan, A., Mihailovic, D.T., 2010a. Calculation of aggregated albedo in rectangular solid geometry on environmental interfaces. In: Mihailović, D.T., Gualtieri, C. (Eds.), Advances in Environmental Fluid Mechanics. World Scientific, Singapore, pp. 145—165.

Kapor, D.V., Mihailović, D.T., Ćirišan, A.M., 2010b. Aggregation of canyon albedo in surface models. In: Mihailović, D.T., Lalić, B. (Eds.), Advances in Environmental Modeling and Measurements. Nova Science Publishers Inc., New York, pp. 57−65.

Kapor, D., Mihailović, D.T., Ćirišan, A., Firanj, A., 2012. Monte Carlo ray-tracing method for the use in the calculation of the albedo for different geometries. In: Mihailovic, D.T. (Ed.), Essays on Fundamental and Applied Environmental Topics. Nova Science Publishers Inc., New York, pp. 59−75.

Kapor, D., Mihailović, D.T., Tošić, T., Rao, S.T., Hogrefe, C., 2002. An approach for the aggregation of albedo in calculating the adiative fluxes over heterogeneous surfaces in atmospheric models integrated assessment and decision support. In: Rizzoli, A., Jakeman, A.J. (Eds.), Proceedings of the Integrated Assessment and Decision Support, 1st Biennial Meeting of the International Environmental Modelling and Software Society (IEMSS), Lugano 24−27. VI 2002, pp. 389−394.

Kreuter, A., Buras, R., Mayer, B., Webb, A., Kift, R., Bais, A., Kouremeti, N., Blumthaler, M., 2014. Solar irradiance in the heterogeneous albedo environment of the Arctic coast: measurements and a 3-D model study. Atmos. Chem. Phys. 14, 5989−6002.

Liou, K.N., 2002. An Introduction to Atmospheric Radiation, second ed. Academic Press, Inc., New York, p. 583.

Marchuk, G.I., Mikhailov, G.A., Nazaraliev, M.A., Darbinjan, R.A., Kargin, B.A., Elepov, B.S., 1980. The Monte Carlo Methods in Atmospheric Optics, vol. viii. Springer-Verlag, Berlin, p. 208.

Mayer, B., Hoch, S.W., Whiteman, C.D., 2010. Validating the MYSTIC three-dimensional radiative transfer model with observations from the complex topography of Arizona Meteor Crater. Atmos. Chem. Phys. Discuss. 10, 13373−21340.

McComiskey, A., Ricchiazzi, P., Gautier, C., Lubin, D., 2006. Assessment of a three dimensional model for atmospheric radiative transfer over heterogeneous land cover. Geophys. Res. Lett. 33, L10813.

McCrea, W.H., 1960. Analytical Geometry of Three Dimensions, second ed. Oliver and Boyd Ltd, Edinburgh and London, p. 160.

Mihailovic, D.T., Balaž, I., 2007. An essay about modelling problems of complex systems in environmental fluid mechanics. Idojaras 111 (2−3), 209−220.

Mihailović, D.T., Kapor, D., Ćirišan, A., Firanj, A., 2012. Parametrization of the albedo over the heterogeneous surfaces for different geometries in a land surface scheme by the Monte Carlo ray-tracing method. Atmos. Res. 107, 51−68.

Mihailović, D.T., Kapor, D., Hogrefe, C., Lazić, J., Tošić, T., 2004. Parametrization of albedo over heterogeneous surfaces in coupled land-atmosphere schemes for environmental modeling. Part I: theoretical background. Environ. Fluid Mech. 4, 57−77.

Mihailovic, D.T., Lazic, J., Leśny, J., Olejnik, J., Lalic, B., Kapor, D.V., Cirisan, A., 2010. A new design of the LAPS land surface scheme for use over and through heterogeneous and non-heterogeneous surfaces: numerical simulations and tests. Theor. Appl. Climatol. 100, 299−323.

Oke, T.R., 1987. Boundary Layer Climates, second ed. Methuen, London, New York, p. 435.

Sanchez, M.C., 1998. Optical Analysis of a Linear-Array Thermal Radiation Detector for Geostationary Earth Radiation Budget Applications. Master of science thesis. Mechanical Engineering Department, Virginia Polytechnic Institute and State University, Blacksburg, VA.

Schwerdtfeger, P., 2002. Interpretation of airborne observation of the albedo. Environ. Model. Softw. 17, 51−60.

Wendisch, M., Pilewskie, P., Jäkel, E., Schmidt, S., Pommier, J., Howard, S., Jonsson, H.H., Guan, H., Schröder, M., Mayer, B., 2004. Airborne measurements of areal spectral surface albedo over different sea and land surfaces. J. Geophys. Res. 109, D08203.

Zhang, K., Huiting, M., Civerolo, K., Berman, S., Ku, J.Y., Rao, S.T., Doddridge, B., Philbric, C.R., Clarck, R., 2001. Numerical investigation of boundary-layer evolution and nocturnal low-level jets: local versus non-local PBL schemes. Environ. Fluid Mech. 1, 171−208.

Complexity measures and time series analysis of the processes at the environmental interfaces

Kolmogorov complexity and the measures based on this complexity

15

15.1 INTRODUCTORY COMMENTS ABOUT COMPLEXITY OF ENVIRONMENTAL INTERFACE SYSTEMS

The systems which we face in modelling the processes on environmental interfaces can be grouped into three categories, i.e., simple, complicated, and complex ones. As a metaphorical illustration of structures of these problems, we will give three examples originating from Brenda Zimmermane (2014), an expert in the strategic management. Her examples, in a simple but essential way, illustrate the structure of the above problems.

Looking at the panels in Fig. 15.1 (left panel), we can see that for making soup we need just the right recipe that is enough to get the same results every time; therefore, they are entitled as simple. In complicated problems, like sending a rocket to the Moon (Fig. 15.1, middle panel), we have to invest much more effort for solving that problem (although the comparison with making soup may be grotesque, it must

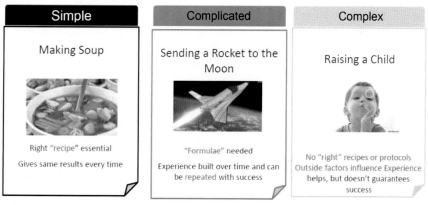

FIGURE 15.1

Types of problems which we face in modelling the processes on environmental interfaces (Zimmerman, 2014).

be admitted that it is deeply obvious). Despite the fact that we have all tools, it does not guarantee that we will be successful in our efforts. To reach the target, we have to build our experience over time and then it can be repeated with success. Finally, let us consider the complex problem through the analysis of raising a child (Fig. 15.1, right panel). Firstly, we have no "right" recipes or protocols, i.e., a set of instructions for recognizing a particular direction we have to follow, including a list of the steps required. Secondly, we have no experience (every child is a special experience regardless of whether the first or succeeding); thus only outside factors influence our experience. It helps but it does not guarantee "solving the problem," i.e., success in this case.

Whereas this book is intended for a wide audience of engineers and scientists, who start from different attitudes and approaches in their work, right now at the beginning of this part of the book and having in mind examples in Fig. 15.1, we will make a distinction between the notion "complicated" and "complex." When we say that something is "complicated," it obviously means that there exists a spectrum of complications but we understand what context that would describe. Complicated matter cannot be broken down into simple parts, for it is made of complicated parts. It is an evincive of something that is problematic, long-winded, difficult, and inconsistent. Complicated usually has to execute with taking time, subsistence hard, or has a lot of limitary factors. Therefore, sending the rocket to the Moon is a good example of a complicated problem. On the other hand, complex problems and systems result from networks of multiple interacting causes that cannot be individually distinguished; must be addressed as entire systems, that is they cannot be addressed in a piecemeal way (Poli, 2013). That kind of problems and systems cannot be controlled like complicated ones, since we know in advance that there is no warranty for success. The best one can do is to have an effect on them, learn to "dance with them," as Donella Meadows ably said (Meadows, 2008). At the end of the story about the types of problems and systems in environmental modelling, let us add the following comments. This question is closely related to the question whether the information we are receiving about the system obeys the same evolution through randomness and determinism, independently of context. "So can a combination of randomness and determinism produce all information and everything else we see around us? It seems that randomness and determinism together can be seen to underlie every aspect of reality" (Verdal, 2010).

The issue of complexity has been touching the scientific community intensively during the last three decades. This is happening on the epistemological as well as the methodological level. The complexity is one of the aspects of self-organization whose fundamental quality is an emergence. Self-organizing systems are complex systems. The term *complexity* has three levels of meaning as concisely elaborated by Arshinov and Fuchs (2003). In the review paper, Crutchfield (2012) has underlined: "Spontaneous organization, as a common phenomenon, reminds us of a more basic, nagging puzzle. If, as Poincaré found, chaos is endemic to dynamics, why is the world not a mass of randomness? The world is, in fact, quite structured, and we now know several of the mechanisms that shape microscopic fluctuations as they are amplified to macroscopic patterns. Critical phenomena in statistical mechanics (Binney et al., 1992) and pattern formation in dynamics (Cross and Hohenberg, 1993; Manneville, 1990) are two arenas that explain in predictive detail how spontaneous organization works. Moreover, everyday experience shows us that nature inherently organizes; it generates pattern. Pattern is as much the fabric of life as life's unpredictability." These sentences are also related to the phenomenon of the complexity of systems in many disciplines, ranging from philosophy and cognitive science to evolutionary and developmental biology and particle astrophysics (Crutchfied, 2012; Wheeler, 1990 and references herein).

There exist a lot of complexity measures in complex system behavior and time series analysis. In particular, in the focus is analysis of time series since the only available evidences about the nature of complex system come through a time series. According to Zunino et al. (2012) and references herein, the recorded signals from experimental measurements give us useful information to establish the deterministic or stochastic character of the system under analysis, but the task to discern between regular, chaotic, and stochastic dynamics from complex time series is a critical issue. In the study of complex system behavior and time series analysis, an important part is played by symbolic sequences, since it is believed that most systems whose complexity we intent to estimate can be reduced to them (Adami and Cerf, 2000). Thus, in searching for an adequate measure for the complexity of sequences, it is difficult to do that consistently. In particular, measures of complexity that are based on the Kolmogorov complexity (Cover and Thomas, 1991), useful in signal analysis, are measures of randomness rather than complexity (Grassberger, 2012). This approach is not capable of discerning between signals with different amplitude variations and similar random components. Note, that the Kolmogorov complexity has some advantages comparing to measures from the theory of nonlinear dynamic systems like the Lyapunov exponent, the correlation dimension, and correlation entropy. This measure does not involve embedding the time series onto a high dimensional, which is necessary for applying measures from the theory of nonlinear dynamic systems (Sen, 2009). Actually, it is easier to use this complexity since it is easily calculated for any type of time series and it does not include any assumption of the probability law of the process generating the time series.

The purpose of this chapter is to elaborate novel complexity measures based on the Kolmogorov complexity to be used in complex systems behavior and time series analysis, offering deeper insights into these issues. Note that in this chapter we will deal specifically with the physical complexity. However, without losing the generality, the methods proposed can be applied as complexity measures for other kind of complexity. We do that through the following steps. In Section 15.2 we consider in what extent Kolmogorov complexity enlightens the physical complexity? We describe in Section 15.3 the measures based on the Kolmogorov complexity [the Kolmogorov complexity spectrum, the Kolmogorov complexity spectrum highest value, and the overall Kolmogorov complexity introduced by Mihailović et al. (2015)]. Finally, in Section 15.4 we apply the Kolmogorov complexities to two different dynamical systems.

15.2 IN WHAT EXTENT KOLMOGOROV COMPLEXITY ENLIGHTENS THE PHYSICAL COMPLEXITY?

In this section, through several steps, we consider the Kolmogorov complexity, which is seen through its *applicability* in illuminating the physical complexity.

Kolmogorov complexity. The Kolmogorov complexity $K(x)$ of an object x is the length, in bits, of the smallest program (in bits) that when run on a Universal Turing Machine (U) outputs $K(x)$ and then stops with the execution. This complexity is maximized for random strings. Thus, $K(x) = |\text{Print}(x)|$. That is, the shortest program to get a U to produce is to just hand the computer a copy and say "print this" (Feldman and Crutchfield, 1998). This measure was developed by Andrey N. Kolmogorov in the late 1960s (Kolmogorov, 1968). A good introduction to the Kolmogorov complexity (in further text KLL) can be found in Cover and Thomas (1991) and with a comprehensive description by Li and Vitanyi (1977). On the basis of Kolmogorov's idea, Lempel and Ziv (1976) developed an algorithm (LZA), which is often used in assessing the randomness of finite sequences as a measure of its disorder.

The Kolmogorov complexity of a time series $\{x_i\}$, $i = 1,2,3,4,...,N$ by the LZA algorithm can be summarized as follows. *Step A*: Encode the time series by constructing a sequence S consisting of the characters 0 and 1 written as $\{s(i)\}$, $i = 1,2,3,4,...,N$, according to the rule

$$s(i) = \begin{cases} 0 & x_i < x_t \\ 1 & x_i \geq x_t \end{cases}. \tag{15.1}$$

Here x_t is a threshold that should be properly chosen. The mean value of the time series has often been used as the threshold (Zhang et al., 2001). Depending on the application, other encoding schemes are also available (Radhakrishnan et al., 2000). *Step B*: Calculate the complexity counter $C(N)$, which is defined as the minimum number of distinct patterns contained in a given character sequence (Ferenets et al., 2006); $c(N)$ is a function of the length of the sequence N. The value of $c(N)$ is approaching an ultimate value $b(N)$ as N approaching infinite, i.e.,

$$c(N) = O(b(N)), \quad b(N) = \frac{N}{\log_2 N}. \tag{15.2}$$

Step C: Calculate the normalized complexity measure $C_k(N)$, which is defined as

$$C_k(N) = \frac{c(N)}{b(N)} = c(N)\frac{\log_2 N}{N}. \tag{15.3}$$

The $C_k(N)$ is a parameter to represent the information quantity contained in a time series, and it is to be a 0 for a periodic or regular time series and a 1 for a random time series, if N is large enough. For a nonlinear time series, $C_k(N)$ is to be between 0 and 1. Let us note that Hu and Gao (2006) derived analytic expression for C_k in the Kolmogorov complexity, for regular and random sequences. In addition they showed that for the shorter length of the time series, the larger C_k value and correspondingly the complexity for a random sequence can be larger than 1.

The above steps are incorporated in codes of different programming languages to estimate the *lower version of the Kolmogorov complexity* (KLL). This version is commonly used by researchers. However, there exists the *upper version of the Kolmogorov complexity* (KLU), which is described by Lempel and Ziv (1976). Note, that in both cases, an extension to a sequence is considered "innovative" in some way, but differently. We describe both of them. The LZA is an algorithm, which calculates the KLL measure of binary sequence complexity. As inputs, it uses a vector S consisting of a binary sequence whose complexity we want to analyze and calculate converting the numeric values to logical values depending on whether (0) or not (1). In this algorithm we can evaluate as a string two types of complexities. One is "exhaustive," i.e., when complexity measurement is based on decomposing S into an exhaustive production process. On the other hand, the so-called "primitive" complexity measurement is based on decomposing S into a primitive production process. Exhaustive complexity can be considered a lower limit of the complexity measurement approach (KLL) and primitive complexity an upper limit (KLU). Let us note that the "exhaustive" is considered as the KLL measure and frequently used in complexity analysis. The KLL calculation is based on finding extensions to a sequence, which are not reproducible from that sequence, using a recursive symbol-copying procedure. The KLU calculation uses the eigenfunction of a sequence. The sequence decomposition occurs at points where the eigenfunction increases in value from the previous one (Mihailović et al., 2014a).

First, we have to find an array consisting of the history components O that were found in the sequence S, while calculating the KLL or KLU (C_k in Eq. (15.3)). Each element in O consists of a vector of logical values (true, false) and represents a history component. Histories are composed by decomposing the sequence S into the following sequence of words

$$O(S) = S(1, h_1)S(h_1 + 1, h_2)S(h_2 + 1, h_3)...S(h_{m-1} + 1, h_m), \qquad (15.4)$$

where the indices $\{h_1, h_2, h_3...h_{m-1}, h_m)$ characterize a history making up the set of "terminals". We do not know how long the histories will be or in other words, how many terminals we need. As a result, we will allocate an array of length equal to the eigenfunction vector length $Es(h)$ (Mihailović et al., 2014a).

For an exhaustive history (i.e., when we calculate the KLL), from Theorem 8 in Lempel and Ziv (1976) the terminal points h_i, $1 \leq i \leq m - 1$ are defined by

$$h_1 = \min\{h|Es(h) > Es(h_m - 1)\}. \qquad (15.5)$$

From the same theorem, for a primitive history (i.e., when we calculate the KLU), the terminal points h_i, $1 \leq i \leq m-1$ are defined by

$$h_1 = \min\{h|Es(h) > Es(h_i - 1)\}, \qquad (15.6)$$

where the eigenfunction, $Es(n)$, is monotonically nondecreasing (Lemma 4 in Lempel and Ziv (1976)). Finally, we use the terminal points to calculate c_{pr}

(primitive, KLU) or c_{ex} (exhaustive, KLL), as the length of the production histories $O_{pr}(S)$ or $O_{ex}(S)$, which are the so-called unnormalized complexities (Eq. 15.2). To get normalized ones we use Eq. (15.3). In this chapter we have designed our own code in FORTRAN90, which partly relies on the MATLAB code by Thai (2012).

Kolmogorov complexity of dynamical systems corrupted with noise. We will make some comments about the KLL complexity: (1) for purely periodic time series and, on the other side, for purely stochastic ones and (2) for dynamical systems corrupted with noise. Since the KLL complexity is a measure of the degree of disorder or irregularity of time series, its value is low for a regular time series like periodic time series with constant periodicity (Hu and Gao, 2006). Therefore, we set focus on the question: How the noise, included in the system, influences the measurement results? The term dynamical noise refers to situations where the output of a dynamical system corrupted with noise is used as an input to the next iteration. The dynamical noise can dramatically change the dynamics of low dimensional chaotic systems. Meaningful analyses of real systems in terms of chaos theory should consider the effect of dynamical noise on the system's dynamics. In fact as Ruelle (1994) put it, real systems can in general be described as deterministic systems with some added noise. This description is sufficiently vague that it appears to cover everything. In economics, for example, such a description is familiar and the noise is called "shocks." A first remark concerning the above picture is that the separation between noise and the deterministic part of the evolution is ambiguous, because one can always interpret "noise" as a deterministic time evolution in infinite dimension (Serletis et al., 2007a). In addition Serletis et al. (2007b) argue that dynamical noise (noise that acts as a driving term in the equations of motion) can dramatically change the dynamics of nonlinear dynamical systems. In fact, dynamical noise can make the detection of chaotic dynamics very difficult as it is possible to lead to rejection of the null hypothesis of chaos.

Now, we will see how noise included in the system can affect the complexity of the system. We consider the effect of noise on complexity of: (1) deterministically generated time series which is contaminated by the adding noise and (2) time series of including measured values of the physical quantity. In many models, there is an interest in cases which occur when parameters of the oscillators have small, random variations due to either internal or external noise. These so-called parametric fluctuations can be simulated by modulating the values of the nonlinearity parameters by uniform random numbers in a small interval. However, in our consideration in both cases this noise enters by an additive "shock," i.e., by the external excitation. The cases we will deal with are (1) logistic map $x_{n+1} = rx_n(1 - x_n)$, where $r = 3.7$ and (2) time series of the measured indoor radon concentration described (Mihailović et al., 2014b). The randomness influence on the logistic equation was analyzed by adding random noise in the logistic equation, i.e., $x_{n+1} = rx_n(1 - x_n) + \Delta\xi_n$. Here $\Delta\xi_n = \mathbf{D}\delta_n$ measures the noise intensity while δ_n is random number uniformly distributed in the interval $[-1,1]$ and $\mathbf{D}$ is the amplitude of the noise as it was done in Chapter 11. Similarly, it was done by adding $\Delta\xi_n$ on the values of the radon time series normalized on its highest value. For generating the random numbers we use the intrinsic subroutine CALL RANDOM NUMBER (arg) from the Microsoft FORTRAN Developer Studio library. To explore the dependence of the Kolmogorov complexity on the amplitude of the noise, we have calculated the KL for each **D** from 0.0001 to 0.05 with step 0.0001, 1500 iterations are applied for an initial state ($x_0 = 0.2$), and the first 500 steps were ignored. The r was taken to have a value of 3.7. The graph representing the KLL of the logistic map as a function of the amplitude **D** of the noise that is introduced by the added noise is depicted in Fig. 15.2. From this figure, it is seen that the KLL complexity is sensitive to the noise introduced to the system in both cases, i.e., when the logistic equation as well as the radon time series are "shocked" by the added noise. However, while the logistic time series, contaminated by the noise, has a trend of an increase of the KLL complexity, the influence of noise on the radon time series is noticeable only after a certain value of amplitude of noise **D** (around **D** = 0.03 in Fig. 15.2).

To demonstrate how dynamical noise changes the dynamics and complexity of chaotic systems, we perform two additional experiments. In the first experiment we explore changes to the KL of a time series which is affected by the size of the number of time iterations N, in the logistic map time simulations. In simulations, for each **D** from 0.0001 to 0.05 with step 0.0001, 2000, 3000, 4000, 5000, 6000, 11,000, and 21,000 iterations were applied for an initial state ($x_0 = 0.2$), and then the first 1000 steps were ignored (the number of iterations included was therefore $N = 1000, 2000, 3000, 4000, 5000, 10,000,$ and 20,000). The r has a value of 3.7

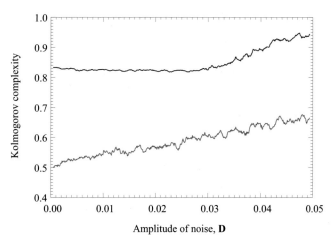

FIGURE 15.2

Kolmogorov complexity of the logistic map $x_{n+1} = rx_n(1 - x_n)$ with $r = 3.7$ (red) and the measured radon time series (blue) (Mihailović et al., 2014b; Mihailović et al., 2015) as a function of the amplitude **D** of the noise that is introduced by the added noise.

Reprinted with permission from Mihailović, D.T., Mimić, G., Nikolić-Đorić, E., Arsenić, I., 2015. Novel mea-
sures based on the Kolmogorov complexity for use in complex system behavior studies and time series analysis.
Open Phys. 13, 1–14.

(Mihailović et al., 2015). The graph representing the Kolmogorov complexity of the logistic map as a function of the amplitude **D** of the noise that is introduced in the logistic equation, for different number of time iterations, is depicted in Fig. 15.3a. From this figure, it is seen that the range of changes in the KLL is about 0.1 for all step sizes. The only differences which occur are in the KLL values which

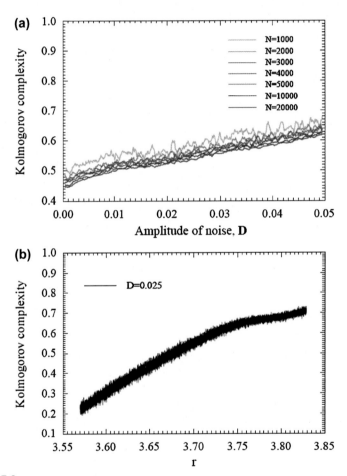

FIGURE 15.3

Kolmogorov complexity (KLL) of the logistic map $x_{n+1} = rx_n(1 - x_n)$ in dependence: (a) of the amplitude **D** of the noise that is introduced in the logistic equation time series, for different number of time iterations N, and (b) of r according to Pomeau-Manneville scenario (3.56995 < < 3.82843).

Reprinted with permission from Mihailović, D.T., Mimić, G., Nikolić-Đorić, E., Arsenić, I., 2015. Novel measures based on the Kolmogorov complexity for use in complex system behavior studies and time series analysis. Open Phys. 13, 1–14.

decrease when N increases. Note that oscillations in the KLL values are more pronounced for $N = 1000$ while they are less present for $N = 20{,}000$. In the second experiment we consider how the logistic map is "shocked" around $r = 3.7$ with a chosen **D**. For this experiment we have used: (1) a Pomeau-Manneville scenario ($3.56995 < r < 3.82843$) (Pomeau and Manneville, 1980) and (2) **D** $= 0.025$, $x_0 = 0.2$ and $N = 5000$. Values of r were changed for an increment of 0.00001. Results of simulations are depicted in Fig. 15.3b. This figure indicates a sharp increase in the KLL of about 0.5 when r is changed in the chosen interval.

Physical complexity. When we use the term complexity in physical systems, we explicitly think that it is a measure of the probability of the state vector of the system. It is a mathematical measure, one in which two distinct states are never combined into a composite whole and considered equal, as is done for the notion of entropy in statistical mechanics. Therefore, this term differs from the *entropy* in statistical mechanics. Note, that here we make a distinction between terms "randomness" and "complexity" in the following sense. The term randomness refers the dissimilarities between amplitudes in a time series whose simplest measure is, for example, the Shannon's entropy. On the other hand, the complexity refers to the sequence appearance disorder of some amplitudes in a time series which will be in the focus of our interest. According to Adami (2002), the physical complexity of a sequence "refers to the amount of information that is stored in that sequence about a particular environment." This should not be confused with mathematical (Kolmogorov) complexity; it is a distinct mathematical complexity, which only deals either with the intrinsic regularity or irregularity of a sequence in this case. Namely, for any two strings $x, y \in P^*$ (P^* is the set of all finite binary strings), the Kolmogorov complexity of given x is $K(x|y) = min_p\{|p| : U(p, y) = x\}$ where $U(p, y)$ is the output of the program p with auxiliary input y when it is run in the machine U. For any time constructible t, the t-time-bounded Kolmogorov complexity of x given y is, $K^t(x|y) = min_p\{|p| : U(p, y) = x$ in at most$(t|p|)steps\}$.

Let us note that the regularity of a sequence we are talking about is just a reflection of the unchanging laws of mathematics. It is not a reflection of the physical world where such a sequence may mean something. The Kolmogorov complexity does not measure pattern or structure or correlation or organization. Structure or pattern is maximized for neither high nor low randomness. If we follow Grassberger (1989), it can be intuitively accepted as something that is placed between uniformity and total randomness (Fig. 15.4). Let us note that the structural complexity versus randomness relation is just one of the *many possible behaviors.* Different systems have different structural complexity versus randomness plots (Grassberger, 2012). There is no "universal" complexity relationship, which is clearly established in the scientific literature.

Adami and Cerf (2000), proposing a measure of physical complexity, cleverly observed that it should closely correspond to our intuition. In addition they stressed that it can consistently be defined within information theory. In studying the complex systems, an important step in using this measure is connected with symbolic sequences. Namely, it is believed that most systems whose complexity we would

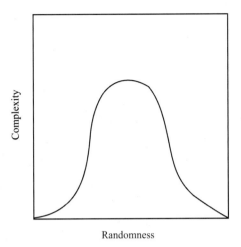

FIGURE 15.4

Complexity versus randomness plotted following the physical intuition (Grassberger, 2012, 1989).

like to estimate can be reduced to them. Contrary to the idea that the regularity of a string is in any way connected to its complexity (as in Kolmogorov theory), it seems that such a classification is, in the absence of an environment within which the string is to be interpreted, thoroughly meaningless. There is no doubt that it is possible to establish a coding system, for example, such that all of Grass' "Tin Drum" (1962) is represented in terms of a uniform (and thus "regular") string of the vanishing Kolmogorov complexity. For example, one possible coding system could be invented in the following way: 1 (one) is assigned to an event when it is described explicitly that Oskar Makowski strikes the drum. Otherwise, it is 0 (zero). Although, this event is presented metaphorically, evidently, in such a case the literature complexity of the string is hidden in the coding rules which relate the string to its environment: the ensemble of books (as mentioned in Adami and Cerf (2000)). Thus, the complexity of a string representing the physical (or any other) complexity can be determined only by analyzing its correlation with a physical or corresponding environment.

15.3 NOVEL MEASURES BASED ON THE KOLMOGOROV COMPLEXITY

The quantification of the complexity of a system is one of the aims of nonlinear time series analysis. Complexity of the system is hidden in the dynamics of the system. However, if there is no recognizable structure in the system, it is considered to be stochastic. Due to noise, spurious experimental results, and artifacts in various forms, it is often not easy to get reliable information from a series of measurements. The time series of some physical quantity is the only information about its physical state, which can be obtained either by measurement or modelling. The time series is the only source for establishing the level of complexity of the physical system. The exact states of an observed physical system are translated into a sequence of symbols via a measurement channel. This process is described by a parameterized partition M_ε of the state space, consisting of cells of size ε that are sampled every τ time units. A measurement sequence consists of the successive elements of M_ε, visited over time by the system's state. Using the instrument $\{M_\varepsilon, \tau\}$ we get information as a sequence of states $\{x_i\}$. Here, we consider a possible way of calculating the physical complexity of the system, i.e., complexity of time series which represents the system passing through different states.

Let us consider *Kolmogorov complexity spectrum* introduced by Mihailović et al. (2015).

Definition 1: The time series $\{x_i\}$, $i = 1,2,3,4,...,N$ we call normalized one (or time series with normalized amplitude) after transformation $x_i = (X_i - X_{min})/(X_{max} - X_{min})$, where $\{X_i\}$ is a time series obtained either by a measuring procedure or as an output from a physical model, while $X_{max} = max\{X_i\}$ and $X_{min} = min\{X_i\}$.

Remark: It follows from *Def.* 1 that all elements in time series $\{X_i\}$ belong to the interval [0,1].

Definition 2: We call the *Kolmogorov complexity spectrum of time series* $\{x_i\}$ the sequence $\{c_i\}$, $i = 1,2,3,4,...,N$ obtained by the LZA algorithm which is applied N times on time series, where thresholds $\{x_{t,i}\}$ are all elements in $\{x_i\}$.

Remark: We transform the time series obtained either by a measuring procedure or as an output from a physical model, into a finite symbol string by comparison with a series of thresholds $\{x_{t,i}\}$, $i = 1,2,3,4,...,N$, where each element is equal to the corresponding element in the considered time series $\{x_i\}$, $i = 1,2,3,4,...,N$, applying the LZA algorithm. The original time series samples are converted into a set of $0-1$ sequences $\{S_i^{(k)}\}$, $i = 1,2,3,4,...,N$, $k = 1,2,3,4,...,N$ defined by comparison with a threshold $x_{t,k}$,

$$S_i^{(k)} = \begin{cases} 0 & x_i < x_{t,k} \\ 1 & x_i \geq x_{t,k} \end{cases}. \tag{15.7}$$

Applying the LZA algorithm on each element of series $\{S_i^{(k)}\}$ we get the KL complexity spectrum $\{c_i\}$, $i = 1,2,3,4,...,N$. We introduce this spectrum to explore the range of amplitudes in a time series representing a process, for which it has highly enhanced stochastic components, i.e., highest complexity.

Definition 3: The highest value K_{max}^C in this series, i.e., $K_{max}^C = max\{c_i\}$, we call the *Kolmogorov complexity spectrum highest value*.

The following examples will help us to demonstrate the meaning of measures we have just introduced: (1) spectrum of the complexity $\{c_i\}$ and (2) Kolmogorov complexity spectrum highest value the spectrum K_{max}^C (in further text KLM), using a time series $\{x_i\}$.

Example 1: In this example we illustrate the meaning of the complexity spectrum $\{c_i\}$. Here we obtain the time series $\{x_i\}$ by the instrument $\{M_\varepsilon, \tau\}$ with $M_\varepsilon = e^{-w\sigma}$, where σ is the random number uniformly distributed in the interval $[0,1]$ while w is the amplitude, which takes values in the interval $[0,1]$; $\{x_i\}$ is sampled every $\tau = 1$ time unit.

Fig. 15.5 shows the KLL complexity spectra for different values of w: $w = 1.0$, 0.75, 0.50, and 0.25, respectively. They are all similar to the curve in Fig. 15.4, which represents just one of the many possible behaviors since different systems have different complexity vs. randomness plots, because there is no "universal" complexity−entropy relationship (Feldman and Crutchfield, 1998).

Example 2. To illustrate the justification for introducing the complexity measure K^C_{max} we deal with a time series $\{x_i\}$, $i = 1,2,3,4,...,500$, which is generated by a generalized logistic map (see Section 10.1). Mathematically, that map has the form

$$\Phi(x) = rx^P(1 - x^P) \tag{15.8}$$

where r is a logistic parameter, $0 < r < 4$. This map expresses the exchange of biochemical substance between cells that is defined by a diffusion-like manner, where the parameter p is the cell affinity. This kind of map is convenient to illustrate the meaning of K^C_{max}. We have calculated K^C_{max} and K^C complexities (K^C, calculated with threshold $x_t = \sum_{i=1}^{N} x_i/N$) for $p = 0.5$ and $p = 1$ ($0 < p \leq 1$). In those computations, for each r from 3.5 to 4.0 and p with step 0.01, 1000 iterations were applied

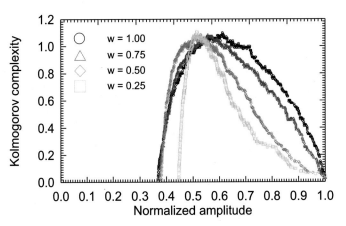

FIGURE 15.5

The Kolmogorov complexity spectra $\{c_i\}$ of time series $\{x_i\}$ obtained by the instrument $\{M_\varepsilon, \tau\}$ with $M_\varepsilon = e^{-w\sigma}$, where w is the amplitude factor, σ is the random number uniformly distributed in the interval $[0,1]$, and sampling $\tau = 1$ time unit.

Reprinted with permission from Mihailović, D.T., Mimić, G., Nikolić-Đorić, E., Arsenić, I., 2015. Novel measures based on the Kolmogorov complexity for use in complex system behavior studies and time series analysis. Open Phys. 13, 1–14.

for an initial state, and then the first 100 steps were abandoned. The results of computations are given in Fig. 15.6.

From this figure it can be seen that in both cases, K^C carries less information about the complexity of the time series than K^C_{max} does. Moreover, for $p = 0.5$, the K^C is recognized only after $r > 3.9$ since it gives us average information about the complexity of the time series. In contrast to that, K^C_{max} carries the information about the highest complexity among all complexities in the spectrum. Therefore, this measure should be included when developing an understanding of a system's randomness and organization (Crutchfield, 2012) and also in the complexity analysis of the time series that an instrument provides. To explore the dependence of (a) the KLL and (b) the KLM on the logistic parameter r and cell affinity p, we have simulated the generalized logistic map defined by Eq. (15.8). In those computations, for each r from 0.0 to 4.0 and p with step 0.01, 1000 iterations were applied for an initial state, and then the first 100 steps were abandoned. Looking at Fig. 15.7a and b, which depicts the KLL and KLM complexities, respectively, we can see regions with different levels of complexity. Further inspection of figures points out that in the region of the KLM (Fig. 15.7b), its values are higher than for the KLL ones (Fig. 15.7a). Apparently, that the KLM is a better indicator about the complexity time series than the commonly used KLL one. This is because the KLL carries average information about a time series. In contrast to that, the KLM carries information about the highest complexity among all complexities in the Kolmogorov complexity spectrum.

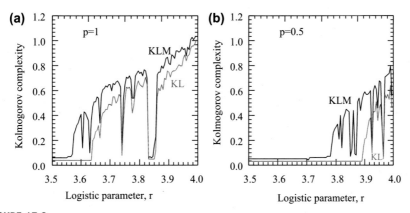

FIGURE 15.6

The dependence on the logistic parameter r of Kolmogorov complexity (KL, red (gray in the print version)) and the KLM (black) of time series generated by the generalized logistic equation $x_{n+1} = rx_n^p(1 - x_n^p)$ for (a) $p = 1$ and (b) $p = 0.5$.

Reprinted with permission from Mihailović, D.T., Mimić, G., Nikolić-Đorić, E., Arsenić, I., 2015. Novel measures based on the Kolmogorov complexity for use in complex system behavior studies and time series analysis. Open Phys. 13, 1–14.

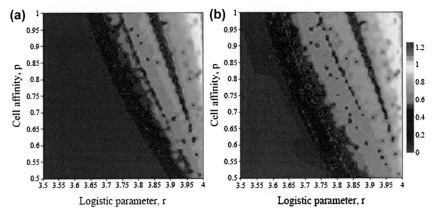

FIGURE 15.7

The dependence on the logistic parameter r and cell affinity p of (a) Kolmogorov complexity (KLL) and (b) Kolmogorov complexity spectrum highest value (KLM), simulated by the generalized logistic equation $\Phi(x) = rx^P(1 - x^P)$.

Reprinted with permission from Mihailović, D.T., Mimić, G., Nikolić-Đorić, E., Arsenić, I., 2015. Novel measures based on the Kolmogorov complexity for use in complex system behavior studies and time series analysis. Open Phys. 13, 1–14).

The overall Kolmogorov complexity measure. Now, we elaborate a new measure based on the Kolmogorov complexity suggested by Mihailović et al. (2015), which can be used for better understanding of physical complexity as well as the complexity of other systems, i.e. their time evolution and predictability. We first briefly consider the term complexity as the possibility for a growth of structural complexity. Many papers regarding this issue have been offered during the last two decades. Among them we underline two contributions about complexity: (1) a comprehensive elaboration from different aspects (epistemological, mathematical as well as physical), summarized in Arshinov and Fuchs (2003) and (2) a recent overview given by Crutchfield (2012), who emphasize the difficulties in perception, which become more problematic when the phenomena of interest arise in systems that spontaneously organize. When a complex system is under observation, only an active subject (a scientist, agent) that creates new communicative parameters of order allows the realization of more complex information about a system that is connected with the idea of constructive chaos and chaos as a space of information. The only available evidence about the nature of a complex physical system is the agent's report written down in the form we call time series. A key question therefore is how we may gather information about complexity expressed through some measure, particularly when the phenomena of interest arise in systems that spontaneously organize. As mentioned above, according to Adami and Cerf (2000), the main aim is to search for a measure of physical complexity that closely corresponds to our intuition but that may also consistently be defined within information theory.

Van der Pol and van der Mark (1927) also noted that much of our positive reception depends on the question of whether or not our minds are "ready" to confront with these intricacies. Further to the psychological observation they concluded that "When confronted by a phenomenon for which we are ill-prepared, we often simply fail to see it, although we may be looking directly at it." With this in mind we consider two points. First, for any complex physical system at any moment, we can establish its entropy either through a measuring procedure or computation. That is only matter that can be written down in the agent's report at the fixed time (we refer to the "white window"). Second, between two successive agent's registration in the record, there exits no information about complexity (except "that it should closely correspond to our intuition" (Adami and Cerf, 2000)—nothing more and nothing less)—this period is behind the window we refer as the "black window." Note that the complexity tells us how the pathway between two states is complex and which corresponding entropies we can measure or compute: the only thing we can do is to anticipate a measure of complexity, which will carry more information. The KLL complexity as a measure is not able to distinguish between time series with different amplitude variations and similar random components. On the other hand, the same feature could also be attached to the suggested KLM measure, although it gives more information about complexity, in a broader context, than the KLL one does. Thus, when we convert a time series into a string, then its complexity is hidden in the coding rules. For example, in the procedure of establishing a threshold for a criterion for coding some information about the structure of the time series can be lost. However, from the spectrum of the KLL complexity $\{c_i\}$ of time series $\{x_i\}$ obtained by the instrument $\{M_\varepsilon, \tau\}$, we do not lose any information since we get N fixed thresholds, each of them contributing to the dynamics of the system, and N calculated complexities, i.e., corresponding spectrum (Fig. 15.5). The shape of this complexity curve depends on variability of time series amplitudes that cannot be captured by the KLL and KLM. From that point of view, the spectrum can be considered as a novel method in quantifying amplitude and complexity variations in the time series. In introducing the way how the spectrum of complexity is computed, we increase the amount of information "that is stored in a sequence about a particular environment," what is according to Adami (2002) a definition of physical complexity of a sequence. The increase of information gives us an opportunity to have better access to insights of the system complexity since the physical or other complexities can be determined only by analyzing its correlation with the corresponding environment. The increase in information gives us opportunity to better understand the system complexity, since the physical or other complexities can be determined only by analyzing its correlation with the corresponding environment. Fig. 15.8 depicts the dependence of the Kolmogorov-based complexities of time series obtained by the instrument $\{M_\varepsilon, \tau\}$ as in Fig. 15.5 in dependence on the amplitude factor. From this figure, it is seen that the KLL values as well as the KLM values are very close for all amplitude factors. Apparently, neither the KLL nor the KLM complexity is able to discern between time series with different Kolmogorov spectra

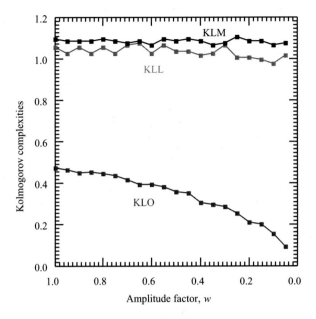

FIGURE 15.8

The dependence on the amplitude factor w of the Kolmogorov complexities (KLL, KLM, and KLO) of the time series obtained by the instrument $\{M_g, \tau\}$ as in Fig. 15.5 (Mihailović et al., 2015).

of complexity. From this reason we introduce *an overall Kolmogorov complexity measure* K_O^C (KLO in further text) defined as

$$K_O^C = \int_X K_s^C dx \tag{15.9}$$

where K_s^C is the spectrum of the Kolmogorov complexity, dx is the differential of the normalized amplitude, while X is a domain of all normalized amplitudes, over which this integral takes values. Since K_s^C is given as the sequence $\{c_i\}$, $i = 1,2,3,4,...,N$ (see *Def. 2*), it is calculated numerically as

$$K_O^C = c_1(x_2 - x_1) + \frac{1}{2}\sum_{i=1}^{N-1} c_i(x_{i+1} - x_{i-1}) + c_N(x_N - x_{N-1}). \tag{15.10}$$

The K_O^C takes value on the interval $(0, K_u)$, where according to Hu and Gao (2006), K_u can take the value up to 1.2. This measure can provide a distinction between different time series having close values of the KL and KLM. It is clearly seen in Fig. 15.8 which shows the descending KLO curve for different values of the amplitude factor w. Thus, if information about the KLO is available we can arrive at a more robust conclusion regarding the Kolmogorov complexity of time series.

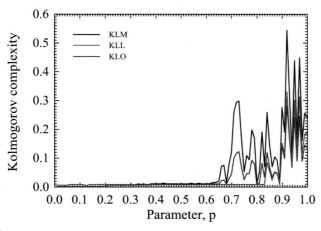

FIGURE 15.9

The dependence on the parameter p of Kolmogorov complexity (KLL, red), the Kolmogorov complexity highest value (KLM, black), and the overall complexity (KLO, blue) of the time series generated by the generalized logistic equation (3.16) for $r = 3.7$.

Reprinted with permission from Mihailović, D.T., Mimić, G., Nikolić-Đorić, E., Arsenić, I., 2015. Novel measures based on the Kolmogorov complexity for use in complex system behavior studies and time series analysis. Open Phys. 13, 1–14.

To see the differences between the KLL and the suggested KLM and KLO measures with respect to the parameter p, we have calculated the KLL, for each p from 0.01 to 1 with step 0.01, 6000 iterations were applied for an initial state ($x_0 = 0.2$), and then the first 1000 steps were ignored ($N = 5000$). The r was taken to have a value of 3.7. Fig. 15.9 depicts all three Kolmogorov complexities of the logistic map as a function of the parameter p. In the region ($0.6 < p < 0.83$), the KLM and KLO are higher than the KLL, which has a negligible value. This means that the KLM and KLO are better indicators of complexity of the time series than the commonly used KLL one. This is because the KLL carries average information about a time series. In contrast to that, the KLM carries the information about the highest complexity among all complexities in the Kolmogorov complexity spectrum, while the KLO gives integral information about complexity for the whole spectrum of complexities.

15.4 APPLICATION TO DIFFERENT DYNAMICAL SYSTEMS

We illustrate the practical performance of the Kolmogorov-based complexity measures for two dynamical systems. Application of the Kolmogorov-based complexity measures to various modeled and natural records of the environmental interface complex systems will be analyzed in the following chapters.

Intracellular concentration dynamics in a multicell system. In the first application we choose to analyze the intracellular concentration dynamics in a multicell system represented by a ring of coupled cells (Fig. 7.3). In our approach, a cell moves locally in its environment without making long pathways. As a generalization of the two-cell system, according to Mihailović and Balaž (2012) and Mihailović et al. (2013), the dynamics of biochemical substance exchange in such a multicell system of cells can be represented by the discrete nonlinear time-invariant dynamical system (Krabs and Pickl, 2010):

$$\mathbf{x}^{(n+1)} = \mathbf{F}\left(\mathbf{x}^{(n)}\right) := C\Phi\left(\mathbf{x}^{(n)}\right) + (I - C)\mathbf{Z}\mathbf{\Psi}\left(\mathbf{x}^{(n)}\right), \tag{15.11}$$

where: $x_k^{(n)}$ is the concentration of the substance in k-th cell in a discrete time step n, $k = 1,2,...,K$, $n=0,1,2...,N$, and $\mathbf{x}^{(n)} := [x_1^{(n)} \, x_2^{(n)} \, ... \, x_K^{(n)}]^T$ is the appropriate vector; $C := diag(c_1, c_2, \leftrightharpoons, c_N)$ is the diagonal matrix of the coupling coefficients for each cell; $\Phi(x^{(n)}) := diag(\varphi(x_1^{(n)}), \varphi(x_2^{(n)}), ..., \varphi(x_K^{(n)}))$ is the diagonal matrix of intracellular behavior modeled by logistic map $\varphi:(0,1) \rightarrow (0,1)$, $\varphi(x) := rx(1 - x)$; $\Psi(x^{(n)}) := diag((x_2^{(n)})^{p_1}, (x_3^{(n)})^{p_2}, ..., (x_N^{(n)})^{p_{K-1}}, (x_1^{(n)})^{p_K})$ is the diagonal matrix of the flow of the substance to each cell, where all the cell's affinities fulfill the constraint $p_1 + p_2 + ...p_K = 1$; $Z \in \{0, 1\}^{K \times K}$ is the upper cyclic permutation matrix, i.e., $Z := [e_K \, e_1, e_2, ... \, e_{K-1}]$, where $e_1, e_2, ..., e_K$ are the standard basis vectors of $\mathbf{R}^N$.

Simulations of biochemical substance exchange in the system represented by a ring of coupled $K = 3$ cells, given by Eq. (15.11), were performed with the values of parameters $r = 4$, $p_1 = p_2 = p_3 = 1/3$. The coupling parameter c cover a broad range of coupling, ranged from weak to strong (0.02, 0.15, 0.19, and 0.50), while the number of iterations was $N = 1000$. The results of simulations are depicted in Fig. 15.10. The curves, describing the Kolmogorov complexity spectrum of time series of concentration, for different values of c, show significant difference in the complexities. Those deference are strongly correlated with the value of c, i.e., the complexity of the concentration dynamics is highest for the weakest coupling ($c = 0.02$) and it takes the lowest values for the strongest one ($c = 0.5$).

The results of the KLL, KLM, and KLO calculations are given in Table 15.1. The order of the KLL complexities (0.957, 0.678, 0.119, 0.109) for c_1, c_2, c_3, and c_4 is pursued by the KLM complexities (0.987, 0.807, 0.478, 0.149) as well as by the KLO complexities (0.711, 0.570, 0.230, 0.105). Here, for these time series the hierarchy of all complexities is clearly enhanced. Therefore, in this case the KLL measure carries enough information about the complexity of this process.

Stock price dynamics. Here we demonstrate the use of Kolmogorov complexities in econophysics and financial econometrics where it is of great importance to measure market efficiency in terms of the patterns contained in price changes relative to the patterns in random sequences. The market is efficient when price changes are unpredictable and random walk hypothesis is satisfied. This means that information is incorporated in prices quickly, eliminating the possibility of market participants

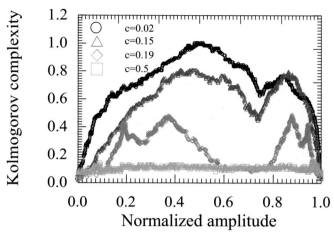

FIGURE 15.10

The Kolmogorov complexity spectrum for the normalized amplitude of the intracellular concentration obtained by the model of biochemical substance exchange in a system represented by a ring of coupled cells for four values of c.

Reprinted with permission from Mihailović, D.T., Mimić, G., Nikolić-Ðorić, E., Arsenić, I., 2015. Novel measures based on the Kolmogorov complexity for use in complex system behavior studies and time series analysis. Open Phys. 13, 1–14.

Table 15.1 Kolmogorov Complexities (Kolmogorov Complexity, KLL; Kolmogorov Complexity Spectrum Highest Value, KLM; Overall Kolmogorov Complexity Measure, KLO) Calculated for Time Series of Different Origin (the Modeled Intracellular Concentration in a Multicell System for Deferent Coupling Parameters and the Stock Prices) (Mihailović et al., 2015)

KLL	KLM	KLO	Origin of Time Series	Time Series
0.957	0.987	0.711	Modeled intracellular concentration in a multicell system	$c = 0.02$ (c1)
0.678	0.807	0.570		$c = 0.15$ (c2)
0.119	0.478	0.230		$c = 0.19$ (c3)
0.109	0.149	0.105		$c = 0.50$ (c4)
0.978	1.013	0.218	Stock Prices	Imlek (IMLK)
1.048	1.062	0.137		Dean Food (DF)

profiting from their information (Mihailović et al., 2015). In addition to the statistical approach to studying market efficiency, information theory can also be applied. Gulko (1999) first applied the concept of entropy to the analysis of financial series. Pincus and Kalman (2004) used approximation entropy to study market stability and Alvarez-Ramirez et al. (2012) applied a multiscale entropy for measuring a time varying structure of market efficiency. Giglio et al. (2008) applied the KLL to rank stock exchanges and exchange rates. To compare the efficiency of stocks from developed and less-developed markets, we have chosen two time series. The first time series is the daily closing stock price of company Imlek (IMLK) from the Belgrade Stock Exchange. The company Imlek is a regional leader of dairy industry. The second time series is related to the daily closing stock price 10 of Dean Foods (DF) from the New York Stock Exchange (NYSE), which is the largest processor and distributor of milk and other dairy products in the United States. The sample period covers $N = 1511$ trading days from January 3, 2011 to December 30, 2015.

Both time series are nonstationary, the impact of which is reduced by converting the original series to returns, taking the logarithm of the ratio of consecutive values of the series $r_i = log(p_i/p_{i-1})$, where p_i are daily closing stock prices. The values of the KLM for both series of returns, given in the corresponding rows of Table 15.1, are greater than 1. That indicates their random behavior. As expected, the value of the KLM is lower for the stock from the less developed market. On the other hand, there is a larger deference between values of KLO, i.e., between areas below the curves describing the Kolmogorov complexity spectrum (Fig. 15.11). The curves

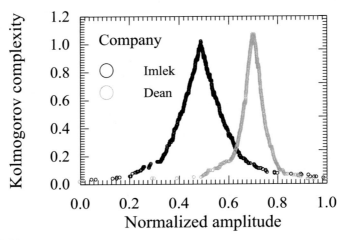

FIGURE 15.11

The Kolmogorov complexity spectrum for normalized daily returns of the companies Imlek from the Belgrade Stock Exchange and Dean Foods from the New York Stock Exchange.

Reprinted with permission from Mihailović, D.T., Mimić, G., Nikolić-Đorić, E., Arsenić, I., 2015. Novel measures based on the Kolmogorov complexity for use in complex system behavior studies and time series analysis. Open Phys. 13, 1–14.

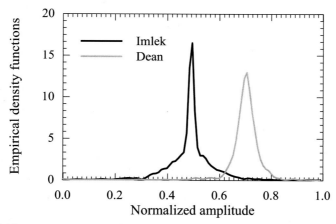

FIGURE 15.12

Empirical density functions for normalized daily returns of the companies Imlek from the Belgrade Stock Exchange and Dean Foods from the New York Stock Exchange.

Reprinted with permission from Mihailović, D.T., Mimić, G., Nikolić-Đorić, E., Arsenić, I., 2015. Novel measures based on the Kolmogorov complexity for use in complex system behavior studies and time series analysis.

Open Phys. 13, 1–14.

may be compared with empirical density functions of normalized returns $r_{iS} = (r_i - min(r_i))/(max(r_i) - min(r_i))$ (Fig. 15.11), estimated using a Gaussian kernel (Härdle and Simar, 2007). Both density curves are approximately symmetrical and have maximum values if normalized returns are close to medians of corresponding distributions $M_e^1 = 0.49$ and $M_e^2 = 0.70$. It should be noted that each complexity curve reaches a maximum value exactly for the median value of normalized amplitude. The distributions of normalized returns (amplitudes) also differ in variability. The calculated values of standard deviations and coefficients of variation are $sd(r_{1S}) = 0.08411851$, $V(r_{1S}) = 17, 8\%$ for the IMLK time series and $sd(r_{2S}) = 0.05572103$, $V(r_{2S}) = 7, 94\%$ for the DF time series. The difference in variability affects the shape of the spectrum of the Kolmogorov complexity curves (Fig. 15.12). So it may be concluded that the spectrum of complexity gives the additional information about differences in amplitudes of time series that was not contained in the KLL and KLM. The deference in spectrum curves is reflected in the values of the KLO. The value of the KLO = 0.218 for IMLK normalized returns is greater than the KLO = 0.137 for DF.

REFERENCES

Adami, C., 2002. What is complexity? Bioessays 24, 1085–1094.

Adami, C., Cerf, N.J., 2000. Physical complexity of symbolic sequences. Phys. D 137, 62–69.

Alvarez-Ramirez, J., Rodriguez, E., Alvarez, J., 2012. A multiscale entropy approach for market efficiency. Int. Rev. Financ. Anal. 21, 64–69.

Arshinov, V., Fuchs, C., 2003. Causality, Emergence, Self-Organisation. NIA-Priroda, Moscow.

Binney, J.J., Dowrick, N.J., Fisher, A.J., Newman, M.E.J., 1992. The Theory of Critical Phenomena. Oxford University Press, Oxford.

Cover, M., Thomas, J.A., 1991. Elements of Information Theory. Wiley, New York.

Cross, M.C., Hohenberg, P.C., 1993. Pattern formation outside of equilibrium. Rev. Mod. Phys. 65, 851–1112.

Crutchfield, J.P., 2012. Between order and chaos. Nat. Phys. 8, 17–24.

Feldman, D.P., Crutchfield, J.P., 1998. Measures of statistical complexity: why? Phys. Lett. A 238, 244–252.

Ferenets, R., Lipping, T., Anier, A., 2006. Comparison of entropy and complexity measures for the assessment of depth of sedation. IEEE Trans. Biomed. Eng. 53, 1067.

Giglio, R., Matsushita, R., Figueiredo, A., Gleria, I., Da Silva, S., 2008. Algorithmic complexity theory and the relative efficiency of financial markets. Europhys. Lett. 84, 48005.

Grass, G., 1962. The Tin Drum, (Transl. By Ralph Manheim). Secker & Warburg, London.

Grassberger, P., 1989. Problems in quantifying self-generated complexity. Helv. Phys. Acta 62, 489–511.

Grassberger, P., 2012. Randomness, Information, and Complexity arXiv:1208.3459 [physics.-data-an].

Gulko, L., 1999. The entropic market hypothesis. Int. J. Theor. Appl. Financ. 2, 293–329.

Härdle, W., Simar, L., 2007. Applied Multivariate Statistical Analysis. Springer-Verlag, Berlin Heidelberg.

Hu, J., Gao, J., Principe, J.C., 2006. Analysis of biomedical signals by the Lempel-Ziv complexity: the effect of finite data size. IEEE Trans. Biomed. Eng. 53, 2606–2609.

Kolmogorov, A.N., 1968. Automata and life. In: Cybernetics Expected and Cybernetics Unexpected (Nauka, Moscow) (In Russian), p. 24.

Krabs, W., Pickl, S., 2010. Dynamical Systems — Stability, Controllability and Chaotic Behavior. Springer.

Lempel, A., Ziv, J., 1976. On the complexity of finite sequences. IEEE Trans. Inform. Theory 22, 75–81.

Li, M., Vitanyi, P., 1997. An Introduction to Kolmogorov Complexity and Its Applications, second ed. Springer Verlag, Berlin.

Manneville, P., 1990. Dissipative Structures and Weak Turbulence. Academic Press, Boston.

Meadows, D.H., 2008. Thinking in Systems: A Primer. Chelsea Green Publishing, Vermont.

Mihailović, D.T., Balaž, I., 2012. Synchronization in biochemical substance exchange between two cells. Mod. Phys. Lett. B 26, 1150031-1.

Mihailović, D.T., Balaž, I., Arsenić, I., 2013. A numerical study of synchronization in the process of biochemical substance exchange in a diffusively coupled ring of cells. Cent. Eur. J. Phys. 11, 440–447.

Mihailović, D.T., Mimić, G., Nikolić-Đorić, E., Arsenić, I., 2015. Novel measures based on the Kolmogorov complexity for use in complex system behavior studies and time series analysis. Open Phys. 13, 1–14.

Mihailović, D.T., Nikolić-Dorić, E., Dresković, N., Mimić, G., 2014a. Complexity analysis of the turbulent environmental fluid flow time series. Phys. A 395, 96–104.

Mihailović, D.T., Udovic, V., Krmar, M., Arsenic, I., 2014b. A complexity measure based method for studying the dependence of 222Rn concentration time series on indoor air temperature and humidity. Appl. Radiat. Isot. 84, 27–32.

Pincus, S., Kalman, R.E., 2004. Irregularity, volatility, risk and financial market time series. Proc. Natl. Acad. Sci. 101, 13709–13714.

Poli, R., 2013. A note on the difference between complicated and complex social systems. CADMUS 2, 142–147.

Pomeau, Y., Manneville, P., 1980. Intermittent transition to turbulence in dissipative dynamical systems. Comm. Math. Phys. 74, 189–197.

Radhakrishnan, N., Wilson, J.D., Loizou, P.C., 2000. An alternative partitioning technique to quantify the regularity of complex time series. Int. J. Bifurc. Chaos 10, 1773–1779.

Ruelle, D., 1994. Where can one hope to profitably apply the ideas of chaos? Phys. Today 47, 24–30.

Sen, A.K., 2009. Complex analysis of riverflow time series. Stoch. Environ. Res. Risk Assess. 23, 361–366.

Serletis, A., Shahmoradi, A., Serletis, D., 2007a. Effect of noise on the bifurcation behavior of nonlinear dynamical systems. Chaos Soliton Fract 33, 914–921.

Serletis, A., Shahmoradi, A., Serletis, D., 2007b. Effect of noise on estimation of Lyapunov exponents from a time series. Chaos Soliton Fract 32, 883–887.

Thai, Q., 2012 http://www.mathworks.com/matlabcentral/fileexchange/38211-calclzcomplexity.

van der Pol, B., van der Mark, J., 1927. Frequency demultiplication. Nature 120, 363–364.

Verdal, V., 2010. Decoding Reality. The Oxford University Press, Oxford.

Wheeler, J.A., 1990. Information flow in causal networks. In: Zurek, W. (Ed.), The Proceedings of the Workshop on Complexity, Entropy, and the Physics of Information, Held may–June, 1989, in Santa Fe, New Mexico. Entropy, Complexity, and the Physics of Information Volume VIII, SFI Studies in the Sciences of Complexity. Addison-Wesley.

Zhang, X.S., Roy, R.J., Jensen, E.W., 2001. EEG complexity as a measure of depth of anesthesia for patients. IEEE Trans. Biomed. Eng. 48, 1424–1431.

Zimmerman, B., 2014. https://thrive.novascotia.ca/sites/default/files/Thrive-Summit-2014-Brenda-Zimmerman-En.pdf.

Zunino, L., Soriano, M.C., Rosso, O.A., 2012. Distinguishing chaotic and stochastic dynamics from time series by using a multiscale symbolic approach. Phys. Rev. E 86, 046210.

Complexity analysis of the ionizing and nonionizing radiation time series

16.1 A COMPLEXITY ANALYSIS OF ^{222}Rn CONCENTRATION VARIATION IN A CAVE

Radon is an odorless and colorless radioactive noble gas. It emanates from soil, rock, sediments, and water and creates decay products in air. It is the greatest source of natural radioactivity, which after prolonged exposure onto humans may cause a negative effect on their health. ^{222}Rn has the most significant impact on the environment due to its relatively long half-life, enabling it to migrate quite significant distances within the geological environment before decaying. It circulates in the external environment. Build-up of radon and elevated radiation exposure levels in the underground places have been observed by researchers worldwide, mostly to assess the radiological hazards to occupational workers and occasional visitors and tourists in mines (Veiga et al., 2004; Lindsay et al., 2004), tunnels (Lam et al., 1988; Abdel-Monem et al., 1996), show caves (Jovanovič, 1996; Chen and Li, 1995; Luo et al., 1996; Lu, 2002; Papastefanou et al., 2003; Lario et al., 2005; Dueñas et al., 2005; Aytekin et al., 2006; Lu et al., 2009; Koltai et al., 2010; Gregorič et al., 2011; Grant et al., 2012; Sánchez et al., 2013), boreholes (Choubey et al., 2011), and underground monuments (Hafez and Hussein, 2001). In these papers, the authors mostly considered the effective doses to visitors comparing them with the recommended one. On the other hand, radon is used as a natural radioactive tracer of air movement in caves to enable better understanding of their microclimate (Fernandez-Cortes et al., 2009). Radon activity concentration in underground environments is usually characterized by the large temporal variations (Eff-Darwich et al., 2002; Perrier et al., 2007; Barbosa et al., 2010). However, no attention has been devoted to analysis of radon concentration time series depending on meteorological factors inside/outside of underground environments, although it is an important step in deriving conclusions about the behavior of ^{222}Rn in a cave environment. Analyses which have been done about this issue do not pursue the number of well-documented measurements, but some of them exist. We have mentioned several of them, which are recently offered. Thus, on the basis of computations, Gregorič et al. (2011) have shown that the effect of the difference between outside and cave air temperatures on ^{222}Rn concentration can be delayed for four days because of the distance between measurement point and cave entrance. Koltai et al. (2010) have considered ^{222}Rn concentration dependence on the temperature and atmospheric pressure data in a cave using correlation and regression analysis, but those methods did not give significant results. Choubey et al. (2011) have found that the high fluctuation in ^{222}Rn concentration is mainly caused by the temperature contrast between the air column inside the borehole and the atmosphere above the Earth's surface.

In this section we analyze ^{222}Rn concentration variation in Domica cave (Slovakia) for the period June 2010–June 2011. For that purpose we have employed a complexity analysis based method which is helpful to get more insight into the complexity of ^{222}Rn concentration time series that cannot be done by traditional mathematical statistics. Our intention is (1) to investigate possible existence of a periodical component in the variation of ^{222}Rn concentration and some environmental parameters, as well as possible correlation in their periodicities, (2) to use complexity measures based on the Kolmogorov complexity (KLL) for establishing the dependence of ^{222}Rn concentration on cave environmental parameters, and (3) to see whether influence of some parameters makes the distribution of measured quantity less or more stochastic (Mihailović et al., 2015).

Monitoring site and methods of measurements. The Domica cave is situated on the south-western edge of the Silická plateau in the Slovak Karst National Park, Southern Slovakia (48°28′40″ N, 20°28′13″ E). The cave is formed in the Middle Triassic lagoon Wetterstein limestone of the Silica Nappe along the tectonic faults by corrosive and erosive activities of Styx River and Domica Brook and smaller underground tributaries draining water mainly from the nonkarst part of the catchment. Limestone is strongly disrupted into blocks by the cracks and mylonite zones (Droppa, 1972; Mello, 2004; Gaal and Vlček, 2011). The cave is mainly horizontal, connected to the Čertova diera cave and they together reach a length of 5358 m. They also form one genetic unit with the Baradla cave in Hungary with a total length of about 25 km, from which almost one-quarter is in the Slovak territory. The cave entrance/exit is on the southern foothill of Domica Hill, at an altitude of 339 m. Three stable monitoring stations equipped with an automatic measuring and registration instruments for continual microclimatic, hydrological, and hydrochemical monitoring are installed in the Domica cave and operated by Microstep-MIS company (Gažík et al., 2009). Radon activity concentration was monitored from June 2010 to July 2011, at the station Virgin passage, situated away from the tourist route (Smetanová et al., 2011). Continual high-resolution monitoring of ^{222}Rn counts was carried out using Barasol probe (Algade, France). Measurement is provided by a passive silicon semiconductor which records alpha particle emissions of the radon by diffusion through a fiber filter, which also eliminates ^{220}Rn. The count sensitivity is 0.02 pulses per hour for 1 Bq m^{-3}. The saturation volumetric activity is 3 MBq m^{-3}. The background count rate is below one event every 24 h. The detection limit for radon is 50 Bq m^{-3} (Papastefanou et al., 2003). Accumulated pulses are recorded automatically at an interval of 10 min and stored in data logger.

Computation of complexity measures and periodograms. For complexity analysis, we use time series of (1) ^{222}Rn concentration and (2) wind speed inside the cave. The length of all the time series used was $N = 51{,}840$. Using the LZA algorithm and procedures outlined in Sections 15.2 and 15.3, we compute all complexity measures. To establish the periodicity in behavior of these quantities, we compute periodogram (Box et al., 2008).

A time series $y_t(t = 1, \ldots, N)$ is observed at equal intervals of time and may be expressed as: $Y_t = \widehat{Y}_t + \varepsilon_t$, where $\widehat{Y}_t$ is unobserved fixed value at time t and $\{\varepsilon_t\}$ is a sequence of random errors identically and independently distributed with expectation 0 and variance σ^2. To determine whether the variability of the time series has periodic components, the series is approximated by finite Fourier series of the following form: if the number of data is even, $N = 2n$,

$$\widehat{Y}_t = A_0 + 2\sum_{m=1}^{n-1}(A_m \cos 2\,\pi m f_1 t + B_m \sin 2\,\pi m f_1 t) + A_n \cos 2\,\pi n f_1 t, \quad \text{or}$$

$$\widehat{Y}_t = A_0 + 2\sum_{m=1}^{n-1}(A_m \cos 2\,\pi m f_1 t + B_m \sin 2\,\pi m f_1 t),$$

if the number of data is odd: $N = 2n - 1$.

Here $R_m = \sqrt{A_m^2 + B_m^2}$ is the amplitude and $\phi_m = arctg(B_m/A_m)$ is the phase of the i-th component. Periodogram is defined as the sum of squared amplitudes of fundamental frequencies $f_m = m/N$, ($j = 0, 1, \ldots, n-1$).

The function $\widehat{Y}_t$ is a linear combination of sine and cosine functions with frequencies proportional to fundamental frequencies $f_1 = 1/N$, so it is linear multiple regression with sine and cosine functions as repressors. Since $\frac{1}{N}\sum_{t=1}^{N} Y_t^2 = R_0^2 + 2\sum_{m=1}^{n-1} R_m^2 + R_n^2$, a contribution of i-th harmonical component to the mean of the total sum of squares of time series is equal to R_i^2. By decomposing the mean of the total sum of squares, it is possible to single out harmonical components which describe the series well.

Analysis of ^{222}Rn *concentration periodicity.* The distribution of frequencies of measured ^{222}Rn data collected in Domica cave is depicted in Fig. 16.1. From this figure, it can be seen that concentrations of ^{222}Rn are measured in relatively broad range, up to 6348 Bq m^{-3}. The shape of distribution of measured values is comparable with a number of other ^{222}Rn measurements.

The concentration of ^{222}Rn in some caves is a result of dynamical steady-state of a number of parameters. It can be expected that stronger than usual influence of some of them or absence of others can result either in higher or lower values of ^{222}Rn concentration in the monitored area.

Parameters which can influence ^{222}Rn concentration (air pressure, temperature, wind speed inside and outside of cave, humidity, etc.) mostly have defined daily and annual course. Thus, their periodicities can cause possible periodicity of collected ^{222}Rn data. We compute periodogram based on time series of 51,840 measurements of ^{222}Rn concentrations. Results obtained are depicted in Fig. 16.2. It is apparent that a dominant periodical component of 27.5 weeks (around half of the year) can describe some periodicity of measured data. Another

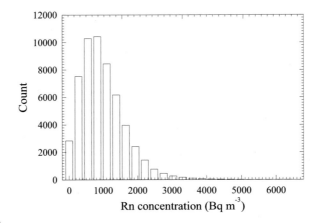

FIGURE 16.1

Distribution of frequencies of measured ^{222}Rn concentrations.

Reprinted with permission from Mihailović, D.T., Krmar, M., Mimic, G., Nikolic-Djoric, E., Smetanova, I., Holy, K., Zelinka, J., Omelka, J., 2015. A complexity analysis of ^{222}Rn concentration variations: a case study for Domica cave, Slovakia for the period June 2010–June 2011. Radiat. Phys. Chem. 106, 88–94.

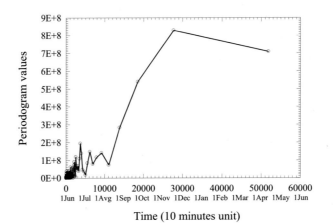

FIGURE 16.2

Periodogram of ^{222}Rn concentration time series.

Reprinted with permission from Mihailović, D.T., Krmar, M., Mimic, G., Nikolic-Djoric, E., Smetanova, I., Holy, K., Zelinka, J., Omelka, J., 2015. A complexity analysis of ^{222}Rn concentration variations: a case study for Domica cave, Slovakia for the period June 2010–June 2011. Radiat. Phys. Chem. 106, 88–94.

prominent periodical component which corresponds to the duration of the full one year time series illustrates existence of the periodicity of 27.5 weeks, seen in Fig. 16.3, where the original ^{222}Rn concentration time series and 27.5 weeks periodic component are presented. Measurements were gathered starting from June

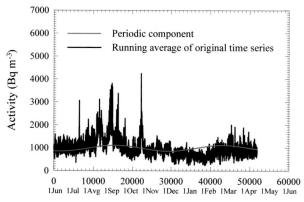

FIGURE 16.3

Time series of ^{222}Rn data fitted by running average method and periodic component in 10 min unit.

Reprinted with permission from Mihailović, D.T., Krmar, M., Mimic, G., Nikolic-Djoric, E., Smetanova, I., Holy, K., Zelinka, J., Omelka, J., 2015. A complexity analysis of ^{222}Rn concentration variations: a case study for Domica cave, Slovakia for the period June 2010–June 2011. Radiat. Phys. Chem. 106, 88–94.

2010, where a slight minimum periodic trend is observed in June and December, and maximum in April and October.

Seasonal changes in natural ventilation often cause large temporal variation in ^{222}Rn levels, mostly characterized by high summer and low winter concentrations (Perrier et al., 2007; Lu et al., 2009; Kowalczk and Froelich, 2010). Because of different structure of caves, there are other atypical patterns which have also been documented, such as maximum concentrations during autumn and minimum during summer in Mammoth Cave, Kentucky (Eheman et al., 1991) or in Moestroff Cave (Luxembourg), where the lowest concentrations are measured in the summer (Kies and Massen, 1997). The present study is likely the first analysis of ^{222}Rn concentration periodicity in a cave (Mihailović et al., 2015).

Dependence of ^{222}Rn *concentration on measured environmental parameters.* Underground ^{222}Rn concentration level, in particular in karstic systems, depends on a complex interrelationship between different external and internal factors like: outside–inside temperature differences, wind speed outside and inside of the cave, atmospheric pressure, humidity, geomorphology, etc. (Kies et al., 1997; Lario et al., 2005). Thus, it is difficult to establish relationships between underground ^{222}Rn concentration level and the mentioned parameters. For this reason we have analyzed other data sets, measured simultaneously with ^{222}Rn concentration (CO$_2$ concentration, temperature inside the cave, external temperature, wind speed in the cave, and external wind speed), to establish a possible correlation of the ^{222}Rn activity and those parameters.

Smetanová et al. (2011) reported that external parameters (atmospheric pressure, external temperature, temperature gradient, rainfall events, etc.) have not affected short-term changes of the observed ^{222}Rn activity. Thus, we have focused on mostly internal parameters. The simplest behavior had the time series of data representing the temperature inside the cave. It was almost constant, oscillating around value of 10°C with variation of 1°C during the considered period but not significantly altering ^{222}Rn activity. Although the time series of the measured CO_2 concentration was much more changeable in time, it did not show any significant correlation with the measured ^{222}Rn concentration time series. However, analysis of the wind speed inside the cave shows very interesting results.

To investigate and compare the lag effect of wind speed inside the cave and external wind speed on ^{222}Rn concentration, a cross-correlation function was used. It was established that there is a negative cross-correlation between ^{222}Rn activity and wind speed inside the cave series for lags 1—20 and that those values are larger than the cross-correlation series of ^{222}Rn activity and external wind speed. According to the Granger test of causality (Asteriou and Hall, 2007), wind speed inside the cave affects a time series of ^{222}Rn concentration significantly, while external wind speed does not.

Additional evidence about possible correlation between ^{222}Rn activity and air circulation inside the cave gives periodogram of the wind speed inside the cave. Fig. 16.4 depicts periodogram based on measured time series of speed of air movement inside the cave, with the values ranging in the interval (0.0—0.7 m s^{-1}). From this figure, it is seen that the time series of wind speed inside the cave has prominent

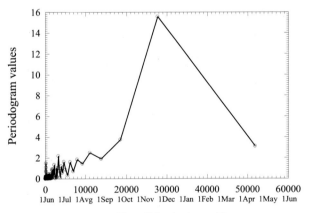

FIGURE 16.4

Periodogram of the time series of the wind speed inside the cave.

Reprinted with permission from Mihailović, D.T., Krmar, M., Mimic, G., Nikolic-Djoric, E., Smetanova, I., Holy, K., Zelinka, J., Omelka, J., 2015. A complexity analysis of ^{222}Rn concentration variations: a case study for Domica cave, Slovakia for the period June 2010—June 2011. Radiat. Phys. Chem. 106, 88—94.

periodic component with a periodicity of 27.5 weeks (half of the year). This periodicity is practically the same as one for the ^{222}Rn concentration time series indicating that both time series has similar dynamics in their annual courses.

A complexity analysis of ^{222}Rn *concentration and wind speed inside the cave.* Analysis of ^{222}Rn concentration time series in a cave is an important step in establishing interrelationship between radon and cave environment. As it is discussed above, these analyses are given either phenomenologically or in the form following relatively simple statistical methods, which just in some segments give us insight of this interaction. However, to get more insight we should apply complexity analysis of time series. Thus, for example, Mihailović et al. (2014) analyzed the dependence of measured ^{222}Rn concentration time series on indoor air temperature and humidity using the product of their KLL complexities (see section 15.2). Here, we consider relationship between ^{222}Rn concentration and wind speed inside the cave, using measures suggested in Section 15.3.

To take into account the KLL variations during the year, we have divided the original time series of the measured ^{222}Rn concentration into 120 three-day time series having the same size, i.e., 432 samples. To compute complexity of each of three-day time series the LZ algorithm has been applied. Identical procedure was used on wind speed inside the cave time series. Then we have computed distributions of frequencies of complexities of the time series for ^{222}Rn concentration and wind speed inside the cave, for the 0.02 interval. Those distributions are given in Fig. 16.5.

From the distribution of frequencies in Fig. 16.5, it is seen that frequencies of ^{222}Rn concentration as well as the wind speed inside the cave are mostly grouped around the highest value of the KLL. It means that changes of both quantities are random and the ^{222}Rn concentration (Fig. 16.5a) and the wind speed inside the

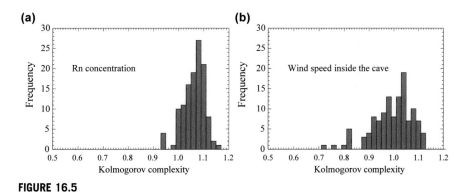

FIGURE 16.5

Distribution of frequencies of the Kolmogorov complexity (KLL) calculated by the LZA algorithm for (a) ^{222}Rn concentration and (b) wind speed inside the cave time series.

Reprinted with permission from Mihailović, D.T., Krmar, M., Mimic, G., Nikolic-Djoric, E., Smetanova, I., Holy, K., Zelinka, J., Omelka, J., 2015. A complexity analysis of ^{222}Rn *concentration variations: a case study for Domica cave, Slovakia for the period June 2010–June 2011. Radiat. Phys. Chem. 106, 88–94.*

cave time series (Fig. 16.5b) both have high values of complexity (greater than 0.7). The mean complexity of 120 three-day series of ^{222}Rn concentration and the wind speed inside the cave is 1.06 ± 0.04 and 0.99 ± 0.08, respectively. ^{222}Rn concentration complexities are distributed in a narrower region than complexities of the wind speed inside the cave. It can be also seen that the number of cases when time series of the wind speed inside the cave are less complex, in the sense of the KL complexity, is greater than for ^{222}Rn concentration time series. Finally, 30 time series of wind speed inside the cave have lower complexity than 0.95 (25% of number of all analyzed time series), while just 3% of ^{222}Rn concentration time series have complexity less than 0.95.

Now, we analyze 120 three-day time series for both ^{222}Rn concentration and wind speed inside the cave, using the overall Kolmogorov complexity (KLO) given by the expression (15.10). Note that using the KLL as a measure, we are not able to distinguish between time series with different amplitude variations and similar random components while with the KLO complexity we can make that distinction, particularly, when they have close values of KLL and KLM, i.e., the Kolmogorov complexity spectrum highest value (see Section 15.2). Computed distributions of frequencies of the KLO complexities of the time series for ^{222}Rn concentration and wind speed inside the cave, for the 0.02 interval are given in Fig. 16.6. From distribution of frequencies of the KLO complexity in Fig. 16.6a, it is seen that frequencies of ^{222}Rn concentration are concentrated in relatively narrow region of the KLO complexities, where the highest obtained value of complexity is just 60% higher than the lowest computed one. The maximal computed overall complexity of the wind speed inside the cave is three times higher than the smallest one (Fig. 16.6b). ^{222}Rn overall

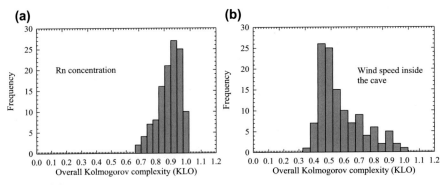

FIGURE 16.6

Distribution of frequencies of the overall Kolmogorov complexity (KLO) computed by the LZA algorithm for (a) ^{222}Rn concentration and (b) wind speed inside the cave time series.

Reprinted with permission from Mihailović, D.T., Krmar, M., Mimic, G., Nikolic-Djoric, E., Smetanova, I., Holy, K., Zelinka, J., Omelka, J., 2015. A complexity analysis of ^{222}Rn concentration variations: a case study for Domica cave, Slovakia for the period June 2010–June 2011. Radiat. Phys. Chem. 106, 88–94.

complexities are grouped around highest values while complexities of wind speed inside the cave have opposite tendency, i.e., they are mostly grouped in the region of low values.

Since the KLO depends on (1) the shape of the Kolmogorov complexity spectrum, (2) the Kolmogorov complexity spectrum highest value (KLM), and (3) the width of the domain of amplitudes of measured quantity in a time series, it is interesting to make a comparison between Figs. 16.5 and 16.6. Before that let us note the following fact: if two time series have very close KLM values and similar shape of the Kolmogorov complexity spectrum then the time series with measured values, which are distributed in a broader domain of amplitudes, will have higher value of the KLO. Otherwise, a time series where measured values vary in relatively narrow interval of amplitudes will have lower value of the KLO. Let us compare results presented in Figs. 16.5a and 16.6a. Apparently, ^{222}Rn three-day time series have high randomness and thus complexity, because all computed values of Kolmogorov complexities have value higher than 0.95 (Fig. 16.5a). Fig. 16.6a shows that the computed KLO values are grouped around high ones. Broadness of distribution depicted in Fig. 16.6a can be considered as a measure of the amplitude growth of the measured ^{222}Rn concentrations in analyzed time series having very similar and very high complexity, as it can be also seen in Fig. 16.5a. Time series of the wind speed inside the cave have high complexity, although some lower values of KL are observed as it is seen in Fig. 16.5b where even 25% of time series have complexity lower than 0.95. However, the KLO of time series of the wind speed inside the cave is distributed in relatively broad region. It is seen from Fig. 16.6b that KLO values are grouped in the region of low values (around 0.5) with distribution showing a decreasing trend. The lowest number of time series has the KLO values around 0.9 and higher. Considering the wind speed inside the cave time series have relatively high values of KLL (Fig. 16.6b), distribution of the KLO depicted in Fig. 16.6b indicates that low values of amplitude in the wind speed inside the cave time series dominate over time series where measured values are spread in the broad region of amplitudes.

To check annual variation of overall complexity of both time series, we have divided the original time series into 12 monthly groups, thus one-month time series had 4320 samples for both quantities. The annual distributions of their KLO complexities are given in Fig. 16.7. Fig. 16.7a indicates that the one month KLO overall complexity of ^{222}Rn concentration is not a constant quantity. The highest complexity is observed in the period from middle of September until middle of October, while lowest complexity is observed in period February–March. The complexity of the wind speed is higher in the period December–March. Data presented in Fig. 16.7b indicate that complexity of ^{222}Rn concentrations could indicate a weak inverse dependence of wind speed ($r = -0.4018$). However, in complexities obtained using three-day intervals, no dependence between complexities of ^{222}Rn concentrations and wind speed data was observed.

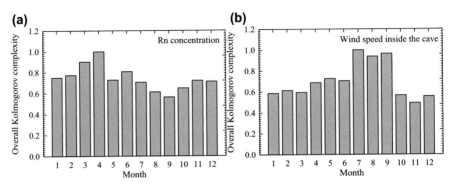

FIGURE 16.7

Annual distribution of the overall complexity (KLO) of (a) ^{222}Rn concentration and (b) wind speed inside the cave time series.

Reprinted with permission from Mihailović, D.T., Krmar, M., Mimic, G., Nikolic-Djoric, E., Smetanova, I., Holy, K., Zelinka, J., Omelka, J., 2015. A complexity analysis of ^{222}Rn concentration variations: a case study for Domica cave, Slovakia for the period June 2010–June 2011. Radiat. Phys. Chem. 106, 88–94.

16.2 USE OF COMPLEXITY ANALYSIS IN ANALYZING THE DEPENDENCE OF ^{222}Rn CONCENTRATION TIME SERIES ON INDOOR AIR TEMPERATURE AND HUMIDITY

Radon (^{222}Rn) is the decay product of radium (^{226}Ra), and both elements are members of the uranium series (^{228}U). After generation from the radioactive decay of (^{226}Ra), mostly in the earth crust, can be transported to the large distances and accumulated indoor due to the fact that radon is a noble gas having no affinity to chemical reactions and relative long half-life of 3.82 days. It is believed that, after smoking, radon is the next most significant source of the lung cancer. The soil and the building materials are the most important source of indoor radon in dwellings. Indoor radon concentrations exceeding the level prescribed can cause possible health hazard of dwelling people (Jelle et al., 2010). Besides in the dwelling control, possible reduction of ^{226}Ra presence is crucial in the area of low-background laboratories. Namely, radioactive gas radon, with its progenies, which emanates from the soil and construction materials, contributes significantly to the background radiation. In a number of experiments where some measurable effects of low-probability process were followed, reduction of background radiation is often the most significant way to improve sensitivity (Antanasijević et al., 1999; Dragić et al., 2011; Udovičić et al., 2009; Garcia et al., 1998; Jovančević and Krmar, 2011). Measured value of ^{222}Rn concentration in some room is a final outcome of plenty of processes including its generation, transport, accumulation, and decay. Therefore, sometimes it is not possible to follow influence of different parameters, through the aforementioned processes, on the final result of radon measurement. Recently, problem is considered through: (1) studies of the dynamic of radon changes in a room with the idea to find correlation with measurable parameters, mostly indoor environmental ones and (2) development of models for predicting the radon concentration and dynamics in dwellings. Let us note that those designed models, even in the cases when affecting processes are treated on a simplified manner, operate with large number of quantities (Collignan et al., 2012; Girault and Perrier, 2012). Analysis of ^{222}Rn concentration time series is an important step in deriving conclusions about interaction between radon and environment. These analyses are mostly based on relatively simple statistical methods, which in some segments give a clear picture about this interaction. However, if we want to get more insight we have to apply the comprehensive mathematical procedures in analysis of those time series. Thus, illustrative examples for that kind of approach are papers by Negarestani et al. (2003) and Seftelis et al. (2008). Negarestani et al. (2003) have proposed a new method based on adaptive linear neuron to estimate the radon concentration in soil associated with the environmental parameters. Seftelis et al. (2008) have developed a mathematical function to describe the diurnal variation of radon progeny. Our intention is to offer a complexity measure-based method for establishing the dependence of ^{222}Rn concentration time series on indoor environmental parameters. A possible field of application of this method is not restricted only on either indoor or outdoor radon time series. Moreover, this mathematical procedure is applicable in analysis of time series, obtained by the measurements, for which we should establish whether influence of some parameter makes the distribution of measured quantity less or more stochastic. In this section we consider the dependence of ^{222}Rn concentration on indoor parameters, in particular on air temperature and humidity, through dynamics of a complex system, which can be analyzed from the signal sent in the form of time series of measured values. In that sense we apply a complexity measure-based method that help us to get an insight into the complexity of the ^{222}Rn concentration time series in dependence on indoor air temperature and humidity.

In this chapter we apply the KLL complexity and KLM complexity (see Section 15.2) for studying the dependence of ^{222}Rn concentration time series on indoor parameters (in particular, air temperature and humidity). For that purpose we use indoor ^{222}Rn concentration time series measured during 2009 in the Low-Background Laboratory for Nuclear Physics at the Institute of Physics in Belgrade. Comparing complexities of ^{222}Rn concentration (R_n), indoor air temperature (T) and humidity (H) time series we establish the dependence of R_n on T and H on indoor air temperature and humidity (Mihailović et al., 2014).

Kolmogorov complexity of the product of two time series. As mentioned above, the complexity of a time series can be lost due to reduction in functioning the system or process represented by that time series. It means that there exists a source, which causes that time series becomes more uniform than random. For example, let us suppose that one physical process (in our case that is detection of indoor ^{222}Rn concentration) is under influence of some parameters (in our case they are indoor air temperature and humidity). We can establish which of these parameters contributes to reducing the complexity of detection indoor ^{222}Rn concentration, by computing the complexity of the product of two or more time series obtained by measurements. Let us suppose that we have two independent and positive definite time series $\{x_i\}$ and $\{y_i\}$, $i = 1, 2, 3, 4, ..., N$, which are generated either computationally or by a measuring procedure. We define product $\{z_i\}$ of two $\{x_i\}$ and $\{y_i\}$ as $\{z_i\} = \{x_i y_i\} = (x_1 y_1, x_2 y_2, x_3 y_3...., x_N y_N)$, where amplitudes in each time series are normalized on its highest value and thus taking the values in the interval $(0, 1)$. Further, with $K^C(x_i)$, $K^C(y_i)$, and $K^C(z_i)$ we define the KLL complexities for corresponding time series $\{x_i\}$, $\{y_i\}$, and $\{z_i\}$, respectively, while $K_m^C(x_i)$, $K_m^C(y_i)$, and $K_m^C(z_i)$ denote their KLM (see section 15.2). It is of interest to explore the KLL and KLM complexities of the product of time series in dependence on the complexity of single ones.

Data and computations. Data we needed for the nonlinear dynamics in this chapter we have obtained from the abovementioned underground low-level laboratory. The special designed system for radon reduction, used in laboratory consists of three stages: (1) The active area of the laboratory is completely lined up with aluminum foil of 1 mm thickness, which is hermetically sealed with a silicon sealant to minimize the diffusion of radon from surrounding soil and concrete used for construction, (2) the laboratory is continuously ventilated with fresh air, filtered through one rough filter for dust elimination followed by the battery of coarse and fine charcoal active filters, and (3) the parameters of the ventilation system are adjusted to give an overpressure of about 2 mbar over the atmospheric pressure. The radon monitor is used to investigate the temporal variations in the radon concentrations. For this type of short-term measurements, the SN1029 radon monitor was used (manufactured by the Sun Nuclear Corporation). The radon monitor device records radon and atmospheric parameter readings every 2 h in the underground laboratory. The data are stored in the internal memory of the device and then transferred to the personal computer. The data obtained from the radon monitor for the temporal variations of the radon concentrations over a long period of time enable the study of the short-term periodical variations (Udovičić et al., 2011). The distribution of frequencies of measured ^{222}Rn concentration values is depicted in Fig. 16.8. It can be seen that the peak of distribution is about 10 Bq m^{-3}. The presence of indoor radon depends on a large number of factors and Maxwell-like distribution of frequencies of measured values indicates probabilistic character of radon appearance in some room. Some authors presented results of similar measurements as lognormal distribution (Bossew, 2010).

For complexity analysis, we use three time series of indoor: (1) ^{222}Rn concentration, (2) air temperature, and (3) air humidity (Fig. 16.9). Using the computation procedure outlined in Sections 15.2 and 15.3, we have computed (1) complexity spectra of ^{222}Rn concentration (R_n); product of ^{222}Rn concentration and indoor air temperature time series ($R_n \times T$); indoor air humidity ($R_n \times H$), and product of ^{222}Rn concentration, indoor air temperature, and air humidity time series ($R_n \times T \times H$) and (2) the KLL and KLM for these series. Before computation procedure the time series were normalized on their highest values. The length of all time

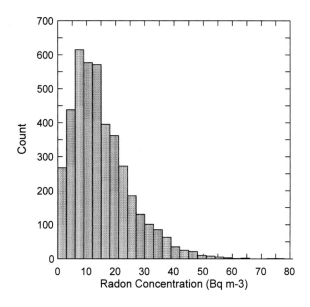

FIGURE 16.8

Distribution of frequencies of measured [222]Rn concentrations.

Reprinted with permission from Mihailović, D.T., Udovičić, V., Krmar, M., Arsenić, I., 2014. A complexity measure based method for studying the dependence of [222]Rn concentration time series on indoor air temperature and humidity. Appl. Radiat. Isot. 84, 27–32.

series used was $N = 4173$. The computations are carried out for the period 1 January–31 December 2009 using the LZA algorithm.

The concentration of [222]Rn in some room is a result of dynamical steady-state of a number of parameters. It can be expected that stronger than usual influence of some of them or absence of another one might result either in higher or lower values of [222]Rn concentration in the monitored area. Thus, [222]Rn concentration strongly depends on parameters of underground environment (Viñas et al., 2007). According to Udovičić et al. (2011) in the long term there exists a clear influence of indoor air temperature and relative humidity on [222]Rn concentration. Further, in the same paper it is underlined that concerning the radon daughters, the relative humidity indoors contributes to the aerosol density and keeps the radon daughters in the indoor air. Although in the last decade a vast number of experimental evidence has been offered about this issue (Choubey et al., 2011; Barbosa et al., 2010; Kamra et al., 2013), we still have no enough knowledge about insights of the influence of the indoor air parameters on the [222]Rn concentration variability. One of the reasons for that is nonlinearity of these dependences, which cannot be elaborated by the traditional mathematical methods (Seftelis et al., 2008). Thus, it seems that the complexity analysis offers more quantitative measures in explanation of this phenomenon.

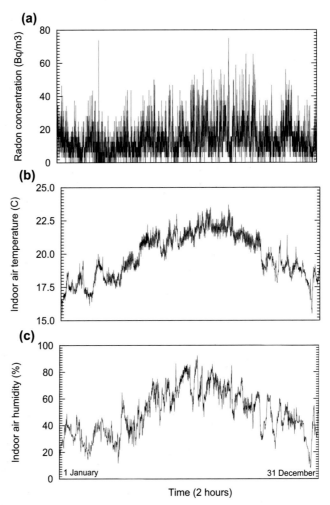

FIGURE 16.9

Time series for (a) ^{222}Rn concentration, (b) indoor air temperature, and (c) indoor air humidity, created from data obtained from the Low-Background Laboratory for Nuclear Physics at the Institute of Physics in Belgrade (Serbia) for the period 1 January–31 December 2009.

Reprinted with permission from Mihailović, D.T., Udovičić, V., Krmar, M., Arsenić, I., 2014. A complexity measure based method for studying the dependence of ^{222}Rn concentration time series on indoor air temperature and humidity. Appl. Radiat. Isot. 84, 27–32.

The results of complexity analysis are given in Fig. 16.10. This figure depicts the KLL spectra of the following time series: (1) ^{222}Rn concentration (R_n), (2) product of ^{222}Rn concentration and indoor air temperature ($R_n \times T$), (3) indoor air humidity ($R_n \times H$), and (4) product of ^{222}Rn concentration, indoor air temperature, and indoor

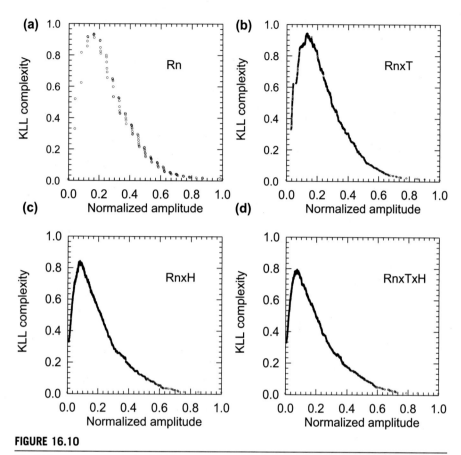

FIGURE 16.10

The Kolmogorov complexity (KLL) sequences of the following time series: (a) ^{222}Rn concentration (R_n), (b) product of ^{222}Rn concentration and indoor air temperature ($R_n \times T$), (c) indoor air humidity ($R_n \times H$), and (d) product of ^{222}Rn concentration, indoor air temperature, and indoor air humidity ($R_n \times T \times H$). All time series are normalized on their highest value.

Reprinted with permission from Mihailović, D.T., Udovičić, V., Krmar, M., Arsenić, I., 2014. A complexity measure based method for studying the dependence of ^{222}Rn concentration time series on indoor air temperature and humidity. Appl. Radiat. Isot. 84, 27–32.

air humidity ($R_n \times T \times H$), where all time series are normalized on their highest value. The peaks in spectra show the KLM. This parameter could be considered as a better indicator of the complexity comparing to the KL, which is not always a suitable measure of the complexity. In particular, this is enhanced in the case of asymmetrical distributions (Nikolić-Đorić, personal communication). Looking at panels it is seen that for the $R_n \times T$ sequence its KLM (Fig. 16.10b) is just slightly different comparing to the R_n one (Fig. 16.10a). Practically, there are no differences in their maximal Kolmogorov complexities. However, in the $R_n \times H$ spectrum (Fig. 16.10c) and the $R_n \times T \times H$ one (Fig. 16.10d), the KLM values are lower than for the R_n spectrum. Note that a process that is least complex has a Kolmogorov complexity value near to zero, whereas a process with highest complexity will have KLL close to one. This measure can be also considered as a measure of randomness. Thus, a value of the KLL near zero is associated with a simple deterministic process like a periodic motion, whereas a value close to one is associated with a stochastic process (Ferreira et al., 2003). Accordingly the KLM values, which are large for the $R_n \times T$ spectrum (0.937), points out the presence of stochastic component in influence of indoor air temperature on ^{222}Rn concentration. The other two computed KLM complexities [$R_n \times H$ (0.865) and $R_n \times T \times H$ (0.850) spectra] indicate that there exists a source of influence, which reduces the complexity of ^{222}Rn concentration. To our opinion it could be attributed to (1) the fact that relative humidity indoors contributes to the aerosol density and keeps the radon daughters in the indoor air and (2) nonlinearities in relation between ^{222}Rn concentration and indoor air humidity (Nikolić-Đorić, personal communication).

Finally, we have plotted the diagram KLL complexity versus Sample Entropy (SempEn) to see behavior of time series. The KLL measure has been often used for evaluation of the randomness present in time series, while entropy is also used to characterize the complexity of a time series. From Fig. 16.11, a strong correlation

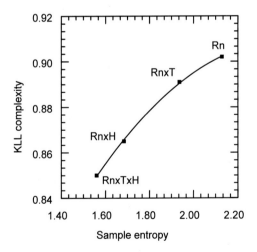

FIGURE 16.11

Kolmogorov complexity (KLL) versus sample entropy (SempEn) for time series used in Fig. 16.9.

Reprinted with permission from Mihailović, D.T., Udovičić, V., Krmar, M., Arsenić, I., 2014. A complexity measure based method for studying the dependence of ^{222}Rn concentration time series on indoor air temperature and humidity. Appl. Radiat. Isot. 84, 27–32.

between these two measures indicating degree of influence of indoor air temperature and relative humidity on ^{222}Rn concentration is clearly seen.

16.3 USE OF THE KOLMOGOROV COMPLEXITY AND ITS SPECTRUM IN ANALYSIS OF THE UV-B RADIATION TIME SERIES

Complexity of the UV-B radiation time series. Influenced by climate, vegetation, geography, and human factors, many meteorological elements including the UV-B radiation in a specific geographic region may range from being relatively simple to complex, which exhibits significant variability in both time and space. Recently, the human factor becomes the most important issue regarding the complexity of the meteorological elements. Namely, actions in the form of different human activities in environment (air, soil, and water) can be either constructive or destructive. They (1) can have positive or negative impact on the human economy and (2) can leave landscape features that are present for a long time. Thus, it is of interest to determine the nature of complexity in the UV-B radiation processes. This approach requires the use of various measures of the complexity of the UV-B radiation, which may provide us: (1) more comprehensive investigation of possible change in UV-B radiation due to human activities and response to climate change and (2) improving the application of the stochastic process concept in radiation its modelling, forecasting, measuring, and other ancillary purposes (Adami, 2002; Boschetti, 2008; Junkermann, 2005; Malinović et al., 2006; Bhattarai et al., 2007; Paulescu et al., 2010; Malinović-Milićević and Mihailović, 2011; Malinović-Milićević et al., 2013). As we mentioned in Section 15.2, the KLL complexity is a measure of the disorder or irregularity in a sequence, while the traditional entropies like approximate entropy and sample entropy (SampEn) (see Section 9.1) quantify only the regularity of time series. Note that these measures have some disadvantages. Therefore, it is of interest to see how these measures can be employed in complexity analysis of the UV-B radiation dose time series for different purposes. Thus, here we investigate the complexity of the UV-B radiation dose time series for places spatially distributed over some area, using the KL and SampEn measures. To reinforce this analysis, we also use the Kolmogorov complexity spectrum and the KLM, i.e. the Kolmogorov spectrum highest value.

For our analysis we use the Vojvodina region (Serbia). UV-B radiation records in the Vojvodina region (Serbia) are of relatively short size. To create the UV radiation time series for seven representative places we include: (1) values measured in Novi Sad using the broadband Yankee UVB-1 biometer, (2) values computed by a parametric numerical model, and (3) values computed by an empirical formula derived on the basis of the linear correlation between the daily sum of the UV-B radiation and the daily sum of the global solar radiation. In the further development we analyze the complexity of the UV-B radiation dose time series from the seven representative places in the Vojvodina region (Serbia) for the period 1990–2007, using the KLL, KLM, and SempEn measures. We also investigate the effect of different human activities, events, and climate change on the UV-B radiation dose complexity by dividing the period 1990–2007 into two equal subintervals: (1) 1990–98 and (2) 1999–2007. Namely, according to Krmar et al. (2012) there was an evident increase in human activity in the Vojvodina region after 1998 (post–civil war period, NATO military activities in air, intensification of economic activity, more intensive traffic, traditional home heating). It has caused high air pollution and further changes in the UV-B radiation dose complexity in the Vojvodina region. We compute the KLL and SempEn values for the various time series in each of the above subintervals to see whether, during the period 1999–2007, there is a decrease in complexity in most of the places in comparison to the period 1990–98. If that is true then the complexity loss may be attributed to (1) human intervention in the post civil war period that caused larger air pollution and (2) increased cloudiness due to climate changes.

Short description of the UV radiation parametric numerical model. We have partly generated time series of the UV-B radiation by a parametric numerical model NEOPLANTA (Malinović et al., 2006). This model computes the solar direct and diffuse UV irradiances on a horizontal surface under cloud-free conditions for the wavelength range 280–400 nm with 1-nm resolution as well as the UV index (UVI). Model simulates the effects of the absorption of the UV radiation by ozone (O_3), sulfur dioxide (SO_2), and nitrogen dioxide (NO_2) and absorption and scattering by aerosol and air molecules in the atmosphere. Atmosphere in model is divided in 40 parallel layers with constant values of meteorological parameters. Its vertical resolution of the model is 1 km for altitudes less than 25 km and above this height 5 km layers were used. The required input parameters are the local geographic coordinates and time or solar zenith angle, altitude, spectral albedo, and the total amount of gases. The NEOPLANTA model includes its own vertical gas profiles (Ruggaber et al., 1994) and extinction cross-sections (Burrows et al., 1999; Bogumil et al., 2000), extraterrestrial solar irradiance shifted to terrestrial wavelength (Koepke et al., 1998), aerosol optical properties for 10 different aerosol types (Hess et al., 1998), and spectral albedo for nine different ground surface types (Ruggaber et al., 1994). The model uses standard atmosphere meteorological profiles although it allows the use of real time meteorological data assimilated from the high-level resolution atmospheric mesoscale models. Output data are spectral direct, diffuse, and global irradiance divided into the UV-A (320–400 nm) and UV-B (280–320 nm) part of the spectrum, erythemally weighted UV irradiance computed using the erythemal action spectrum by McKinley and Diffey (1987), the UVI, spectral optical depth, and spectral transmittance for each atmospheric component. All outputs are computed at the lower boundary of each layer.

The UV irradiance is computed as the sum of the direct and the diffuse components. Computation of the direct part of radiation is carried out by the Beer–Lambert law. The direct irradiance $I_{dir}(\lambda)$ at wavelength λ received at ground level by unit area is given by

$$I_{dir}(\lambda) = I_0(\lambda)T(\lambda) \tag{16.1}$$

where $I_0(\lambda)$ is the extraterrestrial irradiance corrected for the actual Sun–Earth distance and $T(\lambda)$ is the total transmittance that includes O_3, SO_2, NO_2, aerosol, and air transmittances. Each of the individual transmittances is computed using optical depth $\tau(\lambda)$ that is the product of extinction coefficient $\beta(\lambda)$ and ray path through the atmosphere s

$$T(\lambda) = exp(-\tau(\lambda)) = exp(-\beta(\lambda)s). \tag{16.2}$$

Extinction coefficient of the UV radiation β is computed by the product of the cross-sectional area σ and layer particle concentration N

$$\beta(\lambda) = \sigma(\lambda)N. \tag{16.3}$$

The starting point for computation of diffuse part of radiation is the set of equations from Bird and Riordan spectral model (Bird and Riordan, 1986), which represents equations from previous parametric models (Leckner, 1978; Justus and Paris, 1985), improved after comparisons with rigorous radiative transfer model and with measured spectra. The diffuse irradiance $I_{dif}(\lambda)$ is divided into three components: (1) the Rayleigh scattering component $I_{ray}(\lambda)$, (2) the aerosol scattering component $I_{aer}(\lambda)$, and (3) the component that accounts for multiple reflection of irradiance between the ground and the air $I_{rf}(\lambda)$

$$I_{dif}(\lambda) = I_{ray}(\lambda) + I_{aer}(\lambda) + I_{rf}(\lambda). \tag{16.4}$$

The Rayleigh scattered component $I_{ray}(\lambda)$ of diffuse part of UV irradiance is computed as

$$I_{ray}(\lambda) = I_0(\lambda)T_{O_3}(\lambda)T_{SO_2}(\lambda)T_{NO_2}(\lambda)T_{aa}(\lambda)\left(1 - T_{ray}^{0.95}(\lambda)\right)\Big/2. \tag{16.5}$$

T_{O_3}, T_{SO_2}, T_{NO_2}, T_{aer}, and T_{ray} are O_3, SO_2, NO_2, aerosol, and air transmittances that have been defined previously. Transmittance of the aerosol absorption process, $T_{aa}(\lambda)$, is defined in Justus and Paris (1985) as

$$T_{aa}(\lambda) = exp[-(1 - \omega(\lambda))\tau_a(\lambda)], \tag{16.6}$$

where $\omega(\lambda)$ is the single-scattering albedo and $\tau_a(\lambda)$ is aerosol optical thickness.

The aerosol-scattered irradiance is computed as

$$I_{aer}(\lambda) = I_0(\lambda)T_{O_3}(\lambda)T_{SO_2}(\lambda)T_{NO_2}(\lambda)T_{aa}(\lambda)T_{ray}^{1.5}(\lambda)[1 - T_{as}(\lambda)]D_s(\lambda), \tag{16.7}$$

where $T_{as}(\lambda)$ is the transmittance for aerosol scattering, such that

$$T_{as}(\lambda) = exp[-\omega(\lambda)\tau_a(\lambda)] \tag{16.8}$$

and $D_s(\lambda)$ is the fraction of the scattered flux that is transmitted downward. The function $D_s(\lambda)$ is dependent on the aerosol asymmetry factor δ and solar zenith angle θ, according to (Bird and Riordan, 1986) and Justus and Paris (1985) as

$$D_s = F_sC_s, \tag{16.9}$$

$$F_s = 1 - 0.5exp[(B_1 + B_2cos\theta)cos\theta], \qquad (16.10)$$

$$B_1 = B_3[1.459 + B_3(0.1595 + B_3 \times 0.4129)], \qquad (16.11)$$

$$B_2 = B_3[0.0783 + B_3(-0.3824 - B_3 \times 0.5874)], \qquad (16.12)$$

$$B_3 = ln(1 - \delta) \qquad (16.13)$$

and

$$C_s(\lambda) = (\lambda + 0.55)^{1.8}. \qquad (16.14)$$

The asymmetry factor is a key optical characteristic of aerosols and it is used from OPAC database (Hess et al., 1998) for each wavelength and humidity.

Backscattered component of multiple reflections between air and ground is computed following Bird and Riordan (1986) as

$$I_{rf}(\lambda) = \frac{[I_{dir}(\lambda) + I_{ray}(\lambda) + I_{aer}(\lambda)]r_s(\lambda)r_g(\lambda)C_s(\lambda)}{1 - r_s(\lambda)r_g(\lambda)}, \qquad (16.15)$$

where $r_g(\lambda)$ is ground albedo and $r_s(\lambda)$ is sky reflectivity. Ground albedo is used from Ruggaber et al. (1994) while sky reflectivity is computed by

$$r_s(\lambda) = T'_{O_3}(\lambda)T'_{aa}(\lambda)\left[0.5\left(1 - T'_{ray}(\lambda)\right) + (1 - F'_s(\lambda))T'_{ray}(\lambda)\left(1 - T'_{as}(\lambda)\right)\right], \qquad (16.16)$$

where the primed transmittance terms are the regular atmospheric transmittance evaluated at optical mass of 1.8. More details about this model are elaborated in Malinović et al. (2006).

Time series and computations. The Vojvodina region (Serbia) is situated in the northern part of Serbia and the southern part of the Pannonian lowland (18°51′−21°33′E, 44°37′−46°11′N and 75−641 m a.s.l.) (Fig. 16.12a). For the complexity analysis of the UV-B radiation dose time series in this section we select the following places: Sombor (SO), Subotica (SU), Novi Sad (NS), Kikinda (KI), Zrenjanin (ZR), Banatski Karlovac (BK), and Sremska Mitrovica (SM) as shown in Fig. 16.12b. The UV-B radiation has a pronounced impact on the human health and some plants in agricultural activities in this region that is the most important food production area in Serbia with surface area of 21,500 km^2 and a population of about 2 million people. Monitoring details of the UV-B radiation in the Vojvodina region are given in Malinović-Milićević and Mihailović (2011).

We have formed the corresponding time series combining three sources because of the lack of measurement places for the UV radiation in the Vojvodina region. We have included (1) values measured in Novi Sad (45.33°N, 19.85°E, 84 m a.s.l.) by the broadband Yankee UVB-1 biometer, (2) values computed by a parametric numerical model, and (3) values computed by an empirical formula based on linear correlation between the daily dose of the UV-B (UVB_d) and the daily sum of the global solar radiation (G_d) in MJ m^{-2} (Malinović-Milicević et al., 2013) The empirical formula, which is derived on the basis of relationship between daily values of UVB_d (measured UV-B data and corresponding calibration factors) and G_d

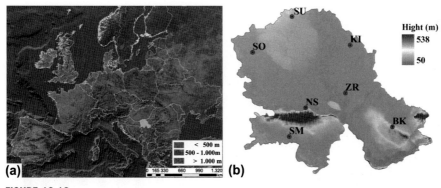

FIGURE 16.12

Location of the Vojvodina region (Serbia) in Europe (a) and places used in study (b); the places are Sombor (SO), Subotica (SU), Novi Sad (NS), Kikinda (KI), Zrenjanin (ZR), Banatski Karlovac (BK), and Sremska Mitrovica (SM).

Reprinted with permission from Mihailović, D.T., Malinović-Milićević, S., Arsenić, I., Drešković, N., Bukosa, B., 2013. Kolmogorov complexity spectrum for use in analysis of UV-B radiation time series. Mod. Phys. Lett. B 27, 1350194.

(computed via an empirical formulae) for the period April 2003—December 2009 in Novi Sad.

Using the computation procedure described in the Sections 15.2, 15.3, and 9.1, we have computed the KLL, KLM, and SampEn values, respectively, for the seven UV-B radiation dose time series (Fig. 16.13). The computations are carried out for the entire time interval 1990—2007 and for two subintervals covering this period: (a) 1990—98 and (b) 1999—2007. All these complexity measures are sensitive to the length of time series, N. For the SampEn, there exists a recommendation for use N that is larger than 200 (Yentes et al., 2012). For the time interval 1990—2007 and two subintervals (1990—98 and 1999—2007), the length of time series was $N = 6574$, 3287, and 3287, respectively. The SampEn is sensitive on input parameters: embedding dimension (m), tolerance (r), and time delay (τ). In this section, it was computed for UV-B radiation dose time series with the following values of parameters: $m = 2$, $r = 0.2$, and $\tau = 1$.

The results of calculations are given in Table 16.1. It is seen from this table that for five places (Sombor, Subotica, Novi Sad, Kikinda, and Zrenjanin) their KL values are close to each other (0.492, 0.498, 0.492, 0.496, and 0.498). However, in contrast to these places for Sremska Mitrovica and Banatski Karlovac have higher values of the KL (0.523 and 0.515). Since Sremska Mitrovica is close to Fruška Gora Mountain while Banatski Karlovac is located in a hilly region (Fig. 16.12), the increase of the complexity in those places can be attributed to enhanced UV-B radiation dose caused by the multiple scattering effects (Kylling et al., 2000; Pfeifer et al., 2006). Following this reason it could be expected that Novi Sad, which is also in the

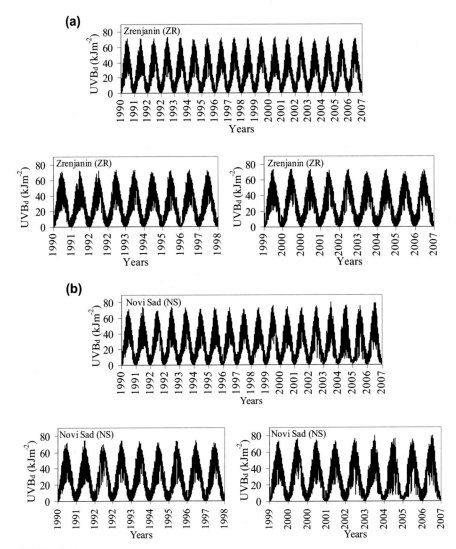

FIGURE 16.13

The UV-B radiation dose time series (1990—2007, 1990—98, 1999—2007 year) for three places in the Vojvodina region (Serbia) analyzed for this section.

Reprinted with permission from Mihailović, D.T., Malinović-Milićević, S., Arsenić, I., Drešković, N., Bukosa, B., 2013. Kolmogorov complexity spectrum for use in analysis of UV-B radiation time series. Mod. Phys. Lett. B 27, 1350194.

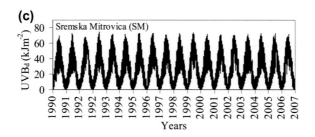

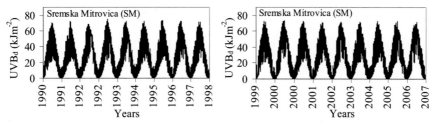

FIGURE 16.13 cont'd.

vicinity of the Fruška Gora mountain, has the higher level of the complexity. However, this place is highly urbanized with more emission sources in comparison with Sremska Mitrovica; thus the urban air pollution reduces the amount of UV-B radiation reaching the ground. Namely, according to Bais et al. (2006) the surface UV-B radiation at locations near the emission sources of O_3, SO_2, or NO_2 in the lower troposphere is attenuated by up to 20%. In result, the complexity of the UV-B radiation dose decreases. Note, if a process is less complex then it has a KLL value close to zero, whereas a process with highest complexity will have the KLL close to one. If we look at the KLM values we reach the same conclusions. To our knowledge, the KLL and KLM measures has not been used for analyzing the complexity of the UV-B radiation dose time series.

In our analysis we have used another complexity measure, i.e., the SampEn. Unlike approximate entropy, SampEn is not often used in the analysis of the complexity of geophysical time series (He et al., 2012). Such analysis was done by Shuangcheng et al. (2006) in measurement of climate complexity using daily temperature time series. The computed values of the SampEn are also listed in Table 16.1 Those values, which are close to each other, indicate a similar behavior of UV-B radiation dose time series for the entire time interval 1990–2007 and all places, i.e., their lower irregularity.

We have also divided the period 1990–2007 into two subintervals: (a) 1990–98 and (b) 1999–2007, and computed the KLL and SampEn values for the various time series in each of these subintervals. These intervals were chosen because we expected a change in the complexity of the UV-B radiation dose after 1999 in the Vojvodina region because of (1) a large increase of air and soil pollution (Krmar et al., 2012) and (2) an increase of cloudiness due to climate change (Rajković et al., 2012)

Table 16.1 Kolmogorov Complexity (KLL), Kolmogorov Complexity Spectrum Highest Value (KLM) and Sample Entropy (SampEn) Values for the UV-B Radiation Dose Time Series of Seven Places in the Vojvodina Region (Serbia) for the Period 1990–2007, and the Subintervals: (a) 1990–98 and (b) 1999–2007. In Computing the Entropy We Have Used the Following Sets of Parameters ($m = 2$, $r = 0.2$, and $\tau = 1$)

Place	Measure	1990–2007	1990–98	1999–2007
SO (45°47′N, 19°05′E)	KL	0.492	0.519	0.505
	KLM	0.511	0.526	0.522
	SampEn	1.206	1.203	1.176
SU (46°06′N, 19°46′E)	KL	0.498	0.498	0.489
	KLM	0.512	0.522	0.522
	SampEn	1.245	1.217	1.202
NS (45°15′N, 19°51′E)	KL	0.492	0.530	0.498
	KLM	0.513	0.547	0.512
	SampEn	1.223	1.262	1.174
KI (45°51′N, 20°28′E)	KL	0.496	0.526	0.501
	KLM	0.509	0.533	0.522
	SampEn	1.238	1.216	1.146
ZR (45°24′N, 20°21′E)	KL	0.498	0.544	0.487
	KLM	0.527	0.565	0.526
	SampEn	1.238	1.252	1.233
SM (44°58′N, 19°38′E)	KL	0.523	0.551	0.508
	KLM	0.536	0.572	0.533
	SampEn	1.252	1.234	1.178
BK (45°03′N, 21°02′E)	KL	0.515	0.530	0.530
	KLM	0.532	0.558	0.530
	SampEn	1.191	1.246	1.243

Reprinted with permission from Mihailović, D.T., Malinović-Milićević, S., Arsenić, I., Dresković, N., Bukosa, B., 2013. Kolmogorov complexity spectrum for use in analysis of UV-B radiation time series. Mod. Phys. Lett. B 27, 1350194.

and corresponding influence on UV radiation dose (Calbo et al., 2005; Rieder et al., 2010). Let us note that the KLL complexity of different kind of biomedical, hydrological, and physical time series may be lost due to different reasons that come from reducing the functionality of some system segments represented by those series. For example, Gómez and Hornero (2010) using entropy and complexity analyses of Alzheimer's disease (AD) have shown that the complexity reduction seems to be associated with the deficiencies in information processing suffered by AD patients. And another example from the river flow time series analysis by Orr and Carling (2006) point out that the complexity loss may be attributed to the extent of human intervention involving land and crop use, urbanization, commercial navigation, and other activity. Thus, decrease of the KLL complexity of some process represented by a time series is an indicator of a simplification of that process caused by some crucial agent.

It is found that during 1999–2007, there was a decrease in complexity in all places (Sombor—0.505; Subotica—0.489; Novi Sad—0.498; Kikinda—0.501; Zrenjanin—0.487; Sremska Mitrovica—0.508 and Banatski Karlovac—0.530) in comparison to the period 1990–1998 (Sombor—0.519; Subotica—0.498; Novi Sad—0.530; Kikinda—0.526; Zrenjanin—0.544; Sremska Mitrovica—0.551, and Banatski Karlovac—0.539) as it presented in Table 16.1. These differences are seen in Fig. 16.14. It shows relative change of the KLL (Fig. 16.14a) and KLM (Fig. 16.14b) from the period 1990–98 comparing to the period 1999–2007 for the seven places. From Fig. 16.14 it is seen that the central and south western parts of the Vojvodina region have the largest decline of the KLL (Fig. 16.14a) and KLM (Fig. 16.14b) complexities. In other parts, that decline is much lower. Among places with the large decline of both complexities, Zrenjanin stands out with the largest one. It is a result of a very large concentration of SO_2 and particles in this place that come from the mentioned human activities. Namely, SO_2 absorbs radiation in the UV-B part of the spectrum, remarkably affecting the reduction of the UV-B radiation through sulfate aerosols. It is estimated that in the industrialized countries on the northern hemisphere sulfate aerosols can reduce the UV-B radiation for 5–18% (Liu et al., 1991). Fig. 16.15 depicts the KLL complexity spectrum of the normalized UV-B radiation dose for three places (Zrenjanin, Novi Sad, and Sremska Mitrovica). From this it is seen that, for all places, the highest differences in spectra of complexity (period 1990–98 versus period 1999–2007) are in the interval (0.3, 0.5) of the normalized UV-B radiation doses.

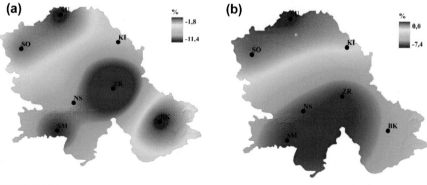

FIGURE 16.14

Relative change of the KLL (a) and KLM (b) from the period 1990–98 comparing to the period 1999–2007 for places in the Vojvodina region (Serbia). Abbreviations are the same as in Fig. 16.12.

Reprinted with permission from Mihailović, D.T., Malinović-Milićević, S., Arsenić, I., Drešković, N., Bukosa, B., 2013. Kolmogorov complexity spectrum for use in analysis of UV-B radiation time series. Mod. Phys. Lett. B 27, 1350194.

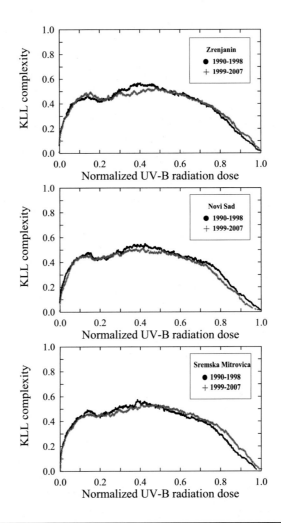

FIGURE 16.15

The Kolmogorov complexity spectrum of the UV-B radiation dose time series for three places in the Vojvodina region (Serbia). On *x* axis are depicted the values of the time series normalized as $x_i = (X_i - X_{min})/(X_{max} - X_{min})$, where $\{X_i\}$ is the time series of the UV-B radiation dose obtained by procedures described in Section 15.3 and $X_{max} = max\{X_i\}$ and $X_{min} = min\{X_i\}$.

Reprinted with permission from Mihailović, D.T., Malinović-Milićević, S., Arsenić, I., Drešković, N., Bukosa, B., 2013. Kolmogorov complexity spectrum for use in analysis of UV-B radiation time series. Mod. Phys. Lett. B 27, 1350194.

REFERENCES

Abdel-Monem, A.A., El Aassy, I.E., El-Naggar, A.M., Attia, K.E., El-Fawy, A.G., 1996. Concentrations of radon gas and daughters in uranium exploration tunnels, Allouga, West Central Sinai, Egypt. Radiat. Phys. Chem. 47, 765−767.

Adami, C., 2002. What is complexity? Bioessays 24, 1085−1094.

Antanasijević, R., Aničin, I., Bikit, I., Banjanac, R., Dragić, A., Joksimović, D., Krmpotić, Đ., Udovičić, V., Vuković, J.B., 1999. Radon measurements during the building of a low-level laboratory. Radiat. Meas. 31, 371.

Asteriou, D., Hall, S.G., 2007. Applied Econometrics: A Modern Approach Using EViews and Microfit, revised ed. Palgrave Macmillan, Basingstoke, Hampshire.

Aytekin, H., Baldik, R., Celebi, N., Ataksor, B., Tasdelen, M., Kopuz, G., 2006. Radon measurements in the caves of Zonguldak (Turkey). Radiat. Protect. Dosimetry 118, 117−121.

Bais, A.F., Lubin, D., Arola, A., Bernhard, G., Blumthaler, M., Chubarova, N., Erlick, C., Gies, H.P., Krotkov, N., Lantz, K., Mayer, B., McKenzie, R.L., Piacentini, R., Seckmeyer, G., Slusser, J.R., Zerefos, C., 2006. Surface Ultraviolet Radiation: Past, Present and Future, in Scientific Assessment of Ozone Depletion. Global Ozone Research and Monitoring Project-Report No. 47, Chap. 7. World Meteorological Organization, Geneva, Switzerland, p. 58, 2007.

Barbosa, S.M., Zafrir, H., Malik, U., Piatibratova, O., 2010. Multi-year to daily Radon variability from continuous monitoring at the Amram tunnel, southern Israel. Geophys. J. Int. 182, 829−842.

Bhattarai, B.K., Kjeldstad, B., Thorseth, T.M., Bagheri, A., 2007. Erythemal dose in Kathmandu, Nepal based on solar UV measurements from multichannel filter radiometer, its deviation from satellite and radiative transfer simulations. Atmos. Res. 85, 112−119.

Bird, E.R., Riordan, C., 1986. Simple solar spectral model for direct and diffuse irradiance on horizontal and tilted planes at the Earth's surface for cloudless atmospheres. J. Clim. Appl. Meteorol. 25, 87−97.

Bogumil, K., Orphal, J., Burrows, J.P., 2000. Temperature dependent absorption cross sections of O_3, NO_2, and other atmospheric trace gases measured with the sciamachy spectrometer. In: Proc. ERS-envistat Symp. 16−20 October 2000. ESA−ESTEC, Gothenburg Sweden, p. 11.

Boschetti, F., 2008. Mapping the complexity of ecological models. Ecol. Complex 5, 37−47.

Bossew, P., 2010. Radon: exploring the log-normal mystery. J. Environ. Radioact. 101, 826−834.

Box, G.E.P., Jenkins, G.M., Reinsel, G.C., 2008. Time Series Analysis, Forecasting and Control, fourth ed. John Wiley & Sons, Inc., Hoboken, New Jersey.

Burrows, J.P., Richter, A., Dehn, A., Deters, B., Himmelmann, S., Voigt, S., Orphal, J., 1999. Atmospheric remote-sensing reference data from GOME—part 2. Temperature-dependent absorption cross sections of O_3 in the 231−794 nm range. J. Quant. Spectrosc. Radiat. Trans. 61, 509−517.

Calbo, J., Pages, D., Gonzalez, J.A., 2005. Empirical studies of cloud effects on UV radiation: a review. Rev. Geophys. 43, RG2002.

Chen, Y., Li, S., 1995. The discussion on the radioprotection of radon and its decay products of tour karst cave. Guizhou Med. J. 19, 251−252 (in Chinese).

Choubey, V.M., Arora, B.R., Barbosa, S.M., Kumar, N., Kamra, L., 2011. Seasonal and daily variation of radon at 10 m depth in borehole, Garhwal Lesser Himalaya, India. Appl. Radiat. Isot. 69, 1070—1078.

Collignan, B., Lorkowski, C., Améon, R., 2012. Development of a methodology to characterize radon entry in dwellings. Build. Environ. 57, 176—183.

Dragić, A., Udovičić, V., Banjanac, R., Joković, D., Maletić, D., Veselinović, N., Savić, M., Puzović, J., Aničin, I., 2011. The new set-up in the Belgrade low-level and cosmic-ray laboratory. Nuclear Technol. Radiat. Protect. 26, 181—192.

Droppa, A., 1972. Contribution to the Development of the Domica Cave, vol. 22. Československý kras, Praha, pp. 65—72 (In Slovak).

Dueñas, C., Fernández, M.C., Cañete, S., 2005. ^{222}Rn concentrations and the radiation exposure levels in the Nerja Cave. Radiat. Meas. 40, 630—632.

Eff-Darwich, A., Martín-Luis, C., Quesada, M., de la Nuez, J., Coello, J., 2002. Variations on the concentration of 222 Rn in the subsurface of the volcanic island of Tenerife, Canary Islands. Geophys. Res. Lett. 29, 26(1)—26(4).

Eheman, C., Carson, B., Rifenburg, J., Hoffman, D., 1991. Occupational exposure to radon daughters in Mammoth-Cave-National-Park. Health Phys. 60, 831—835.

Fernandez-Cortes, A., Sanchez-Moral, S., Cuezva, S., Cañaveras, J.C., Abella, R., 2009. Annual and transient signatures of gas exchange and transport in the Castañar de Ibor cave (Spain). Int. J. Speleol. 38, 153—162.

Ferreira, F.F., Francisco, G., Machado, B.S., Murugnandam, P., 2003. Time series analysis for minority game simulations of financial markets. Phys. A 321, 619—632.

Gaal, L., Vlček, L., 2011. Tectonics of the Domica cave (Slovak karst). Aragonit 16, 3—11 (In Slovak).

Garcia, E., Gonzalez, D., Morales, A., Morales, J., Ortiz, A., Solorzano, De, Puimedon, J., Saenz, C., Salinas, A., Sarsa, M.L., Villar, J.A., 1998. Analysis of airborne radon in an ultra-low background experiment. Appl. Radiat. Isot. 49, 1749—1754.

Gažík, P., Haviarová, D., Zelinka, J., 2009. Integrovaný monitorovací systém jaskýň/Integrated monitoring system of caves. Aragonit 14, 109—112 (In Slovak).

Girault, F., Perrier, F., 2012. Estimating the importance of factors influencing the radon-222 flux from building walls. Sci. Total Environ. 433, 247—263.

Gómez, C., Hornero, R., 2010. Entropy and complexity analyses in Alzheimer's disease: an MEG study. Open. Biomed. Eng. J. 4, 223—235.

Grant, C.N., Lalor, G.C., Balczar, M., 2012. Radon monitoring in sites of economical importance in Jamaica. Appl. Radiat. Isot. 71, 96—101.

Gregorič, A., Zidanšek, A., Vaupotič, J., 2011. Dependence of radon levels in Postojna Cave on outside air temperature. Nat. Hazards Earth Sys. Sci. 11, 1523—1528.

Hafez, A.F., Hussein, A.S., 2001. Radon activity concentrations and effective doses in ancient Egyptian tombs of the Valley of the Kings. Appl. Radiat. Isot. 55, 355—362.

He, W., Feng, G., Wu, Q., He, T., Wand, S., Choue, J., 2012. A new method for abrupt dynamic change detection of correlated time series. Int. J. Climatol. 32, 1604—1614.

Hess, M., Koepke, P., Schult, I., 1998. Optical properties of aerosols and clouds: the software package OPAC. Bull. Am. Meteor. Soc. 79, 831—844.

Jelle, B.P., Noreng, K., Erichsen, T.H., Strand, T., 2010. Implementation of radon barriers, model development and calculation of radon concentration in indoor air. J. Build. Phys. 34, 195—222.

Jovancević, J., Krmar, M., 2011. Neutrons in the low-background Ge-detector vicinity estimated from different activation reactions. Appl. Radiat. Isot. 69, 629—635.

Jovanovič, P., 1996. Radon measurements in karst caves in Slovenia. Environ. Int. 22 (S1), 429—432.

Junkermann, W., 2005. The actinic UV-radiation budget during the ESCOMPTE campaign 2001: results of airborne measurements with the microlight research aircraft D-MIFU. Atmos. Res. 74, 461—475.

Justus, C.G., Paris, M.V., 1985. A model for solar spectral irradiance and radiance at the bottom and top of a cloudless atmosphere. J. Clim. Appl. Meteorol. 24, 193—205.

Kamra, L., Choubey, V.M., Kumar, N., Rawat, G., Khandelwal, D.D., 2013. Radon variability in borehole from multi-parametric geophysical observatory of NW Himalaya in relation to meteorological parameters. Appl. Radiat. Isot. 72, 137—144.

Kies, A., Massen, F., 1997. Radon generation and transport in rocks and soil. In: Massen, F. (Ed.), The Moestroff Cave — a Study on the Geology and Climate of Luxemburg's Largest Maze Cave. Centre de Recherche Public — Centre Universitaire, Luxemburg, pp. 159—183.

Kies, A., Massen, F., Feider, M., 1997. Measuring radon in underground locations. In: Virk, H.S. (Ed.), Rare Gas Geochemistry. Guru Nanak Dev University, Amritsar, pp. 1—8.

Koepke, P., Bais, A., Balis, D., Buchwitz, M., De Backer, H., de Cabo, X., Eckert, P., Eriksen, P., Gillotay, D., Heikkilä, A., Koskela, T., Lapeta, B., Litynska, Z., Lorente, J., Mayer, B., Renaud, A., Ruggaber, A., Schauberger, G., Seckmeyer, G., Seifert, P., Schmalwieser, A., Schwander, H., Vanicek, K., Weber, M., 1998. Comparison of models used for UV index calculations. Photochem. Photobiol. 67, 657—662.

Koltai, K., Ország, J., Tegzes, Z., Bárány-Kevei, I., 2010. Comprehensive radon concentration measurements in caves located in the area of Mecsek mountains. Acta Carsol. 39, 513—522.

Kowalczk, A.J., Froelich, P.N., 2010. Cave air ventilation and CO_2 outgassing by radon-222 modeling: how fast do caves breathe? Earth Planet. Sci. Lett. 289, 209—219.

Krmar, M., Radnović, D., Frontasyeva, M.V., 2012. Moss biomonitoring technique used to study the spatial and temporal atmospheric deposition of heavy metals and airborne radionuclides in Serbia. In: Mihailović, D.T. (Ed.), Essays on Fundamental and Applied Environmental Topics. Nova Science Publishers, New York, pp. 253—276.

Kylling, A., Dahlback, A., Mayer, B., 2000. Effect of clouds and surface albedo on UV irradiances at a high latitude site. Geophys. Res. Lett. 27, 1411—1414.

Lam, W.K., Tsin, T.W., Ng, T.P., 1988. Radon hazard from Caisson and tunnel construction in Hong Kong. Ann. Occup. Hyg. 32, 317—323.

Lario, J., Sánchez-Moralb, S., Cañaverasc, J.C., Cuezvab, S., Soler, V., 2005. Radon continuous monitoring in Altamira Cave (northern Spain) to assess user's annual effective dose. J. Environ. Radioact. 80, 161—174.

Leckner, B., 1978. The spectral distribution of solar radiation at the earth's surface-elements of a model. Sol. Energy 20, 143—150.

Lindsaya, R., de Meijera, R.J., Josepha, A.D., Motlhabanec, T.G.K., Newmand, R.T., Tselac, S.A., Speelmana, W.J., 2004. Measurement of radon exhalation from a gold-mine tailings dam by g-ray mapping. Radiat. Phys. Chem. 71, 797—798.

Liu, S., McKeen, S.A., Madronich, S., 1991. Effects of anthropogenic aerosols on biological ultraviolet radiation. Geophys. Res. Lett. 18, 2265—2268.

Lu, H., 2002. Study on radon pollution in tourist karst caves and protection from the radiation. Carsol. Sin. 21, 137—139 (In Chinese).

Lu, X., Li, L.Y., Zhang, X., 2009. An environmental risk assessment of radon in Lantian karst cave of Shaanxi, China. Water Air Soil Pollut. 198, 307–316.

Luo, K., Gao, Y., Zhou, C., Tang, S., Yang, F., Zhang, D., 1996. Radon and its decay products concentrations in karst cave of Hunan province and effective dose assessment. Chinese J. Radiol. Health 5, 36–37 (In Chinese).

Malinović, S., Mihailović, D.T., Kapor, D., Mijatović, Z., Arsenic, I., 2006. NEOPLANTA: a short description of the first Serbian UV index model. J. Appl. Meteorol. Climatol. 45, 1171–1177.

Malinović-Milićević, S., Mihailović, D.T., 2011. The use of NEOPLANTA model for evaluating the UV index in the Vojvodina region (Serbia). Atmos. Res. 101, 621–630.

Malinović-Milićević, S., Mihailović, D.T., Lalic, B., Dresković, N., 2013. Thermal environment and UV-B radiation indices in the Vojvodina region, Serbia. Climate Res. 57, 111–121.

McKinley, A.F., Diffey, B.L., 1987. A reference action spectrum for ultraviolet induced erythema in human skin. CIE J. 6, 17–22.

Mello, J., 2004. Geological Setting of the Domica Cave and Its Surrounding, vol. 9. Aragonit, Liptovský Mikuláš, pp. 3–8 (In Slovak).

Mihailović, D.T., Udovičić, V., Krmar, M., Arsenić, I., 2014. A complexity measure based method for studying the dependence of ^{222}Rn concentration time series on indoor air temperature and humidity. Appl. Radiat. Isot. 84, 27–32.

Mihailović, D.T., Krmar, M., Mimic, G., Nikolic-Djoric, E., Smetanova, I., Holy, K., Zelinka, J., Omelka, J., 2015. A complexity analysis of ^{222}Rn concentration variations: a case study for Domica cave, Slovakia for the period June 2010–June 2011. Radiat. Physics Chem. 106, 88–94.

Mihailović, D.T., Malinović-Milićević, S., Arsenić, I., Dresković, N., Bukosa, B., 2013. Kolmogorov complexity spectrum for use in analysis of UV-B radiation time series. Mod. Phys. Lett. B 27, 1350194.

Negarestani, A., Setayeshi, S., Ghannadi-Maragheh, M., Akashe, B., 2003. Estimation of the radon concentration in soil related to the environmental parameters by a modified Adaline neural network. Appl. Radiat. Isot. 58, 269–273.

Orr, H.G., Carling, P.A., 2006. Hydro-climatic and land use changes in the river Lune catchment, North West England, implications for catchment management. River. Res. Appl. 22, 239–255.

Papastefanou, C., Manolopoulou, M., Stoulos, S., Ioannidou, A., Gerasopoulos, E., 2003. Radon concentrations and absorbed dose measurements in a Pleistocenic cave. J. Radioanal. Nucl. Chem. 258, 205–208.

Paulescu, M., Stefu, N., Tulcan-Paulescu, E., Calinoiu, D., Neculae, A., Gravila, P., 2010. UV solar irradiance from broadband radiation and other meteorological data. Atmos. Res. 96, 141–148.

Perrier, F., Richon, P., Gautam, U., Tiwari, D.R., Shrestha, P., Sapkota, S.N., 2007. Seasonal variation of natural ventilation and radon-222 exhalation in a slightly rising dead-end tunnel. J. Environ. Radioact. 97, 220–235.

Pfeifer, M.T., Koepke, P., Reuder, J., 2006. Effects of altitude and aerosol on UV radiation. J. Geophys. Res. 111, 1–11.

Rajković, B., Veljović, K., Djurdjević, V., 2012. Dynamical downscaling: monthly, seasonal and climate case studies. In: Mihailoviić, D.T. (Ed.), Essays on Fundamental and Applied Environmental Topics. Nova Science Publishers, New York, pp. 135–158.

Rieder, H.E., Staehelin, J., Weihs, P., Vuilleumier, L., Maeder, J.A., Holawe, F., Blumthaler, M., Lindfors, A., Peter, T., Simic, S., Spichtinger, P., Wagner, J.E., Walker, D., Ribatet, M., 2010. Relationship between high daily erythemal UV doses, total ozone, surface albedo and cloudiness: an analysis of 30 years of data from Switzerland and Austria. Atmos. Res. 98, 9—20.

Ruggaber, A., Dlugi, R., Nakajima, T., 1994. Modelling radiation quantities and photolysis frequencies in the troposphere. J Atmos. Chem. 18, 171—210.

Sánchez, A.M., de la Torre Pérez, J., Ruano Sánchez, A.B., Naranjo Correa, F.L., 2013. Additional contamination when radon is in excess. Appl. Radiat. Isot. 81, 212—215.

Seftelis, G., Nicolaou, N.F., Tsagas, 2008. A mathematical description of the diurnal variation of radon progeny. Appl. Radiat. Isot. 66, 75—79.

Shuangcheng, L., Qiaofu, Z., Shaohong, W., Erfu, D., 2006. Measurement of climate complexity using sample entropy. Int. J. Climatol. 26, 2131—2139.

Smetanová, I., Holy, K., Jurčak, D., Omelka, J., Zelinka, J., 2011. Radon monitoring in Domica cave, Slovakia— A preliminary results. In: Bella, P., Gazik, P. (Eds.), ISCA 6th Congress Proceedings.

Udovičić, V., Aničin, I., Joković, D., Dragić, A., Banjanac, R., Grabež, B., Veselinović, N., 2011. Radon time-series analysis in the underground low-level laboratory in Belgrade, Serbia. Radiat. Protect. Dosimetry 145, 155—158.

Udovičić, V., Grabež, B., Dragić, A., Banjanac, R., Joković, D., Panić, B., Joksimović, D., Puzović, J., Aničin, I., 2009. Radon problem in an underground low-level laboratory. Radiat. Meas. 44, 1009—1012.

Veiga, L.H.S., Melo, V., Koifman, S., Amaral, E.C.S., 2004. High radon exposure in a Brazilian underground coal mine. J. Radiol. Protect. 24, 295—305.

Viñas, R., Eff-Darwich, A., Soler, V., Martín-Luis, M.C., Quesada, M.L., de la Nuez, J., 2007. Processing of radon time series in underground environments: implications for volcanic surveillance in the island of Tenerife, Canary Islands. Spain. Radiat. Meas. 42, 101—115.

Yentes, J.M., Hunt, N., Schmid, K.K., Kaipust, J.P., McGrath, D., Stergiou, M., 2012. The appropriate use of approximate entropy and sample entropy with short data sets. Ann. Biomed. Eng. 41, 349—365.

Complexity analysis of the environmental fluid flow time series

17

17.1 COMPLEXITY ANALYSIS OF THE MOUNTAIN RIVER FLOW TIME SERIES

Scientists in different fields (physicists, meteorologists, geologists, hydrologists, engineers etc., among others) study environmental fluid motion. Behavior of these fluids are significantly influenced by (1) human activities, (2) climatic change, and (2) increasing water pollution, changing mass and energy balance of the fluid. Understanding their complexity can help us to learn how to improve our systems by understanding how complexity underlies and affects the environments and the systems. Influenced by the aforementioned factors, the river flow in different geographic regions may range from being simple to complex, varying in both time and space. For turbulent environmental fluids, like mountain rivers, the speed of the water flow can vary within a system and is subject to chaotic turbulence. This turbulence results in divergences of flow from the mean downslope flow vector as typified by eddy currents. The mean flow rate vector is based on variability of friction with the bottom or lateral sides of the channel, sinuosity, obstructions, and the incline gradient (Allan, 1995). Over the last decade, controversial results have been obtained about the hypothetical chaotic nature of river flow dynamics (Schertzer et al., 2002; Salas et al., 2005 Zunino et al., 2012; Sivakumar and Singh, 2012; Mihailović et al., 2014). For example, Zunino et al. (2012) analyzed the streamflow data corresponding to the Grand River at Lansing (Michigan) trying to provide new insights regarding this issue, while Hajian and Sadegh Movahed (2010) have used the detrended cross-correlation analysis to investigate the influence of sun activity represented by sunspot numbers on river flow fluctuation as one of the climate indicators. The river flow fluctuations have been also analyzed using the formalism of the fractal analysis by Sadegh Movahed and Hermanis (2008). Therefore, it is of interest to determine the nature of complexity in mountain river flow processes, which requires the use of different measures of complexity, which cannot be done by commonly used mathematical statistics. These measures help us to get an insight into the complexity of the environmental fluid flow; i.e., the mountain river flow in this section. Using them, we can more comprehensively investigate possible changes in (1) river flow due to human activities, (2) response to climate changes, and (3) nonlinear dynamic concepts for a catchments classification framework. Also, we shall be able to improve application of the stochastic process concept in hydrology for its modelling, forecasting, and other ancillary purposes (Porporato and Ridolfi, 2001; Schertzer et al., 2002; Stoop et al., 2004; Otache et al., 2011).

In this section we consider the complexity of the river flow dynamics of two mountain rivers in Bosnia and Herzegovina for the period 1926—90, using the lower Kolmogorov complexity (KLL), upper Kolmogorov complexity (KLU), sample entropy (SampEn), and permutation entropy (PermEn) measures, which are described in Section 15.2 (KLL and KLU) and Section 9.1 (SampEn and PermEn). We will do that through (1) sensitivity tests for all considered measures which is dependent on data length and (2) their application on two mountain river flow time series.

Description of river locations and time series. The River Bosnia and the River Miljacka flow through the Sarajevo Valley, which is located between mountain depressions and between the massive Bjelasnica and Igman mountains on the southwest as well as the low mountains and middle mountains on the northeast. The valley generally stretches in the NW—SE direction and there are low mountains and middle mountain areas on the southeastern slopes of the Trebevic Mountain and on the northwestern slopes between valley peaks (Fig. 17.1).

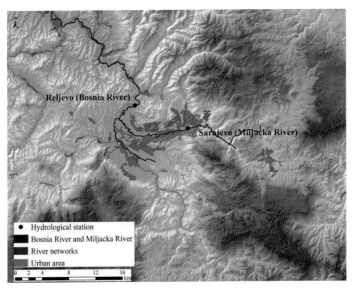

FIGURE 17.1

Topological location of the Sarajevo Valley with hydrological stations Reljevo (the Bosnia River) and Sarajevo (the Miljacka River) used in this study (designed by N. Drešković).

Reprinted with permission from Mihailović, D.T., Nikolić-Đorić, E., Drešković, N., Mimić, G., 2014. Complexity analysis of the turbulent environmental fluid flow time series. Physica A 395, 96—104.

The mean altitude of the bottom of the valley is approximately 515 m. The valley is a hydrological input for the source area of the Bosnia River with seven tributaries including the Miljacka River. In this part of their flow both of them fully represent mountain rivers. In this study for time series we used the monthly mean values (Fig. 17.2) from hydrological stations Reljevo (the Bosnia River) and Sarajevo (the Miljacka River) since they have representative and reliable instruments for hydrological monitoring since 1926 (Hadžić and Drešković, 2012; Mihailović et al., 2014).

The Bosnia River has the mean annual river flow about 8.0 m^3 s^{-1}, except during the precipitation season when it takes a value of 24.0 m^3 s^{-1}. The hydrological station Reljevo is located 11.6 km away from its source. Usually the mean annual river flow of this river is 28.7 m^3 s^{-1}, with a maximum of 44.9 m^3 s^{-1} (in 1937) and a minimum value of 17.9 m^3 s^{-1} (in 1990) during the period 1926−90. The entire Miljacka River system upstream has a very steep and wavy longitudinal profile. Downstream from this site, it flows through the alluvial plateau with a very small drop (3−5%) passing the highly urbanized Sarajevo Valley with over 400,000 inhabitants. The hydrological station Sarajevo is located on the bridge in the central part of Sarajevo. Usually the mean annual river flow of the Miljacka River is 5.5 m^3 s^{-1}, with a maximum of 9.1 m^3 s^{-1} (in 1937) and a minimum value of 3.0 m^3 s^{-1} (in 1990) during the period indicated. The river flow time series for the Miljacka River and the Bosnia River for the period 1926−90 are depicted in Fig. 17.2 (Mihailović et al., 2014).

An example of changes in complexity of the mountain river flow fluid time series. The mountain river is a typical example of the turbulent environmental fluid for which the changes in complexity of its flow rate primarily depends on human activities and climate change. These process and phenomena can contribute to the loss of the complexity, which leads to reducing the stochastic component in the river flow. However, the nature of its complexity may be investigated by the complexity measures that give more insights into the complexity of its flow rate. In an example that follows, we shall illustrate the impact of the mentioned factors on mountain river flow complexity. In these experiments we use the time series for the Bosnia

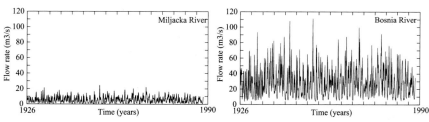

FIGURE 17.2

River flow time series for the Miljacka River and the Bosnia River for the period 1926−90.

Reprinted with permission from Mihailović, D.T., Nikolić-Đorić, E., Drešković, N., Mimić, G., 2014. Complexity analysis of the turbulent environmental fluid flow time series. Physica A 395, 96−104.

River (the right panel in Fig. 17.2) to simulate loss of flow complexity of this river as a result of the anticipated (1) human activities and (2) projected climate changes in the region from Fig. 17.1.

We simulate the influence of the human activity (for example, urbanization and building capacities for the water consumption, etc.) on the mountain river flow complexity. Namely, when a value of the KLL is close to zero then it is associated with a simple deterministic process like a periodic motion, whereas a value close to one is associated with a stochastic process (Mihailović et al., 2014). Thus, by human activities, many stochastic components can disappear from the flow of the mountain river depending on the level and intensity of those activities (Mihailović et al., 2014; Gordon et al., 2004; Orr and Carling, 2006). We illustrate the influence of the human activity on the mountain river flow complexity in the following way. First, depending on the intensity of activity (symbolically depicted in percentage on x axis in Fig. 17.3): (1) we have removed amplitudes in the time series setting them to be zero and (2) we have kept those samples in the time series always having the size N ($N = 780$ in this experiment). Then, using the procedure described in Section 15.2 we have calculated the KLL for each created time series. Changes in the KLL complexity of a mountain river flow rate which is dependent on simulated human activity are depicted in Fig. 17.3. From this figure is seen a descending trend of this curve, which is finished by a straight line on the lowest level of complexity depicting the absence of turbulent eddies as a result of regularization of the river flow. The descending curve is rather wavy than linear because of the nonlinearity of the river flow.

In the context of climate change, significant perturbations can be expected in natural systems in different regions including mountain ones. Because mountains are the source region for over 50% global rivers, the impact of climate change on hydrology is likely to have significant repercussions. Expected changes in mountain river energy exchange processes under a changing climate can be listed in the following declining order: short-wave radiation, long-wave radiation, latent heat flux, and sensible heat flux. Therefore, some expectations about flow under a

FIGURE 17.3

Changes in the KLL complexity of a mountain river flow which is dependent on the level of human activity.

Reprinted with permission from Mihailović, D.T., Nikolić-Đorić, E., Drešković, N., Mimić, G., 2014. Complexity analysis of the turbulent environmental fluid flow time series. Physica A 395, 96–104.

changing climate can be summarized as follows: more variable and severe precipitations, higher evapotranspiration, difficulties in forecast of how annual and seasonal balance between precipitation and evapotranspiration will change, more frequent floods, more frequent droughts, changing flow regimes from snowmelt to winter rainy, etc (Caissie, 2006). We simulate the climate change impact on the mountain river flow complexity in the following way. The time series of the flow rate was divided into three subintervals: (1) 1−280th, (2) 280−520th, and (3) 580−780th month. The impact of climate change on the river flow complexity was introduced during the period (2) of simulation following the results of regional climate simulations by Djurdjevic and Rajkovic (2008) that include the area depicted in Fig. 17.1. According to them, projections for the year 2030 indicate an evident increase of air temperature and evaporation (about 20%) as well as the decrease of precipitation. For the periods (1) and (3) we have calculated the KLL. For the period (2), first we have recalculated the monthly river flow rates by changing their values, according to values of evaporation and precipitation obtained by the regional climate model (Djurdjevic and Rajkovic, 2008), and then we have applied the same procedure for the KLL calculations as in the previous experiment. As a consequence, an evident decrease of the complexity of the river flow time series is seen from Fig. 17.4, which is visualized through the fitting curve.

Complexity analysis of two mountain river flow time series. Using the calculation procedure outlined in Sections 15.2 and 9.1, we have computed the KLL, KLU, SampEn, and PermEn values for the two mountain river flow time series. The calculations are carried out for the entire time interval 1926−90 and for three subintervals covering this period: (a) 1926−45, (b) 1946−65, and (c) 1966−90 obtained by sensitivity tests which is dependent on length of time series. Since all measures are sensitive to the length of time series, N we perform some sensitivity tests. For PermEn, the length of the time series must be larger than the factorial of the embedding dimension (Frank et al., 2006).

To explore the sensitivity of these measures which is dependent on the length of time series we calculated the KLL, KLU, SampEn, and PermEn values for $N = 200$

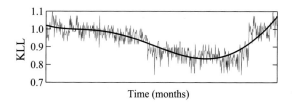

FIGURE 17.4

Changes in the KLL complexity of a mountain river flow rate which is dependent on simulated climate changes. Heavy *solid line* is a fitting curve, which depicts the trend of the complexity change.

Reprinted with permission from Mihailović, D.T., Nikolić-Đorić, E., Drešković, N., Mimić, G., 2014. Complexity analysis of the turbulent environmental fluid flow time series. Physica A 395, 96–104.

up to $N = 780$ (Fig. 17.5). In these experiments we have had in mind the following facts. In this section the SampEn was calculated for the mountain river flow time series with the following values of parameters: $m = 2$, $r = 0.2$, and $\tau = 1$ (see Section 16.3). Besides N, the embedding dimension (m), also called the permutation order, is an input parameter for PermEn. Therefore we have considered its sensitivity on the PermEn outputs. Due to the length of time series ($N = 780$) we chose the embedding dimension to be less than 6 (Fig. 17.6). Our results indicate that the KLL and SampEn decrease and the KLU and PermEn slightly increase when the number of observations increases. All considered measures are sensitive to random component and may be considered as indicators of randomness, but they do not give information about amplitude variations. In particular, we have calculated

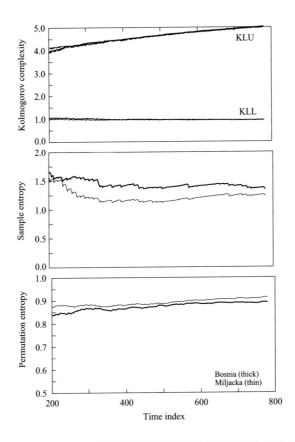

FIGURE 17.5

Sensitivity of the KLL (lower), KLU (upper), SampEn (middle), and PermEn (lower) panel which is dependent on the length of the mountain river flow time series for the Miljacka River and the Bosnia River.

Reprinted with permission from Mihailović, D.T., Nikolić-Đorić, E., Drešković, N., Mimić, G., 2014. Complexity analysis of the turbulent environmental fluid flow time series. Physica A 395, 96–104.

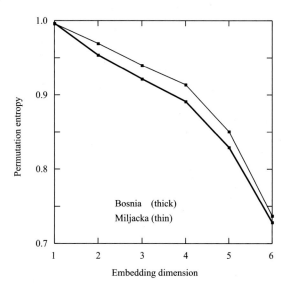

FIGURE 17.6

Permutation entropy as a function of embedding dimension for river flow time series for the Miljacka River and the Bosnia River for the period 1926–90.

Reprinted with permission from Mihailović, D.T., Nikolić-Đorić, E., Drešković, N., Mimić, G., 2014. Complexity analysis of the turbulent environmental fluid flow time series. Physica A 395, 96–104.

the frequencies of the mountain river flow time series. They have the same dominant frequencies (1/12 and 1/6 for the Miljacka River and the Bosnia River, respectively) as well as the similar distribution of the random component. Thus the values of complexities, calculated for the whole time series and subintervals for both rivers, are close to each other.

The results of computations are given in the corresponding rows of Table 17.1. It is seen from this table that the KLL values in both rivers are close while the KLU ones are practically the same. Note that a process that is the least complex has a KLL value near to zero, whereas a process with the highest complexity will have KLL close to one. As we said the KLL measure can be also considered as a measure of randomness. Thus, a value of the KLL near zero is associated with a simple deterministic process like a periodic motion, whereas a value close to one is associated with a stochastic process (Ferreira et al., 2003; Mihailović et al., 2014). Accordingly, the KLL values, which are large for both rivers (0.936), point out the presence of stochastic influence in these typically mountain rivers. The other two calculated measures indicate on a similar behavior of time series for both rivers, i.e., their increased irregularity. The SampEn values are slightly different (1.240 for Mil and 1.357 for Bos) while the PermEn values are very close to each other (0.914 for Mil and 0.891 for Bos).

Table 17.1 Kolmogorov Complexities (Lower—KLL and Upper—KLU), Sample Entropy (SampEn), and Permutation Entropy (PermEn) Values for the River Flow Time Series of Two Mountain Rivers for the Period 1926–90, and the Subintervals: (a) 1926–45, (b) 1946–65, (c) 1966–90. In Computing the Entropies We Have Used the Following Sets of Parameters ($m = 2$, $r = 0.2$, and $\tau = 1$) and ($m = 5$) for the SampEn and PermEn, Respectively.

River	Measure	1926–90	1926–45	1946–65	1966–90
Miljacka (Mil)	KLL	0.936	0.988	0.955	0.988
	KLU	5.002	4.210	3.944	4.557
	SampEn	1.240	1.438	0.903	1.478
	PermEn	0.914	0.879	0.832	0.903
Bosnia (Bos)	KLL	0.936	1.054	0.977	0.988
	KLU	5.024	4.103	4.031	4.471
	SampEn	1.357	1.526	1.214	1.367
	PermEn	0.891	0.843	0.847	0.869

Reprinted with permission from Mihailović, D.T., Nikolić-Đorić, E., Drešković, N., Mimić, G., 2014. Complexity analysis of the turbulent environmental fluid flow time series. Physica A 395, 96–104.

We have chosen the above time intervals for two reasons. Firstly, a change in the complexity of both rivers in the period 1945 (end of the Second World War)–65 (end of the most intensive human intervention, in particular, urbanization and building capacities for the water consumption) was expected. Secondly, we have performed the sensitivity tests to check the reliability of the chosen time series of subintervals. On the basis of those tests, in the computing procedure we have used the following parameters: (1) embedding dimension ($m = 2$), tolerance ($r = 0.2$) and time delay ($\tau = 1$) for the SampEn and (2) embedding dimension ($m = 5$) for the PermEn. As a result the time series for periods (a), (b), and (c) were 240, 240, and 300, respectively. It is found that during 1946–65, there is a decrease in complexity in Mil and Bos rivers (0.955 and 0.977, respectively) in comparison to the other subintervals. This complexity loss may be interpreted as results of intensive different human activities on those rivers after the Second World War. The same result is found for the KLU complexity, i.e., 3.944 for Mil and 4.031 for Bos, which are the lowest of their values in comparison to the other subintervals. Lower values of both entropies for both rivers: (1) the SampEn (Mil—0.903; Bos—1.214) and (2) the PermEn (Mil—0.832), support the conclusion about more regular river flow time series in this period. Only in the case of PermEn, there is minor decrement of the regularity for the period 1946–65. In the case of PermEn, the same conclusion holds for other considered values of embedding dimension.

17.2 RANDOMNESS REPRESENTATION IN TURBULENT FLOWS WITH BED ROUGHNESS ELEMENTS USING THE KOLMOGOROV COMPLEXITY SPECTRUM

The influence of bed roughness elements on turbulent flow is a crucial topic in many different fields, for instance, modelling the transport of pollutants carried by a turbulent flow in the atmosphere, the study of the hydraulic effects of fully/partially submerged vegetation in experimental flumes or in natural rivers, or the parameterizing of the deposition velocity over vegetative or urban area in chemical transport models (Raupach et a., 1996; Jimenez, 2004; Poggi et al., 2004; Flack et al., 2005; Gioia and Chakraborty, 2006; Allen et al., 2007; Nezu and Sanjou, 2008; Mihailović et al., 2009; Gualtieri, 2010; Cushman-Roisin et al., 2012). It is recognized that aquatic plants in rivers have considerable effects on the velocity distributions, turbulence structure, and also on the mass and momentum exchanges between the zones with and without vegetation (Raupach et a., 1996; Nezu and Sanjou, 2008). In the presence of submerged bed roughness elements of height k (e.g., canopy, submerged aquatic vegetation, cylindrical obstacles in laboratory channels) a roughness sublayer (RSL) is formed (Raupach et a., 1996; Poggi et al., 2004; Hussain, 1983; Nepf, 2012; Thoraval et al., 2012). Within the RSL there exist three distinct zones (Fig. 17.7a): (1) the deep zone (RS1 when $z < k$) which is mainly driven by von Kármán vortex streets (Thoraval et al., 2012), but from time to time disrupted by strong sweep episodes with features which are influenced by bed roughness elements density (Poggi et al., 2004); the second zone (RS2), extending upward from $z = k$ to (typically) about $2k$, is a superposition of attached eddies and Kelvin−Helmholtz waves produced around the inflection point on the mean velocity profile (Fig. 17.7a), which develops between two coflowing streams having different velocities (Raupach et a., 1996; Poggi et al., 2004; Attili and Bisetti, 2013; Goncharov and Pavlov, 2015). In this turbulent mixing region are formed coherent turbulent structures caused by Kelvin−Helmholtz instability that travel downstream in the environmental fluids, which are often used to identify the extent of the mixing layer thickness (Raupach et a., 1996; Brown and Roshko, 1974; Rogers and Moser, 1994). The RS2 zone is a superposition of all three constituents, while the RS3 zone is shifted rough wall boundary layer.

For a long time, there was an interest in the environmental fluid science for a better understanding of the physical processes involved in flow-roughness element space interaction, which includes interaction between zones described above, e.g., the turbulent boundary layer and outer laminar region, wall- and free-shear turbulent flows exhibiting coherent structure (Ichimiya and Nakamura, 2013; Tsuji and Nakamura, 1974). However, to our opinion regarding the turbulent flow, some problems remain unanswered because a more complete definition of turbulence is not yet proposed as it is emphasized by Ichimiya and Nakamura (2013).

Namely, one of the fundamental properties in the definition of the turbulence is not clearly included and it is usually expressed verbally by sentences written in

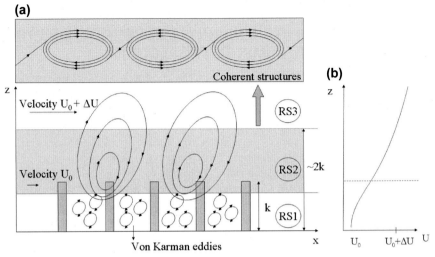

FIGURE 17.7

(a) A schematic diagram of eddies structure over and within the rough elements: (1) RS1 zone ($z/k < 1$) where the flow field is primarily dominated by small eddies associated with the von Kármán streets; (2) RS2 zone straddles the top portion of the bed roughness element space, and is dominated by a mixing layer; and (3) RS3 zone ($z/k > 2$) is the classical boundary-layer region dominated by eddies with length scales proportional to ($z - d$), where d is the displacement height. (b) The mean velocity profile within the bed roughness space $u(z) = u(k)\sqrt{-\beta_1 e^{-\beta_3 z} + \beta_2 e^{\beta_3 z}}$ is obtained from the solution of the partial differential equation $\partial u^2/\partial z = [(2C_d\lambda_d k^2)/(\sigma_s P_s)]u^2$ where $u(k)$ is the velocity at the height k; β_1, β_2 are parameters depending on the morphological and aerodynamic characteristics of the bed roughness element space and $\beta_3 = [(2C_d\lambda_d k^2)/(\sigma_s P_s)]$; C_d, the drag coefficient; σ_s, the parameter of proportionality between the turbulent transport coefficient and velocity within the bed roughness element space; λ_d, the roughness density; and P_s, shelter factor (Rogers and Moser, 1994; Sellers et al., 1986; Mihailović et al., 2016).

terms of irregular or random fluid flows but without its quantification (Reynolds, 1883; Prandtl, 1965; Batchelor, 1956; Hinze, 1975; Frish, 1995). An overview of such definitions is offered by Ichimiya and Nakamura (2013). Only exception is Pope's definition of randomness related to turbulence (Pope, 2000). On the other hand, this definition cannot be used as a measure of the randomness as it is in measure developed by Kolmogorov, on the basis of which is developed the LZA algorithm for calculating the measure of the randomness (see Section 15.2). This algorithm we use for evaluation of the randomness present in time series.

The goal of this section is to quantify randomness of turbulent flows that develop from passing over bed roughness elements, using the Kolmogorov complexity spectrum. For that purpose, we use the results from an experimental study carried out in a laboratory channel with variable bed slope at the University of Naples Federico II (Naples, Italy). First, we describe the experimental setup, channel, and instrumentation. Second, we use the complexity measures based on the Kolmogorov complexity (the Kolmogorov complexity spectrum and the Kolmogorov complexity spectrum highest value), KLM in the analysis of the turbulence randomness. Then, we perform analysis based on a classical turbulence statistics including a simple empirical model for the estimation of the relative sizes of mixing lengths representing the typical scale of an eddy in the corresponding part of the surface layer, to nearly quantify the randomness of the turbulence, for velocity profiles and for turbulence intensity. Finally, we discuss the suggested measures of the randomness of turbulent flow in the surface layer.

Experimental setup. Chanel and instrumentation. The experiments were performed in a laboratory channel with variable bed slope at the University of Naples Federico II (Naples, Italy). The channel was 8 m long and 0.4 m wide with a variable bed slope (Fig. 17.8a). Vegetation covered the bed of the channel and consisted of rigid cylinder rods of the same height and diameter ($k = 0.015$ m, $d_c = 0.004$ m—Fig. 17.8b), set in different aligned arrangements (rectangles or squares), with three different densities, $\lambda_{d,1} = 0.024$ m^2 m^{-3}, $\lambda_{d,2} = 0.048$ m^2 m^{-3}, and $\lambda_{d,3} = 0.096$ m^2 m^{-3}, corresponding, respectively, to 400, 800, and 1600 cylinders per unit area. Vegetation density was evaluated as the total roughness frontal area per unit area. The vegetation was always fully submerged with submergence h_u/k, of about 4, where h_u is the uniform flow depth.

The experimental conditions are listed in Table 17.2, where u_b is the bulk velocity or depth-averaged velocity, u_* is the shear velocity, and u_k is the velocity at the top of the bed roughness element. Instantaneous values of streamwise velocities were recorded in uniform flow in a vertical cross-section located at the mid-length of a square or rectangular array. The velocity measurements were carried out in about 25 vertical locations using a laser Doppler velocimeter (LDV) equipped with a frequency shifter and a frequency tracker. The sampling rate was 2000 Hz for 135 s. The turbulent data were postprocessed using the LabView software to derive the distribution of time-averaged velocity and standard deviation related to turbulence intensity.

(a)　　　　　　　　　　　　　　　　　　**(b)**

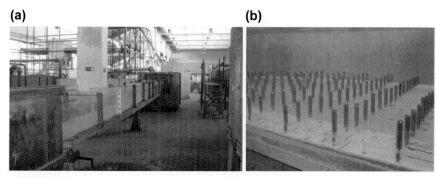

FIGURE 17.8

(a) Channel and (b) vegetated bed (Mihailović et al., 2016).

Table 17.2 Experimental Conditions (described in Mihailović et al., 2016)

Test	λ_d (m² m⁻³)	Slope (°)	Q (l s⁻¹)	h_u (cm)	Cylinders per Unit Area	u_b (m s⁻¹)	u_* (m s⁻¹)	u_k (m s⁻¹)
D1	0.024	0.03	33	6.35	400	1.128	0.119	1.192
D2	0.048	0.02	22	6.44	800	0.735	0.098	0.763
D3	0.096	0.03	22	6.29	1600	0.715	0.119	0.726

Classical turbulence statistics. Fig. 17.9a–c shows the lateral distributions of the streamwise component of time-averaged mean velocity U ($u = U + u'$, where u and u' are instantaneous streamwise measured velocity and velocity fluctuation, respectively) at several relative heights z/k within and above the bed with bed roughness elements space for all the density cases (D1, D2, and D3).

From these figures, it is seen that vertical velocity profiles over a bed with bed roughness elements of different densities do not follow standard logarithmic profile. For $z/k = 1$ and densities D1 and D3, it is seen that the inflection points and shape of velocity profiles follow the theoretical curve in Fig. 17.9b. According to the vertical velocity distribution, the flow could be separated into two zones: (1) a lower zone within the bed roughness elements space ($z/k < 1$) and (2) upper zone ($z/k > 1$). However, these comments can be just partly addressed to the velocity profile for the density D2 (Fig. 17.9b). Let us point out that the most essential difference between surfaces with bed roughness elements of different densities is the magnitude of the inflection in the mean velocity profile, which is a necessary condition for the occurrence of Kelvin–Helmholtz instabilities. However, according to Poggi et al. (2004), its magnitude assigns a framework of the relative importance of that mechanism on the whole turbulence structure.

The turbulence statistics applied on measured values of velocity quantify how the flow within and just above the bed roughness elements space behaves as a perturbed mixing layer. According to Ruelle (1978), the average properties of a turbulent flow can be described by a measure, which is invariant under time evolution. Here, we quantify the turbulence intensity as $\bar{u} = \sqrt{\sum_{i=1}^{N} u_i^2 / N}$ normalized by the friction velocity u_*, i.e., $\sigma_u = \bar{u}/u_*$. Fig. 17.10a depicts the measured profiles of σ_u for all three densities D1, D2, and D3. From this figure it is seen that with increasing λ_d, turbulence intensity σ_u is remarkably damped for $z/k < 1$. It is seen that σ_u changes from 1.69 (D1), for sparse density, which is typical for rough wall layers, to about

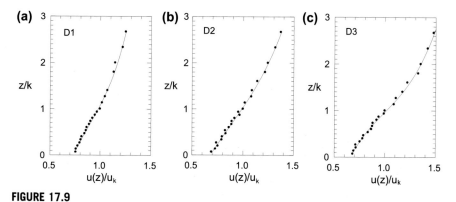

FIGURE 17.9

Time-averaged mean vertical velocity profiles for all the density cases normalized by the velocity at the canopy top (u_k) (Mihailović et al., 2016).

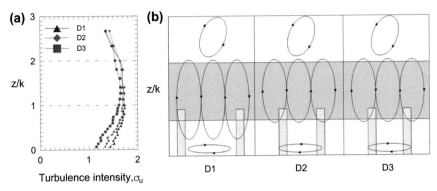

FIGURE 17.10

Turbulence intensity $\bar{u}$ normalized by u_* (a) and mixing lengths l_v, l_{ml}, and l_{bl} normalized by k (b) against relative depth ratio of vertical distance from bed z to bed roughness elements k for all the density cases. The relative sizes of mixing lengths l_v/k, l_{ml}/k, and l_{bl}/k which are calculated represent the typical scale of an eddy in the corresponding layer (Mihailović et al., 2016).

1.49 (D3) for dense canopies, which is typical for mixing layers. Those values are close to the ones reported by Raupach et al. (1996). Further inspection of this figure shows that for all densities, σ_u increases from the bed to the level near the top of the canopy ($z/k < 1$). Above this, σ_u weakly decreases in the RS2 zone and toward the free surface (i.e., in the RS3 zone). These measures of the traditional turbulence statistics provide us an insight in the structure of turbulence within the bed roughness elements space and coherent motions near the top of that space. To explain the behavior of σ_u, we calculate empirically the three basic length scales l_v, l_{ml}, and l_{bl} for zones RS1, RS2, and RS3, respectively, partly following parameterization by Poggi et al. (2004). In RS1, where the flow field is primarily dominated by small vortices associated with the von Kármán streets, the mixing length is evaluated as $l_v = d_r/0.21$. The mixing length l_{ml} in the RS2, i.e., the mixing layer, is parameterized as $l_{ml} = \bar{u}_k/(d\bar{u}/dz)_{z=k}$. Finally, the region RS3 is a classical boundary-layer region dominated by eddies with length scales $l_{bl} = \kappa(z - d)$, where κ is the von Kármán constant. The relative mixing lengths l_v, l_{ml}, and l_{bl} normalized by k, calculated for all the density cases, with zero plane displacement d taken to be $2/3k$, are depicted in Fig. 17.10b. The calculated values of the relative sizes of mixing lengths have the following values: (1) $l_v/k = 0.85$ and $l_{bl}/k = 0.82$ in the zones RS1 and RS3, respectively, for all the density cases and (2) $l_{ml}/k = 2.20$ (D2), 1.47 (D1, D3) in the zone RS2, respectively. Here, the relative sizes of mixing lengths l_v/k, l_{ml}/k, and l_{bl}/k represent the typical scale of an eddy in the corresponding layer. The eddies in all the three zones are seen in Fig. 17.10b. In the RS1 zone, von Kármán eddies are smaller with sizes, which are proportional to d_r and independent of λ_d and local velocity Poggi et al. (2004). In the RS2 zone the eddies are larger, organized in a coherent structure, carrying the highest amount of energy, while in the RS3 zone they are again of smaller size. This simple consideration empirically explains the behavior of σ_u in Fig. 17.10a. However, we still have no available measure of randomness, which is a crucial property of the turbulence as we already mentioned.

When stable laminar flows evolve toward the turbulence, they become high order and complex, exhibiting irregular-like motions with organized dissipative arrangements. To precisely specify their fields (velocity and displacement), more parameters are required than for description of laminar flows, i.e., topological measures that quantify the order or disorder of the flow. One such measure is the Shannon entropy, which has been already used in the analyses of geophysical fluids (Wijesekera and Dillion, 1997; Wesson et al., 2003). The Shannon entropy (*SH*) is defined as $SH = -\sum p_i \ln p_i$ (Shannon, 1948), where p_i is a discrete probability distribution satisfying the following conditions: $p_i \geq 0$; $\sum p_i = 1$ and $p_{i \cup j \cup ...} = p_i + p_j + ...$ In our calculations, p_i is defined as a probability that velocity amplitude occurs within the interval $u_i + du$, where du is obtained when the entire interval of velocity amplitudes is divided into N intervals. From Fig. 17.11a, it is seen that the *SH* is the highest in the mixing layer ($1 < z/k < 2$) where the turbulence intensity σ_u is the highest. However, it decreases going to top of the water jet. This behavior of the *SH* coincides with the conclusion in Wijesekera and Dillion (1997). A decrease of the *SH* going to the rough wall can be addressed to the occurrence of smaller eddies carrying smaller amount of the energy.

The randomness of turbulent flow in the surface layer. To avoid confusion in the following discussion we will make some comments. Namely, the term complexity in physical systems has the connotation of an explicit measure of the probability of the state of the system. It is a mathematical measure which should not be equalized

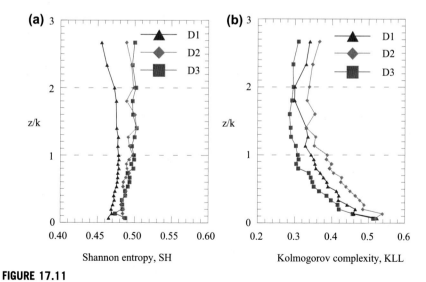

FIGURE 17.11

Shannon entropy *SH* (a) and Kolmogorov complexity KLL (b) against relative depth ratio of vertical distance from bed *z* to bed roughness elements *k* for all the density cases.

with entropy in statistical mechanics (Mihailović et al., 2015). Thus, by the Shannon entropy, we can describe dissimilarities between amplitudes in a time series, while the Kolmogorov complexity refers to the apparent sequence disorder of some amplitudes in a time series. This complexity is of interest in this section, which we intuitively understand as a measure that is ranged between uniformity and total randomness (Grassberger, 1989; Mihailović et al., 2015) as we noted in Section 15.2. Comparing Fig. 17.11a and b, they seem to have overall symmetrical trends. In Fig. 17.11a, the randomness weakly increases in the RS1 zone, it has a constant value in the RS2 zone, and then decreases in the RS3 zone. This trend is clearer for sparse bed roughness elements (D1) but, anyway, the density of the bed roughness elements seems to affect randomness, in fact lower randomness corresponds to sparse density (D1). In Fig. 17.11b it is seen that the value of the KL complexity decreases with height from the rough wall to the mixing layer ($1 < z/k < 2$). This can be explained by the fact that in the RS1 zone, flow is dominated by smaller eddies (see Fig. 17.10b) contributing to the higher randomness which becomes lower in the mixing layer having a constant value in this zone. This is because the eddies in this zone are larger and coherently organized, without the possibility of introducing more randomness in the flow. Above the mixing layer, the KL slightly increases since the eddies become smaller providing conditions for higher complexity in flow.

Using the LZA algorithm we have calculated the Kolmogorov spectra for all the density cases and selected relative depth ratio z/k, which are depicted in Fig. 17.12. From these figures, it is seen that the KLM values (when the randomness is the highest) are very close to the KLL at the corresponding relative heights (Fig. 17.11b). Moreover, as the density increases, Kolmogorov spectra in the mixing layer, where complexity is constant, tend to be very close. However, the Kolmogorov complexity spectrum provides us additional information. Namely, the area below this spectrum (overall complexity in Mihailović et al. (2015), which is not considered in this section) gives integral information about the complexity for the whole spectrum of complexities, i.e., it comprises both the (1) dissimilarities between amplitudes (SH) and (2) disorder of some amplitudes (KLL). Thus, the Kolmogorov-based complexities measures allow the quantification of the degree of turbulence in relation to the randomness rather than it being expressed verbally in terms of irregular or random.

17.3 APPLICATION OF THE COMPLEXITY MEASURES BASED ON THE KOLMOGOROV COMPLEXITY ON THE ANALYSIS OF DIFFERENT RIVER FLOW REGIMES

Influenced by the factors mentioned in Section 17.1, the river flow may range from being simple to complex, fluctuating in both time and space. In the last decade, many authors have devoted their attention to the chaotic nature of river flow dynamics by analyzing their daily, monthly and annual time series (Lange et al., 2013; Mihailović et al., 2014; Serinaldi et al., 2014). Here, we quantify the randomness of the river flow dynamics of seven rivers, in Bosnia and Herzegovina for the period

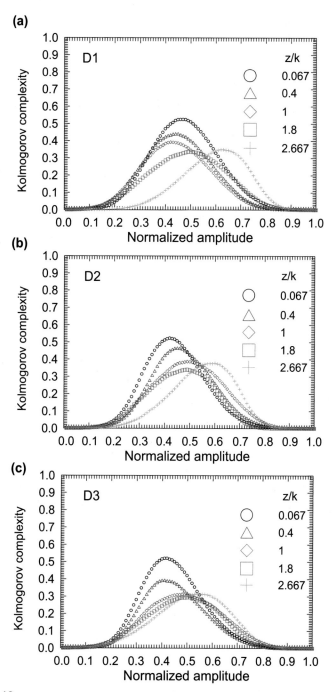

FIGURE 17.12

Kolmogorov spectra for all the density cases and selected relative depth ratio of vertical distance from bed z to bed roughness elements k.

1965—86, using the KLL, KLU, KLM, and KLO complexity measures and the Kolmogorov complexity spectrum.

Datasets and computations. River flow records in the Bosnia and Herzegovina are, in general, of relatively short duration. Except for several rivers (Mihailović et al., 2014) many measurements began in the late 1950s. For this study we have selected seven rivers placed in the territory of Bosnia and Herzegovina, which is located in the western Balkans, surrounded by Croatia to the north and southwest, Serbia to the east, and Montenegro to the southeast. It lies between latitudes 42° and 46°N, and longitudes 15° and 20°E. The country is mostly mountainous, encompassing the central Dinaric Alps. The northeastern parts reach into the Pannonian basin, while in the south it borders the Adriatic Sea. Dinaric Alps generally run in the east-west direction and get higher toward the south. The highest point of the country is peak Maglić at 2386 m, at the Montenegrin border, while the major mountains include Kozara, Grmeč, Vlašić, Čvrsnica, Prenj, Romanija, Jahorina, Bjelašnica, and Treskavica (Fig. 17.13).

Since we want to quantify the randomness degree in the river flow time series of rivers in Bosnia and Herzegovina, in different parts of their courses, we have made a selection of hydrological stations on the basis of classification of typology for mountains and other relief classes according to Meybeck et al. (2001). This classification can be summarized in the following way: (1) lowlands (0—200 m mean altitude—in further text L type), (2) platforms and hills (200—500 m—H type), and (3) mountains with mean elevations between 500 and 6000 m (M type). Thus, we have analyzed: the lower and upper course (the rivers Neretva, Drina and Bosnia), the upper course (the rivers Una and Miljacka), and the lower course (the river Vrbas and Ukrina). These catchments are listed in Table 17.3. They are spread across the country and are representative for catchments in these regions. Datasets of monthly river flow rates for the period 1965—86 were taken from the Annual Report of the Hydrometeorological Institute of Bosnia and Herzegovina, consisting of 252 data points in each time series.

Computation of complexity measures for seven river flow time series and analysis. Using the calculation procedure outlined in Section 15.3, we have computed the KLL, KLU, the Kolmogorov complexity spectrum, and the KLO values for the 10 river flow time series of seven rivers (Fig. 17.14) in Bosnia and Herzegovina. The calculations are carried out for the entire time interval 1965—86.

Results for Kolmogorov complexities (lower—KLL, upper—KLU, Kolmogorov complexity spectrum highest value—KLM) and overall Kolmogorov complexity measure (KLO) are given in the corresponding rows of Table 17.4. From this table it is seen that the KLL values for seven rivers can be classified into two intervals, i.e., (0.948, 1.076) and (0.791, 0.918), which corresponds to the upper (H and M regimes) and the lower river course (L regime), respectively.

As we said, the process that is less complex has a KLL value near to zero, whereas a process with highest randomness will have the KLL close to one. Accordingly, the KLL values, which are large for rivers from the first interval, i.e., (0.948, 1.076), indicate the presence of stochastic influence in their upper courses, where

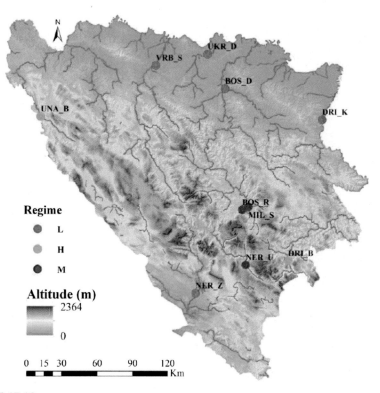

FIGURE 17.13

Relief of Bosnia and Herzegovina with location of 10 hydrological stations on seven rivers used in the study (their abbreviations for rivers and letters indicating the river regime are given in Table 17.3) (Mihailovic et al., 2015).

these rivers show behavior that is typical for mountain rivers. Inversely, the KLL complexities are smaller for the lower river courses (0.791, 0.918). The only exception is the Ukrina River (UKR_D) having greater randomness which is closer to the KLL of mountain rivers (0.981), which could be attributed to the fact that the KLL measure neglects variability in time series amplitudes. Similar results are observed in analysis of the KLU measure. However, now except besides the Ukrina River (UKR_D) and also the Vrbas River (VRB_S) has the KLU closer to the H and M regimes (4.324). Fig. 17.15 depicts the changes in the KLL and KLM complexities of river flow rate for 10 time series of seven rivers in Bosnia and Herzegovina for the period 1965–86, depending on the altitude of hydrological station. There is a positive trend in changes of the KLL and KLM with coefficients of correlation r, which are close to each other 0.649 and 0.602, respectively (Mihalović et al., 2015).

We analyze the monthly river flow time series of the seven rivers in Bosnia and Herzegovina for the period 1965–86, with $N = 252$ data points. The curves

Table 17.3 Rivers in Bosnia and Herzegovina Used in the Study With the Corresponding Flow Rates (FR—Mean; FR_{max}—Maximal; FR_{min}—Minimal for the Period 1965–86) and Their Classification Following a Classification of Typology for Mountains and Other Relief Classes by (Meybeck et al., 2001): Lowland (alt < 200 m)—(L Regime), Platforms and Hills (200 < alt < 500 m)—(H Regime), and Mountains (500 < alt < 6000 m)—(M Regime) (Mihailovic et al., 2015)

Catchment	Number	Abb.	Long (°E)	Lat (°N)	Alt (m)	FR (m³/s)	FR_{max} (m³/s)	FR_{min} (m³/s)	Regime
River Neretva to Zitomislić	1	NER_Z	17°47'	43°12'	16	252.0	734.0	53.0	L
River Neretva to Ulog	2	NER_U	18°14'	43°25'	641	8.0	35.3	0.5	M
River Bosna to Doboj	3	BOS_D	18°16'	43°49'	137	172.0	650.0	30.0	L
River Bosna to Reljevo	4	BOS_R	18°06'	44°44'	500	29.0	99.5	4.8	M
River Drina to Kozluk	5	DRI_K	19°07'	44°30'	121	380.0	1160.0	66.0	L
River Drina to Bastasi	6	DRI_B	18°46'	43°27'	425	155.0	497.0	32.7	H
River Miljacka to Sarajevo	7	MIL_S	18°21'	43°12'	530	5.0	21.9	0.6	M
River Una to Martin Brod	8	UNA_B	16°07'	43°30'	310	54.0	188.0	9.2	H
River Ukrina to Derventa	9	UKR_D	17°55'	44°50'	105	17.0	92.3	1.1	L
River Vrbas to Delibašino Selo	10	VRB_S	18°16'	43°49'	141	111.0	358.0	38.2	L

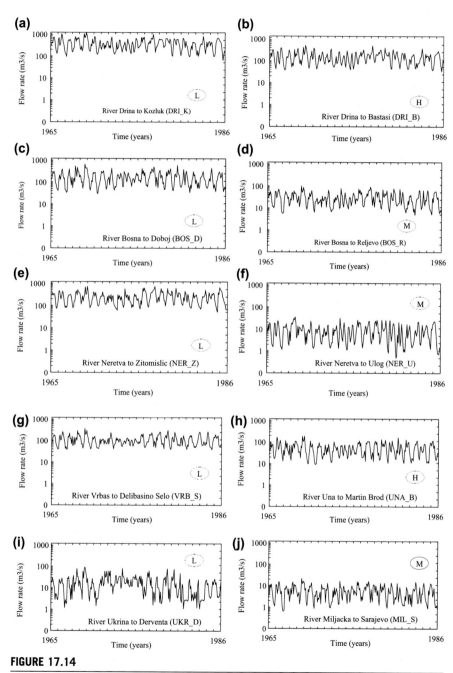

FIGURE 17.14

Ten river flow time series of seven rivers in Bosnia and Herzegovina for the period 1965–86 (Mihailovic et al., 2015).

Table 17.4 Kolmogorov Complexities (Lower—KLL, Upper—KLU, Kolmogorov Complexity Spectrum Highest Value—KLM) and Overall Kolmogorov Complexity Measure (KLO) Values of the Flow Rate for Ten Time Series of Seven Rivers in Bosnia and Herzegovina for the Period 1965—86 (Mihailovic et al., 2015)

Catchment	Number	Regime	Abb.	KLL	KLU	KLM	KLO
River Neretva to Zitomislić	1	L	NER_Z	0.918	3.799	0.948	0.506
River Neretva to Ulog	2	M	NER_U	1.013	3.309	1.013	0.529
River Bosna to Doboj	3	L	BOS_D	0.791	3.277	0.886	0.470
River Bosna to Reljevo	4	M	BOS_R	0.948	4.092	0.981	0.538
River Drina to Kozluk	5	L	DRI_K	0.823	3.213	0.981	0.502
River Drina to Bastasi	6	H	DRI_B	0.948	3.605	1.045	0.529
River Miljacka to Sarajevo	7	M	MIL_S	1.076	4.194	1.077	0.558
River Una to Martin Brod	8	H	UNA_B	0.948	4.227	1.045	0.540
River Ukrina to Derventa	9	L	UKR_D	0.981	4.294	1.013	0.479
River Vrbas to Delibašino Selo	10	L	VRB_S	0.918	4.324	0.918	0.486

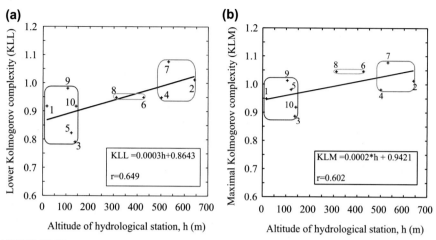

FIGURE 17.15

The dependence of KLL and KLM complexities of river flow rate on altitude, for 10 time series of the seven rivers in Bosnia and Herzegovina for the period 1965–86. Closed contours indicate the river regime: L (blue (dark gray in print versions)), H (green (light gray in print versions)), and M (red (gray in print versions)) (Mihailovic et al., 2015).

describing the Kolmogorov complexity spectra for the time series of flow rate of seven rivers and 10 stations are depicted in Fig. 17.16. From this figure, it is seen that flow dynamics of rivers are different in sense of the position and value of the maximum of the Kolmogorov complexity spectrum highest value (KLM). A simple inspection of this figure indicates the following facts for the same river: (1) when the river is in the M regime, i.e., with a pronounced presence of stochastic components in the river flow, then the KLM is greater in comparison with its KLM in the L regime; (2) the position of the maximum in the M regime is shifted toward the smaller normalized amplitudes (Fig. 17.16a and b); (3) when the river is in the H regime, i.e., when presence of stochastic components in the river flow is less pronounced than in M regime, then the KLM is slightly greater in comparison with its KLM in the L regime, while the position of the maximum in the H regime is still shifted toward the smaller values in the spectrum (Fig. 17.16c). Note that these conclusions could not be obtained if we compare the regimes of different rivers. From Fig. 17.16d—g, it is seen that the maximum of the Kolmogorov complexity spectrum highest value of the river flow time series is more shifted to the smaller values for the regimes where river flow has the higher randomness.

Analysis of Table 17.4 indicates that the KLM values in seven rivers are classified into two intervals, i.e., (1.013, 1.077) and (0.886, 0.948) corresponding to H and M river regimes and L river regime, respectively, while the UKR_D river is an exception with the KLM value of 1.013.

The KLL as a measure does not "see" a difference between time series which have different amplitude variations but similar random components. This could

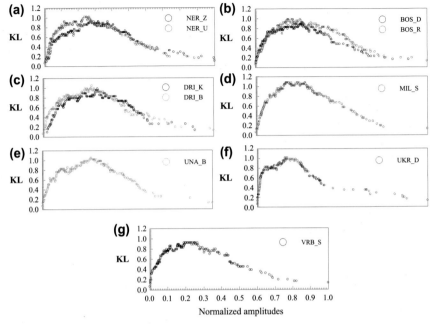

FIGURE 17.16

The Kolmogorov complexity spectra for the normalized amplitude of the monthly flow rate time series of the seven rivers in Bosnia and Herzegovina for the period 1965–86. The *circles* indicate the regime of river course: (blue (dark gray in print versions)), H (green (light gray in print versions)), and M (red (gray in print versions)) (Mihailovic et al., 2015).

also be said for the KLM measure, although it gives more information about complexity, in a broader context, than the KLL one does. It is seen from the above analysis of KLL and KLM values. However, it seems that the KLO measure better takes into account both, i.e., the amplitude and the place of the components in a time series. A detailed inspection of column KLO in Table 17.4 and left panel of Fig. 17.17 shows that there exist two intervals of this measure, which are clearly separated: (1) (0.470, 0.506) and (2) (0.529, 0.558). The first interval includes KLO of the flow rate of rivers in the L regime while another one refers to the H and M regimes. Now, the complexities of the UKR_D and VRB_S rivers, described by the KLO measure, correspond to the less stochastic time series. It seems that this is a more realistic measure than when their time series are described by the KLL, KLU, and KLM measures.

The right panel of Fig. 17.17 depicts a spatial distribution of the KLO measure of flow rate for 10 stations of seven rivers for the period 1965–86. This map enhances two regions: (1) with higher KLO, which is strongly influenced by the high mountain relief in the northwestern and eastern part of Bosnia and Herzegovina including the H

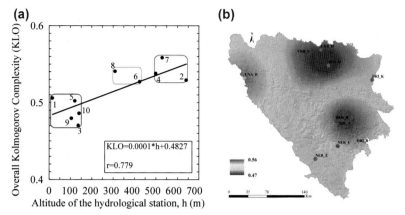

FIGURE 17.17

The KLO measure: dependence on altitude (left) and spatial distribution (right) of the monthly flow rate for 10 time series of seven rivers in Bosnia and Herzegovina for the period 1965–86 (Mihailovic et al., 2015). For correspondence between number of the station and its abbreviation see columns (1) and (2) in Table 17.4.

(three stations) and M (two stations) river regimes and (2) with lower KLO, which corresponds to the lowland regions (the northern, northeastern, western and southeastern parts with L river regimes (five stations). In the central part of Bosnia and Herzegovina there exists a transition belt with the KLO values indicating the mixed influences of the relief.

REFERENCES

Allan, J.D., 1995. Stream Ecology: Structure and Function of Running Waters. Chapman and Hall, London, p. 388.

Allen, J.J., Shockling, M.A., Kunkel, G.J., Smits, A.J., 2007. Turbulent flow in smooth and rough pipes. Phil. Trans. R. Soc. A 365, 699–714.

Attili, A., Bisetti, F., 2013. Fluctuations of a passive scalar in a turbulent mixing layer. Phys. Rev. E 88, 033013.

Batchelor, G.K., 1956. The Theory of Homogeneous Turbulence. Cambridge University Press, Cambridge.

Brown, G.L., Roshko, A., 1974. On density effects and large structure in turbulent mixing layers. J. Fluid Mech. 64, 775–816.

Caissie, D., 2006. The thermal regime of rivers: a review. Freshwater Biol. 51, 1389–1406.

Cushman-Roisin, B., Gualtieri, C., Mihailović, D.T., 2012. In: Gualtieri, C., Mihailović, D.T. (Eds.), Environmental Fluid Mechanics: Current Issues and Future Outlook in Fluid Mechanics of Environmental Interfaces, second ed. CRC Press/Balkema, Leiden.

Djurdjevic, V., Rajkovic, B., 2008. Verification of a coupled atmosphere-ocean model using satellite observations over the Adriatic Sea. Ann. Geophys. 26, 1935–1954.

Ferreira, F.F., Francisco, G., Machado, B.S., Murugnandam, P., 2003. Time series analysis for minority game simulations of financial markets. Physica A 32, 619–632.

Flack, K.A., Schultz, M.P., Shapiro, T., 2005. Experimental support for Townsend's Reynolds number similarity hypothesis on rough walls. Phys. Fluids 17, 035102.

Frank, L.D., Sallis, J.F., Conway, T.L., Chapman, J.E., Saelens, B.E., Bachman, W., 2006. Many pathways from land use to health: associations between neighborhood walk ability and active transportation, body mass index, and air quality. J. Am. Plann. Assoc. 72, 75–87.

Frish, U., 1995. Turbulence, The Legacy of A. N. Kolmogorov. Cambridge University Press, Cambridge.

Gioia, G., Chakraborty, P., 2006. Turbulent friction in rough pipes and the energy spectrum of the phenomenological theory. Phys. Rev. Lett. 96, 044 502.

Goncharov, V.P., Pavlov, V.I., 2015. Algebraic instability in shallow water flows with horizontally nonuniform density. Phys. Rev. E 91, 043004.

Gordon, N.D., McMahon, T.A., Finlayson, B.L., Gipel, C.J., 2004. Stream Hydrology: An Introduction for Ecologists. Wiley, New York.

Grassberger, P., 1989. Problems in quantifying self-generated complexity. Helv. Phys. Acta 62, 489–511.

Gualtieri, C., 2010. RANS-based simulation of transverse turbulent mixing in a 2D geometry. Environ. Fluid Mech. 10, 137–156.

Hadžić, E., Dresković, N., 2012. Climate change impact on water river flow: a case study for Sarajevo valley (Bosnia and Herzegovina). In: Mihailović, D.T. (Ed.), Essays on Fundamental and Applied Environmental Topics. Nova Science Publishers, New York, pp. 307–332.

Hajian, S., Sadegh Movahed, M., 2010. Multifractal Detrended Cross- Correlation Analysis of sunspot numbers and river flow fluctuations. Physica A 389, 4942–4957.

Hinze, J.O., 1975. Turbulence, second ed. McGraw-Hill College Division, New York.

Hussain, A.K.M.F., 1983. Coherent structures—reality and myth. Phys. Fluids 26, 2816–2850.

Ichimiya, M., Nakamura, I., 2013. Randomness representation in turbulent flows with Kolmogorov complexity (In laminar-turbulent transition due to periodic injection in an inlet boundary layer in a circular pipe). J. Fluid Sci. Technol. 8, 407–422.

Jimenez, J., 2004. Turbulent flows over rough walls. Annu. Rev. Fluid Mech. 36, 173–196.

Lange, H., Rosso, O.A., Hauhs, M., 2013. Ordinal patterns and statistical complexity of daily stream flow time series. Eur. Phys. J. 222, 535–552.

Meybeck, M., Green, P., Vorosmarty, C., 2001. A new typology for mountains and other relief classes: an application to global continental water resources and population distribution. Mt. Res. Dev. 21, 34–45.

Mihailovic, D.T., Alapaty, K., Podrascanin, Z., 2009. The combined non-local diffusion and mixing schemes, and calculation of in-canopy resistance for dry deposition fluxes. Environ. Sci. Pollut. Res. 16, 144–151.

Mihailović, D.T., Mimić, G., Gualtieri, P., Arsenić, I., Gualtieri, C., 2016. Randomness representation in turbulent flows with bed roughness elements using the spectrum of the Kolmogorov complexity, Proceedings of the 8th International Congress on Environmental Modelling and Software (iEMSs), July 10-14 2016, Toulouse, France.

Mihailović, D.T., Mimić, G., Dresković, N., Arsenić, I., 2015. Kolmogorov complexity based information measures applied to the analysis of different river flow regimes. Entropy 17, 2973–2987.

Mihailović, D.T., Nikolić-Đorić, E., Drešković, N., Mimić, G., 2014. Complexity analysis of the turbulent environmental fluid flow time series. Physica A 395, 96–104.

Nepf, H.M., 2012. Flow and transport in regions with aquatic vegetation. Annu. Rev. Fluid Mech. 44, 123–142.

Nezu, I., Sanjou, M., 2008. Turbulence structure and coherent motion in vegetated canopy open-channel flows. J. Hydro Environ. Res. 2, 62–90.

Orr, H.G., Carling, P.A., 2006. Hydro-climatic and land use changes in the river Lune catchment, North West England, implications for catchment management. River Res. Appl. 22, 239–255.

Otache, Y.M., Sadeeq, M.A., Ahaneku, I.E., 2011. ARMA modelling of Benue River flow dynamics: comparative study of PAR model. Open J. Mod. Hydrol. 1, 1–9.

Poggi, D., Porporato, A., Ridolfi, L., Albertson, J.D., Katul, G.G., 2004. The effect of vegetation density on canopy sub-layer turbulence. Bound Lay Meteorol. 111, 565–587.

Pope, S.B., 2000. Turbulent Flows. Cambridge University Press, Cambridge.

Porporato, A., Ridolfi, L., 2001. Multivariate nonlinear prediction of river flows. J. Hydrol. 248, 109–122.

Prandtl, L., 1965. Führer durch die Strömungslehre, 6. Auflage. Friedr. Vieweg & Sohn, Braunschweig.

Raupach, M.R., Finnigan, J.J., Brunet, Y., 1996. Coherent eddies in vegetation canopies – the mixing layer analogy. Bound Lay. Meteorol. 78, 351–382.

Reynolds, O., 1883. An experimental investigation of the circumstances which determine whether the motion of water shall be direct or sinuous and of the law of resistance in parallel channels. Philos. T. R. Soc. Lond 174, 935–982.

Rogers, M.M., Moser, R.D., 1994. Direct simulation of a self-similar turbulent mixing layer. Phys. Fluids 6, 903.

Ruelle, D., 1978. What are the measures describing turbulence? Supp. Prog. Theor. Phys. 64, 339–345.

Sadegh Movahed, M., Hermanis, E., 2008. Fractal analysis of river flow fluctuations. Physica A 387, 915.

Salas, J.D., Kim, H.S., Eykholt, R., Burlando, P., Green, T.R., 2005. Aggregation and sampling in deterministic chaos: implications for chaos identification in hydrological processes. Nonlinear Proc. Geophys. 12, 557–567.

Schertzer, D., Tchiguirinskaia, I., Lovejoy, S., Hubert, P., Bendjoudi, H., Larchevesque, M., 2002. Discussion of "Evidence of chaos in rainfall-runoff process" Which chaos in rainfall-runoff process? Hydrol. Sci. J. 47, 139–149.

Sellers, P., Mintz, Y., Sud, Y.C., Dachler, A., 1986. A simple biosphere model (SiB) for use within general circulation models. J. Atmos. Sci. 43, 505–531.

Serinaldi, F., Zunino, L., Rosso, O.A., 2014. Complexity–entropy analysis of daily stream flow time series in the continental United States. Stoch. Env. Res. Risk A 28, 1685–1708.

Shannon, C.E., 1948. A mathematical theory of communication. Bell Syst. Tech. J. 27, 379–623.

Sivakumar, B., Singh, V.P., 2012. Hydrologic system complexity and nonlinear dynamic concepts for a catchment classification framework. Hydrol. Earth Syst. Sci. 16, 4119–4131.

Stoop, R., Stoop, N., Bunimovich, L., 2004. Complexity of dynamics as variability of predictability. J. Stat. Phys. 114, 1127–1137.

Thoraval, M.J., Takehara, K., Etoh, T.G., Popinet, S., Ray, P., Josserand, C., Zaleski, S., Thoroddsen, S.T., 2012. von Kármán Vortex Street within an impacting drop. Phys. Rev. Lett. 108, 264506.

Tsuji, Y., Nakamura, I., 1994. The fractal aspect of an isovelocity set and its relationship to bursting phenomena in the turbulent boundary layer. Phys. Fluids 6, 3429–3441.

Wesson, K.H., Katul, G.G., Siqueira, M., 2003. Quantifying organization of atmospheric turbulent eddy motion using nonlinear time series analysis. Bound Lay. Meteorol. 106, 507–525.

Wijesekera, H.W., Dillion, T.M., 1997. Shannon entropy as an indicator of age for turbulent overturns in the oceanic thermocline. J. Geophys. Res. 102, 3279–3291.

Zunino, L., Soriano, M.C., Rosso, O.A., 2012. Distinguishing chaotic and stochastic dynamics from time series by using a multiscale symbolic approach. Phys. Rev. E 86, 046210.

How to face the complexity of climate models?

18

18.1 COMPLEXITY OF THE OBSERVED CLIMATE TIME SERIES

In the study of the climate system, an important role plays the complexity of the observed time series as well as the complexity of the climate models used to get projection of the climate in the future. In this chapter we shall deal with the complexity of the observed time series while the next chapter will be devoted to the complexity of climate models. The huge diversity and complexity of climate elements, including their intensity, and temporal and spatial distribution, do not allow easy descriptions. However, as we already emphasized, the traditional statistical models and methods may not be appropriate for some climate elements, which exhibit odd behaviors. This challenged the development of some of the finest nonlinear dynamics methodologies to describe these behaviors.

In Section 16.3 we have already applied some of those methods. Namely, we have used the Kolmogorov complexity (KLL) and the Kolmogorov complexity spectrum highest value (KLM) of the UV-B radiation time series relying on the paper by Mihailović et al. (2013). It was found that the complexity loss of the observed UV-B radiation time series may be attributed to (1) the increased human intervention in the post–civil war period causing increase of the air pollution after 1999 and (2) the increased cloudiness due to climate changes.

First, we demonstrate the complexity of the spatial distribution of the monthly air temperature for 46 places in Bosnia and Herzegovina for the period 1960–90, using the overall Kolmogorov complexity (KLO) (see Section 15.3), which is a measure of disorder in a time series. Fig. 18.1 depicts a spatial distribution of the monthly air temperature for selected places in Bosnia and Herzegovina for the period indicated. This map depicts two regions: (1) with higher KLO values which is strongly influenced by the vicinity of the Pannonian Basin (northern and partially central area) and (2) with lower KLO values, i.e., lower disorder in time series, which correspond with the vicinity of the Adriatic Sea (southern and south-western area). Between them is a transition belt (belt between them) with the KLO values indicating on the mixed influences of the above-mentioned geographical factors and relief.

The study of precipitation up till now continues to be an exciting research area, since quite a few aspects of precipitation generation and evolution have not been

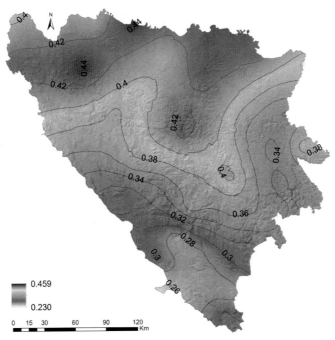

FIGURE 18.1

Spatial distribution of the overall Kolmogorov complexity (KLO) of the monthly air temperature time series in Bosnia and Herzegovina for the period 1960—90.

understood, clarified, and described satisfactorily. There are still several problems related to the perception and modelling of precipitation. It is difficult to describe easily precipitation because of its huge complexity, comprising its forms, extent, intermittency, intensity, and temporal and spatial distribution (Koutsoyiannis, 2005). As an example, we analyze the long-term precipitation dynamics for two locations with different geographies in Bosnia and Herzegovina. The pluviographic regime of one location is influenced by the vicinity of the Adriatic Sea (Mostar— Mo), while the other is surrounded by mountain and flat areas (Bihac—Bi). The time series were updated for the time interval 1960—84 having the length of $N = 300$ (Drešković and Djug, 2012). The curves in Fig. 18.2 that describe the Kolmogorov complexity spectrum for time series of both places suggest that their long-term precipitation dynamics differs. The values of the KLM are slightly different (1.152—Bi and 1.097—Mo), while the difference in the calculated KLO values is more pronounced (0.511 for Bi and 0.558 for Mo). The differences in the KLL (1.152—Bi and 1.097—Mo) are negligible, but the differences in the KLM are larger. This is additional information about complexity that is not contained in KL and KLM. This allows us to conclude that the Mo time series is

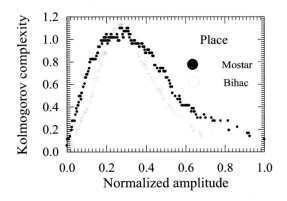

FIGURE 18.2

The Kolmogorov complexity spectrum for the normalized amplitude of the monthly precipitation time series for two locations, Mostar and Bihac (Bosnia and Herzegovina) for the period 1960–84.

Reprinted with permission from Mihailović, D.T., Mimić, G., Nikolić-Đorić, E., Arsenić, I., 2015b. Novel measures based on the Kolmogorov complexity for use in complex system behavior studies and time series analysis. Open Phys. 13, 1–14.

more complex than Bi time series having larger variability of amplitudes as it is seen from the shape of its spectrum in Fig. 18.2.

Here, we demonstrate the complexity of the spatial distribution of the monthly precipitation amount time series for 23 places in Bosnia and Herzegovina for the period 1960–84, using the Precipitation Complexity Index (PCI) introduced by Mihailović et al. (2015a) that is practically the KLO measure (Mihailović et al., 2015b). Fig. 18.3 depicts a spatial distribution of the PCI index of the selected places. This map depicts two regions: (1) with higher PCI index which is strongly influenced by the presence of the Adriatic Sea and (2) with lower PCI index, which corresponds with the vicinity of the Pannonian Basin (northern and partially central area). Between them is a transition belt (southern and south-western area, up to central part) with the PCI values indicating the mixed influences of the above-mentioned geographical factors and relief. This index quantitatively indicates that the difference in complexity of spatial distribution of precipitation in Bosnia and Herzegovina is caused by the influence of (1) Adriatic Sea (close to the Mediterranean Sea), (2) relief, and (3) the Pannonian Basin (Fig. 18.2) on its climatic regime and accordingly to spatial and temporal distribution of precipitation.

In extension of this chapter we will analyze the time series of daily values for three meteorological elements, two continuous and a discontinuous one, i.e., the maximum and minimum air temperature and the amount of precipitation. The analysis is based on the observations from seven weather stations in Serbia in the period 1951–2010 (Fig. 18.4 and Table 18.1), to quantify the complexity of the annual values for the above time series and to calculate the rate of its change.

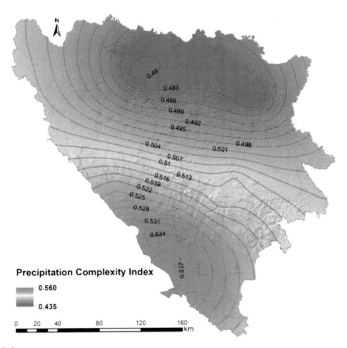

FIGURE 18.3

Spatial distribution of the overall Kolmogorov complexity (KLO) of the monthly precipitation amount time series in Bosnia and Herzegovina for the period 1960—84.

Reprinted with permission from Mihailović, D.T., Drešković, N., Mimić, G., 2015a. Complexity analysis of spatial distribution of precipitation: an application to Bosnia and Herzegovina. Atmos. Sci. Lett. 16, 324—330.

For that purpose, we have used the sample entropy (SampEn) and the Kolmogorov complexity (KLL) as the measures which can indicate the variability and irregularity of a given time series.

Many scientists try to estimate the effects of climate changes in various regions of the world. In those efforts they often use various statistical methods to analyze spatial and temporal variability of the precipitations. For example, Serbia, which is a very interesting region in climate projections of precipitation (Mihailović et al., 2015c), was a subject of the study of the precipitation trend resulting with several papers in which are used methods of classical statistics. Thus, in analysis of the variability of the summer and winter precipitations Tošić (2004) used the empirical orthogonal function. Unkašević and Tošić (2011) performed the statistical analysis of daily precipitations over Serbia calculating the trends and indices such as the number of days exceeding different threshold values, the fraction of the total annual rainfall due to events above the 95th percentile, or the number of very wet days. Gocić and Trajković (2014) analyzed the spatial and temporal patterns of precipitation in Serbia with the precipitation concentration index and performed the principal component analysis for the same time series. The spatial analysis of annual

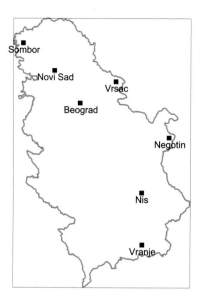

FIGURE 18.4

The location of the weather stations in Serbia used in the complexity analysis.

Reprinted with permission from Mimić, G., Mihailović, D.T., Kapor, D., 2015. Complexity analysis of the air temperature and the precipitation time series in Serbia. Theor. Appl. Climatol. http://dx.doi.org/10.1007/s00704-015-1677-6.

and seasonal air temperature trends in Serbia was also in the focus. Thus, Bajat et al. (2015) have analyzed the magnitude of the trends, which were derived from the slopes of linear trends using the least square method based on the mean monthly data for the period 1961−2010. These methods give just a partial insight into the behavior of the climate since the Earth's atmosphere has evolved into a complex system in which life and climate are intricately interwoven. The interface between

Table 18.1 The List of the Stations With Name, Abbreviation, Latitude, Longitude, and Altitude

Station	Abbreviation	Latitude	Longitude	Altitude (m)
Sombor	SO	45°47′	19°05′	87
Novi Sad	NS	45°20′	19°51′	86
Belgrade	BG	44°48′	20°28′	132
Vršac	VS	45°06′	21°18′	94
Negotin	NE	44°14′	22°33′	42
Niš	NI	43°20′	21°54′	204
Vranje	VR	42°33′	21°55′	432

Reprinted with permission from Mimić, G., Mihailović, D.T., Kapor, D., 2015. Complexity analysis of the air temperature and the precipitation time series in Serbia. Theor. Appl. Climatol. http://dx.doi.org/10.1007/s00704-015-1677-6.

Earth and atmosphere as a "pulsating biophysical organism" is a complex system itself (Mihailović et al., 2014). Thus, discovering the complexity of climate change process direct us toward the complexity measures of the climate time series, which can indicate to the irregularity and randomness of these time series. This is an essential step since this quantification is the crucial step in understanding complexity of the climate, which has no single universal definition.

Here, we partially review the results of the complexity analysis of the maximum and minimum air temperature and the precipitation time series based on the observations from seven stations in Serbia, during the period 1951–2010, which is performed by Mimić et al. (2015). For that purpose they used the sample entropy (SampEn) and the KLL complexity, calculating the annual values of the measures estimating the trends during the considered period.

The Kolmogorov complexity of the maximum temperature has statistically significant increasing trend for all of the seven stations. That trend is seen in Fig. 18.5b for station Sombor. Higher KLL complexity means lower predictability of the time series owing to the increase of randomness and irregularity of time series that could be a potential problem in climate modelling in the future. The sample entropy of the maximum temperature has also increasing but statistically insignificant trends (Fig. 18.5a for the station Sombor). The trends of the complexity measures of the minimum temperature are nonsignificant (Fig. 18.5c and d for the station Sombor); also some of the trends are negative while the rest are positive, depending on the station. On the other hand, the precipitation time series have decreasing trends of both the sample entropy and the Kolmogorov complexity for all of the seven stations (Figs. 18.5e and f for the station Sombor).

18.2 COMPLEXITY OF THE MODELED CLIMATE TIME SERIES

The question of the weather and climate modelling and predictability has been initiated in the early sixties of the 20th century, which was elaborated in pioneering works by Lorenz (1960, 1963a, 1963b, 1964). He was the first person in the scientific world who explicitly pointed out importance of the following points in the nonlinear dynamics in atmospheric motion: (1) question of prediction and predictability, (2) importance of understanding the nonlinearity in modelling procedure, (3) demand for discovery of chaos, and (4) careful consideration of sensitivity of differential equations in modelling system on initial conditions. Here, we will not consider them more comprehensively, since their detailed elaboration can be found in several papers, for example in Mihailović et al. (2014). Undoubtedly, the complex ocean/atmosphere/land dynamical system, called weather and its long time average climate, can be considered as a complex one. This system is modeled by climate models having different levels of sophistication.

An important concept in climate system modelling is that of a spectrum of models of differing levels of complexity, each being optimum for answering specific

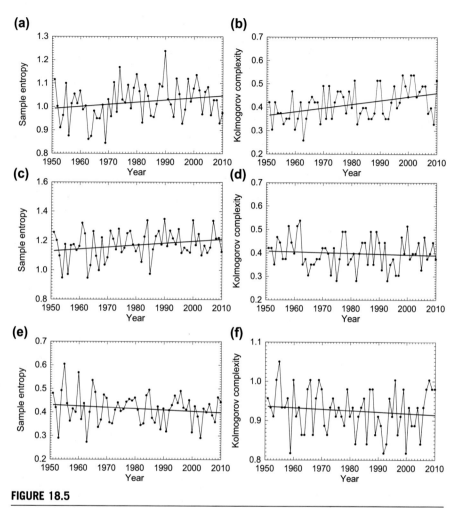

FIGURE 18.5

The trends of the complexity measures for the station Sombor compared for two periods 1951–2010 and 1961–2010, for the maximum temperature (a and b); the minimum temperature (c and d) and the precipitation (e and f).

questions. It is not meaningful to judge one level as being better or worse than another independently of the context of analysis (Mihailović et al., 2014). What is important is that each model be appropriately questioned for its level of complexity and quality of its simulation as emphasized by Randall et al. (2007). In that paper they comprehensively considered the following points: (1) Earth system models of intermediate complexity, that is, reduced-resolution models that incorporate most of the processes represented by AOGCMs (Atmosphere-Ocean General Circulation Models) and models of reduced complexity. However, in this section we consider the model complexity on the basis of calculation of the maximum complexity that can be generated by a model as it was done by Mihailović et al. (2014). Given a time series and the problem of choosing among a number of climate models to study it, we suggest that models whose maximum complexity is lower than the time series complexity should be disregarded because of being unable to reconstruct some of the structures contained in the data. The increasing complexity of those models is a growing concern in the modelling community. They are used to integrate and process knowledge from different parts of the system and in doing so allow us to test system understanding and create hypotheses about how the system will respond to the virtual numerical experiments. However, if we strive to design our models to be more "realistic," we have to include more and more parameters and processes. Then, within this approach the model complexity increases, and thus we are less able to manage and understand the model behavior (Mihailović et al., 2014). Obviously, the question about model complexity could be considered from the standpoint of a practitioner who sees it as a compromise between complexity and manageability. His/her question is basically very simple: "How can I check if this model is appropriate to study this problem with this data set?" According to Boschetti (2008): "As a result, the ability of a model to simulate complex dynamics is no more an absolute value in itself, rather a relative one: we need enough complexity to realistically model a process, but not so much that we ourselves cannot handle."

Clearly, an answer to the above question requires (1) a definition and a measure of complexity and (2) that this measure is equally applicable to the model and to the data, because some sort of comparison is necessary. It is a hard task to find that measure even approximately. However, intuitively we can put an accent on a view of complexity which is more related to a model's dynamical properties rather than its architecture. Thus, we can say that, in developing tools, an advantage will be given to a tool which gives answers to the following questions: (1) what is the maximal dynamical complexity a given model can generate? and (2) what kind of different dynamical behaviors can a given model generate? as it is underlined by Boschetti (2008). For our consideration we will rely on Boschetti (2008) who defined the complexity of an ecological model as the statistical complexity of the output it produces that allows a direct comparison between data and model complexity. Among the many different measures of complexity available in the literature, for that purpose, he adopted the statistical complexity defined by Lempel and Ziv (1976).

An example of comparison between complexities of a global and regional model. In this section we will illustrate an example of comparison between complexities of global and regional model. Here, we do not deal with statistical complexity of the global and regional models. Our intention is just to show possible differences in complexities of time series of precipitation as well as air temperature for both models, applying the algorithm for calculating the KLL complexity (Mihailović et al., 2014).

In order to calculate complexities of model time series we use (1) air temperature and (2) precipitation time series which are outputs from climate simulations for

Belgrade and Novi Sad in Serbia (Djurdjević and Rajković, 2008; Arsenić et al., 2013). The Belgrade data set, for the period 2071–2100, was derived from (1) the SINTEX-G which is a coupled Atmosphere-Ocean General Circulation Model (Gualdi et al., 2003) and (2) the Eta-POM regional model (Djurdjevic and Rajkovic, 2008). The Novi Sad data set, for the period 2021–50, was derived from (1) the ECHAM5 which is the fifth generation of the ECHAM5 general circulation model (Liu et al., 2013) and (2) the RegCM regional model (Giorgi, 2011).

We have calculated the KLL for each time series obtained when each sample, in the original time series, is used as a threshold ($N = 10,800$ for Belgrade and $N = 11,323$ for Novi Sad). The results are depicted in Fig. 18.6. We also have calculated Kolmogorov complexity (KLL) and its maximal value KLM of time series from Fig. 18.6. Results of those calculations are given in Table 18.2. Hereinafter, we will rely on the analysis from Mihailović et al. (2014).

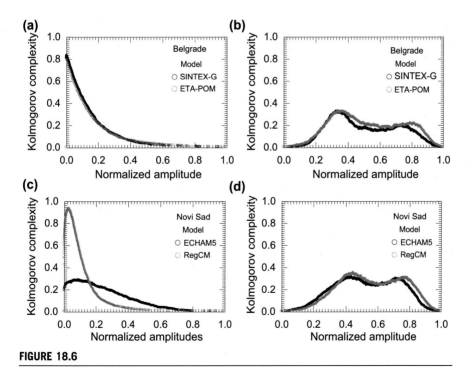

FIGURE 18.6

The Kolmogorov complexity spectrum of the (a) precipitation and (b) air temperature time series for Belgrade and (c) precipitation and (d) temperature for Novi Sad, in Serbia, obtained from climate simulations using different models. On x axis are depicted the values of the time series normalized as in Fig. 16.15.

Table 18.2 Kolmogorov Complexities (KLL and Its Maximum—KLM) Values for the Precipitation and Air Temperature Time Series for Belgrade and Novi Sad, in Serbia, Obtained From Climate Simulations Using Different Models.

| | | Model | | | |
| | | Global | | Regional | |
Quantity	Measure	SINTEX-G	ECHAM5	ETA-POM	RegCM
Temperature (Belgrade)	KLL	0.176		0.207	
	KLM	0.326		0.331	
Temperature (Novi Sad)	KLL		0.241		0.251
	KLM		0.318		0.354
Precipitation (Belgrade)	KLL	0.705		0.671	
	KLM	0.834		0.793	
Precipitation (Novi Sad)	KLL		0.265		0.871
	KLM		0.289		0.935

Reprinted with permission from Mihailovic, D.T., Mimic, G., Arsenic, I., April 2014. Climate predictions: the chaos and complexity in climate models. Adv. Meteorol. 1–14, 878249.

From Fig. 18.6a it is seen that there is no difference between complexities of the precipitation time series for Belgrade obtained by both models (global SINTEX-G and regional Eta-POM) over all amplitudes in time series.

Moreover, the SINTEX-G model has slightly higher complexity. In contrast to that, 18.6b depicts that the Eta-POM model mostly has the higher complexity than the SINTEX-G model for the air temperature time series. From Table 18.2 we can see that for air temperature time series the KLL for the Eta-POM model (0.207) is higher than for the SINTEX-G model (0.176), while the KLM values are practically the same (0.331 and 0.326). Note that all of these complexities are pronouncedly low. Further inspection of this table clearly shows that the precipitation time series obtained by the SINTEX-G model has higher complexities (KLL: 0.705 and KLM: 0.834) than those obtained by the Eta-POM model (KLL: 0.671 and KLM: 0.793). This analysis indicates that the SINTEX-G and Eta-POM models, in particular for precipitation, have approximately the same level of complexity. Now, we analyze the air temperature and precipitation time series for Novi Sad obtained by the global ECHAM5 and regional RegCM models. From Fig. 18.6c it is seen that there is a large difference between complexities of the precipitation time series over all amplitudes in time series. Moreover, the RegCM model has pronouncedly higher complexity. Fig. 18.6d depicts that the RegCM and ECHAM5 models mostly have very similar complexities for the air temperature time series. From Table 18.2 we can see that for air temperature time series the KLL for the RegCM model (0.251) is higher than for the ECHAM5 model (0.241) and also for the KLM values, 0.354 and 0.318, respectively. Similarly, as for the above-analyzed models, these values of complexity are still low. Further inspection of this table clearly shows that the precipitation time series obtained by the ECHAM5 model has lower complexities (KLL: 0.265 and KLM: 0.289) than those obtained by the RegCM model (KLL: 0.871 and KLM: 0.935). This analysis indicates the ECHAM5 and RegCM models have approximately the same level of complexity in simulation of the air temperature. In contrast to that, there is a large difference in capabilities of these models to simulate the precipitation. Note that a higher value of the KLL points out the presence of stochastic influence of different factors on a time series. In this paper (Mihailovic et al., 2014) we suggest that climate models whose maximum complexity is lower than the time series complexity should be disregarded because of being unable to reconstruct some of the structures contained in the data. To our knowledge this complexity analysis has not been used for analyzing the complexity of climate models. However, for more reliable conclusion that could be reached we need to test outputs of many different GCM and RegCM models.

An example of comparison between complexities of a regional climate model and observed time series. In the previous example we have considered complexity through the optics of the relationship between global and climate regional model. The fact is that the increasing complexity of climate models is a growing concern in the modelling community. However, with increasing model complexity, we are less able to manage and understand the model behavior (Mihailović et al., 2015c).

We consider the complexity of the EBU-POM model using the observed and modeled time series of temperature and precipitation. Therefore, we have computed the KLL spectrum, KLL, KLM, and SampEn values for temperature and precipitation. The calculations are performed for the time interval 1961−1990: (1) on a daily basis with a size of $NN = 10,958$ samples for temperature and (2) on a monthly basis with a size $NN = 360$ for the precipitation. The simulated time series of temperature and precipitation were obtained by the EBU-POM model for the given period. The observed time series of temperature and precipitations for two stations in Serbia, Zlatibor (ZL) (1028 m) and Sombor (SO) (88 m), taken from daily meteorological reports of the Republic Hydrometeorological Service of Serbia. The results for both sites are given and seen in Figs. 18.7 and 18.8. Based on Fig. 18.7, the KLL and KLM values for ZL are higher than for SO in both the observed and modeled temperatures (0.361 vs. 0.273 for the KL and 0.391 vs. 0.347 for the KLM values). Thus, for both sites, the modeled (M) complexity is lower than the observed (O) one. This finding means that the models with a KLL (and KLM) complexity lower than the measured time series complexity are unable to reconstruct some of the structures contained in the observed data. This difference is more pronounced for the ZL site than for SO. As we noted, the KLL measure is also used as a measure of randomness taking values between 0 (for deterministic process such as, for example, a periodic motion) and 1 indicating processes that can be considered as the stochastic ones. However, for both sites, the KLL complexity is lower, and its value for ZL indicates the presence of stochastic influences that are not present at the SO site.

We have calculated the following parameters: embedding dimension ($m = 2$), tolerance ($r = 0.2$), and time delay ($\tau = 1$) following Mihailović et al. (2013). Lower values of the SE for both sites (ZL: observed—1.060, modeled—1.007; SO: observed—0.910, modeled—0.849) support the conclusion regarding a more regular

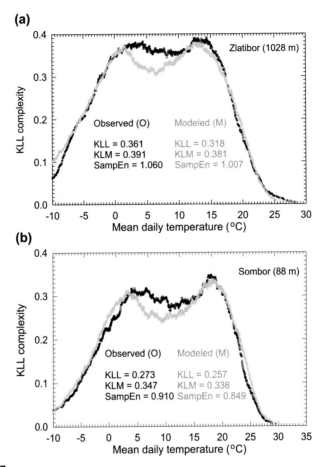

FIGURE 18.7

The Kolmogorov complexity spectrum of the observed (O) and modeled (M) mean daily temperatures for the sites (a) ZL and (b) SO for the period 1961—90. Modeled values are obtained from the EBU-POM model for the same period.

Reprinted with permission from Mihailović, D.T., Lalić, B., Drešković, N., Mimić, G., Djurdjević, V., Jančić, M., 2015c. Climate change effects on crop yields in Serbia and related shifts of Köppen climate zones under the SRES-A1B and SRES-A2. Int. J. Clim. 35, 3320—3334.

mean daily temperature time series in this period. As might be expected, the SampEn values (observed and modeled) for SO are lower than for ZL. This is because the complexities of SO's mean daily temperature time series are lower than those of ZL. Note that a similar use of the SampEn values as a measure in studying the climate complexity was used by Shuangcheng et al. (2006).

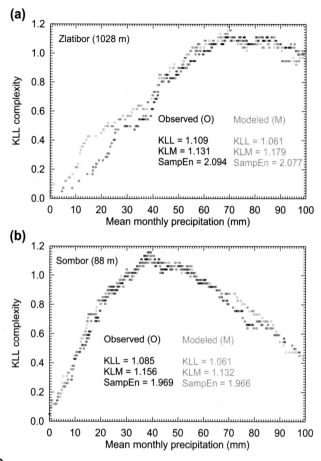

FIGURE 18.8

The Kolmogorov complexity spectrum of the observed (O) and modeled (M) mean monthly precipitation amounts for the sites (a) ZL and (b) SO for the period 1961–90. Modeled values are obtained from the EBU-POM model for the same period.

Reprinted with permission from Mihailović, D.T., Lalić, B., Drešković, N., Mimić, G., Djurdjević, V., Jančić, M.,
2015c. Climate change effects on crop yields in Serbia and related shifts of Köppen climate zones under the
SRES-A1B and SRES-A2. Int. J. Clim. 35, 3320–3334.

Based on Fig. 18.8, the complexities of the mean monthly precipitation amount time series are higher than that of temperature because of the greater stochastic nature of the former's time series. Furthermore, the KLL value of the observed values for ZL is higher than for the SO (1.109 vs. 1.085), whereas for modeled values, the complexities are the same (1.061). The KLM value of the observed time series for ZL is lower than for SO. In contrast, the KLM value of the modeled

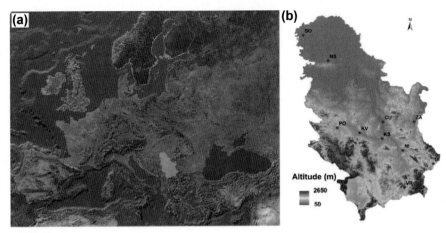

FIGURE 18.9

(a) Serbia in Europe and (b) the 10 sites used in calculating the Kolmogorov complexity spectrum highest value (KLM) and sample entropy (SampEn) values in Table 18.3—CU, Ćuprija; DI, Dimitrovgrad; KS, Kruševac; KV, Kraljevo; NI, Niš; NS, Novi Sad; PO, Požega; SO, Sombor; VR, Vranje; ZA, Zaječar.

Reprinted with permission from Mihailović, D.T., Lalić, B., Drešković, N., Mimić, G., Djurdjević, V., Jančić, M., 2015c. Climate change effects on crop yields in Serbia and related shifts of Köppen climate zones under the SRES-A1B and SRES-A2. Int. J. Clim. 35, 3320–3334.

Table 18.3 Kolmogorov Complexity Spectrum Highest Value (KLM) and Sample Entropy (SampEn) Values for the Precipitation Amount Time Series Obtained by the EBU-POM Regional Climate Model Under the A1B Scenario, for the Periods: (a) 2001–30 and (b) 2071–2100. In computing the SampEn we have used the following sets of parameters ($m = 2$, $r = 0.2$, and $\tau = 1$).

Site	A1B (2001–2030)		A1B (2071–2100)		A2 (2001–2030)		A2 (2071–2100)	
	KCM	SE	KCM	SE	KCM	SE	KCM	SE
SO	1.109	1.920	1.109	1.839	1.156	2.030	1.203	1.911
NS	1.156	1.950	1.109	1.913	1.156	1.983	1.156	2.011
PO	1.203	2.005	1.156	1.944	1.156	1.844	1.132	1.889
CU	1.156	2.059	1.132	1.895	1.109	1.907	1.132	1.892
KR	1.156	2.006	1.132	1.966	1.109	1.901	1.132	1.776
KS	1.132	2.039	1.179	1.907	1.179	1.907	1.155	1.729
ZA	1.132	1.914	1.132	1.780	1.156	1.885	1.156	1.709
NI	1.132	2.050	1.156	1.971	1.156	2.034	1.179	1.841
DI	1.156	2.034	1.132	2.086	1.132	2.008	1.132	1.881
VR	1.109	2.032	1.156	1.981	1.156	1.893	1.156	1.914

Reprinted with permission from Mihailović, D.T., Lalić, B., Drešković, N., Mimić, G., Djurdjević, V., Jančić, M., 2015c. Climate change effects on crop yields in Serbia and related shifts of Köppen climate zones under the SRES-A1B and SRES-A2. Int. J. Clim. 35, 3320–3334.

time series for ZL (1.179) is higher in comparison with that of SO (1.132). In comparison with temperature, slightly higher values of the SE of the precipitation amount time series for both sites (ZL: observed—2.094, modeled—2.077; SO: observed—1.969; modeled—1.966) indicate the presence of a more stochastic component in the mean monthly precipitation amount time series.

Additionally, we have computed the KLM and SampEn values, for 10 sites in Serbia (Fig. 18.9), for the precipitation amount time series, which have a high level of stochastic behavior. This time series was obtained from the EBU-POM regional climate model under the A1B and A2 scenarios for the periods: (1) 2001—30 and (2) 2071—2100. In computing the SE values, we used $m = 2$, $r = 0.2$, and $\tau = 1$. The results are given in the corresponding rows of Table 18.3. Based on this table, the values of both scenarios and periods are in the following ranges: (1) the KLM: 1.109—1.203 and (2) the SampEn: 1.709—2.086. These ranges indicate that the EBU-POM model has a level of complexity that can provide reliable climate time series, i.e., precipitation amount in this case.

REFERENCES

Arsenic, I., Podrascanin, Z., Mihailovic, D.T., Lalic, B., Firanj, A., 2013. Projection of Vojvodina climate using the regional climate model: preliminary results. In: Proceedings of the 6th Scientific-Technical Meeting (InterRegioSci '13), Novi Sad, Serbia.

Bajat, B., Blagojević, D., Kilibarda, M., Luković, J., Tošić, I., 2015. Spatial analysis of the temperature trends in Serbia during the period 1961—2010. Theor. Appl. Climatol. 121, 289—301.

Boschetti, F., 2008. Mapping the complexity of ecological models. Ecol. Complex. 5 (1), 37—47.

Djurdjević, V., Rajković, B., 2008. Verification of a coupled atmosphere-ocean model using satellite observations over the Adriatic Sea. Ann. Geophys. 26, 1935—1954.

Dreškovic, N., Djug, S., 2012. Applying the inverse distance weighting and kriging methods of the spatial interpolation on the mapping the annual precipitation in Bosnia and Herzegovina. In: International Environmental Modelling and Software Society (IEMSS) International Congress on Environmental Modelling and Software Managing Resources of a Limited Planet, Sixth Biennial Meeting, Leipzig, Germany.

Giorgi, F., 2011. Regcm Version 4.1 Reference Manual. Tech. Rep., ICTP, Trieste, Italy.

Gocić, M., Trajković, S., 2014. Spatio-temporal patterns of precipitation in Serbia. Theor. Appl. Climatol. 117, 419—431.

Gualdi, S., Navarra, A., Guilyardi, E., Delecluse, P., 2003. Assessment of the tropical Indo-Pacific climate in the SINTEX CGCM. Ann. Geophys. 46, 1—26.

Koutsoyiannis, D., 2005. Uncertainty, entropy, scaling and hydrological stochastics. 2. Time dependence of hydrological processes and time scaling. Hydrol. Sci. J. 50 (3), 405—426.

Lempel, A., Ziv, J., 1976. On the complexity of finite sequences. IEEE Trans. Inf. Theory 22, 75—81.

Liu, P., Li, T., Wang, B., Zhang, M., Luo, J.-J., Masumoto, Y., Wang, X., Roeckner, E., 2013. MJO change with A1B global warming estimated by the 40-km ECHAM5. Clim. Dyn. 41, 1009−1023.

Lorenz, E.N., 1960. The statistical prediction of solutions of dynamic equations. In: Proceedings of the International Symposium on Numerical Weather Prediction, Tokyo, Japan, pp. 629−635.

Lorenz, E.N., 1963a. The predictability of hydrodynamic flows. Trans. N. Y. Acad. Sci. 25, 409−432.

Lorenz, E.N., 1963b. Deterministic nonperiodic flow. J. Atmos. Sci. 20, 130−141.

Lorenz, E.N., 1964. The problem of deducing the climate from the governing equations. Tellus 16, 1−11.

Mihailovic, D.T., Mimic, G., Arsenic, I., April 2014. Climate predictions: the chaos and complexity in climate models. Adv. Meteorol. 1−14, 878249.

Mihailović, D.T., Drešković, N., Mimić, G., 2015a. Complexity analysis of spatial distribution of precipitation: an application to Bosnia and Herzegovina. Atmos. Sci. Lett. 16, 324−330.

Mihailović, D.T., Mimić, G., Nikolić-Đorić, E., Arsenić, I., 2015b. Novel measures based on the Kolmogorov complexity for use in complex system behavior studies and time series analysis. Open Phys. 13, 1−14.

Mihailović, D.T., Lalić, B., Drešković, N., Mimić, G., Djurdjević, V., Jančić, M., 2015c. Climate change effects on crop yields in Serbia and related shifts of Köppen climate zones under the SRES-A1B and SRES-A2. Int. J. Clim. 35, 3320−3334.

Mihailović, D.T., Malinović-Milićević, S., Arsenić, I., Drešković, N., Bukosa, B., 2013. Kolmogorov complexity spectrum for use in analysis of UV-B radiation time series. Mod. Phys. Lett. B 27, 1350194.

Mimić, G., Mihailović, D.T., Kapor, D., 2015. Complexity analysis of the air temperature and the precipitation time series in Serbia. Theor. Appl. Climatol. http://dx.doi.org/10.1007/s00704-015-1677-6.

Randall, D.A., Wood, R.A., Bony, S., Colman, R., Fichefet, T., Fyfe, J., Kattsov, V., Pitman, A., Shukla, A., Srinivasan, J., Stouffer, R.J., Sumi, A., Taylor, K.E., 2007. Climate models and their evaluation. In: Solomon, S., Qin, D., Manning, M., Chen, Z., Marquis, M., Averyt, K.B., Tignor, M., Miller, H.L. (Eds.), Climate Change 2007: The Physical Science Basis. Contribution of Working Group I to the Fourth Assessment Report of the Intergovernmental Panel on Climate Change. Cambridge University Press, Cambridge, United Kingdom and New York, NY, USA.

Shuangcheng, L., Qiaofu, Z., Shaohong, W., Erfu, D., 2006. Measurement of climate complexity using sample entropy. Int. J. Climatol. 26, 2131−2139.

Tošić, I., 2004. Spatial and temporal variability of winter and summer precipitation over Serbia and Montenegro. Theor. Appl. Climatol. 77, 47−56.

Unkašević, M., Tošić, I., 2011. A statistical analysis of the daily precipitation over Serbia: trends and indices. Theor. Appl. Climatol. 106, 69−78.

Phenomenon of chaos in computing the environmental interface variables

VI

Interrelations between mathematics and environmental sciences

19.1 THE ROLE OF MATHEMATICS IN ENVIRONMENTAL SCIENCES

In this chapter we will consider several issues related to the interrelations between mathematics and environmental sciences. They mainly emerge as a result of dealing with different mathematical formalisms in representing various environmental phenomena. It turned out that the mathematical as well as the physical pathways have converged to one point discovering many inconsistencies in the use of mathematical formalisms and physical (also biological, chemical, etc.) approaches in modelling the environmental interfaces. In both cases, problems have emerged either on an epistemological level or due to the lack of careful treatment of nonlinearity and complexity of the real world. Even in the case when the physical picture was quite clear, the mathematical formalization used has not always pursued that picture. In summary, we can say that during our work, two issues have always been present explicitly or latently: (1) nonlinearities in the real world and (2) relationship between mathematics and environmental sciences. Regarding the first issue, our experience can be summarized as that "the nonlinearity can uncover a lot but it does not allow much to be discovered" (D.M.). The second issue seen through physicist–mathematician's optics has been interestingly elaborated in paper by Wigner (1960), in which he was talking about "unreasonable effectiveness of mathematics in the natural sciences," deeply touching the interrelation between mathematics and environmental sciences but mainly physics. In this section, we will shortly make several comments about this interrelationship as well as about our personal experience.

Probably the right person who was able to see the both sides of this interrelationship is Eugene Wigner a Hungarian American theoretical physicist and mathematician. He shared the Nobel Prize in Physics in 1963 "for his contributions to the theory of the atomic nucleus and the elementary particles, particularly through the discovery and application of fundamental symmetry principles" (Nobel Media AB, 2014). Wigner and Hermann Weyl introduced the group theory into physics, particularly the theory of symmetry in physics. He performed ground-breaking work in pure mathematics, in which he was the author of a number of mathematical theorems. In particular, Wigner's theorem is a cornerstone in the mathematical formulation of quantum mechanics. He also left a trail in the research of the structure of the atomic nucleus. In the above-mentioned paper he reviewed his huge scientific experience about the role of mathematics in physical theories. We briefly elaborate his reasoning.

The mathematics is used in physics for evaluating the results of the laws of nature. In order to make it possible, the laws of nature must previously be formulated in mathematical language. However, this is not the most important role of mathematics in physics since it is just evaluating the consequences of already established theories. In this situation mathematics (rather, applied mathematics as noted by E.W.), is merely a serving tool for physics. Certainly, mathematics also plays a more paramount role in physics. This fact is already implied in the statement (attributed to Galileo Galilei) that the laws of nature must have been formulated in the mathematical language which means that each of them is an object for the use of applied mathematics. The importance which mathematical concepts possess in the formulation of the laws of physics can be illustratively seen, for example, through the axioms of quantum mechanics as formulated, explicitly, by Paul Dirac. In quantum mechanics there are two basic concepts: states and observables. The states are vectors in Hilbert space, the observables self-adjoint operators acting on these vectors. The possible values of the observations are the characteristic values of the operators. Now, the question is how physics chooses certain mathematical concepts for the formulation of the laws of nature, which are certainly only a fraction of all those concepts used in physics. One way we can follow to reach an answer is that the concepts which were chosen were not selected arbitrarily from a list of mathematical terms. Namely, in many if not most cases, those terms were developed independently by the physicists (E.W.) and then they have been conceived by the mathematicians. According to Wigner (1960), "The concepts of mathematics are not chosen for their conceptual simplicity even sequences of pairs of numbers are far from being the simplest concepts but for their amenability to clever manipulations and to striking, brilliant arguments. The Hilbert space of quantum mechanics is the complex Hilbert space, with a Hermitean scalar product. Surely, complex numbers are far from natural or simple and they cannot be suggested by physical observations. The use of complex numbers in this case is not a trick of applied mathematics, but comes from a necessity in the formulation of the laws of quantum mechanics." Now, we will make some personal reminiscences.

At the beginning of this century, one of us, D.M. started to cooperate with the second author (I.B.) who has expressed his intention to be his Ph.D. student. Such an idea, of connection between a physicist and a biologist, came from the third author (D.K.), who is also a physicist. The original idea was to initiate the work on a problem that could be subsumed under the most general topic—modelling of living systems. There was a risk of being lost in a field unknown for us since none of us does have the necessary experience, especially mathematical, in such a field. The first author intuitively perceived the fact that for physicists (and other scientists from the environmental sciences), it is hard to get used to the level of mathematical abstraction from which sometimes they should start to solve a problem they deal with. In other words when they are closer to the point when the interrelation between mathematics and environmental sciences is not quite "clear," then this issue becomes more pronounced. Having in mind this anticipation, the first author asked his fellow professor S. Crvenković whether he could present him and his PhD student (the second author), a course in set theory, selected chapters of algebra and Category Theory. This was the first time at the University of Novi Sad that somebody gave such a course, which includes a comprehensive elaboration of the Category Theory. The course was very interesting. The situation was similar to the situation when three persons, mathematician, biologist, and physicist, find themselves in one room where each of them is not entirely sure how he can cooperate with the other two. Occasionally, professor Crvenković was asking: "I do not understand for what you need this abstract mathematics?" The first author always was answering: "Don't worry about that, just teach us as much as possible consistent!" He was a good teacher, in particular, in elaboration of the proofs of Gödel's incompleteness theorems and scientific and educational explanation of the Category Theory. From this course the first author carries two impressions: (1) about the level of abstraction in mathematics, that is much higher than in environmental sciences and (2) the power of the Category Theory as the most general theory of modelling.

After a short view as well as personal insights about the interrelations between mathematics and physics we are going to deal with another aspect of that relationship. This aspect appears when mathematical concept has already been established covering the environmental phenomenon and also including its evolution in time. The part of that aspect has already been touched upon in Chapter 3, where we stated our opinion about dilemma whether environmental interface systems' models should be built in the form of differential or difference equations, i.e., whether we should deal either with the continuous-time or discrete-time, where time is considered as a continuous or discrete variable, respectively. This choice is a dilemma that occurs in interrelations between classical continuum mathematics and reality. Here, the aforementioned aspect, seen from our point of view, we complement with two more elements: complexity and chaos.

Traditional mathematical analysis of physical systems tacitly assumes that integers and all real numbers, no matter how large or how small, are physically possible and all mathematically possible trajectories are physically possible (Kreinovich and Kunin, 2003). Traditionally, this approach has worked well in physics, engineering, and environmental sciences, but it does not lead to a very good understanding of chaotic systems, which, as is now known, are extremely important in the study of real-world phenomena ranging from weather to biological systems. In this chapter, we deal with some issues in modelling pathways in environmental sciences in their broadest context (in particular, in autonomous dynamical systems, which are common subject under consideration in modern environmental sciences (Mihailović and Mimic, 2012). They are (1) already mentioned issue of how to replace given differential equations by appropriate difference equations in the modelling of phenomena in the environmental world; (2) whether a mathematically correct solution to the corresponding differential equation or system of equations are always physically possible (Kreinovich and Kunin, 2003); and (3) phenomenon of chaos in autonomous dynamical systems, in particular in computing the environmental variables from the corresponding equations that describe processes at the environmental interfaces (Mihailović and Mimic, 2012).

19.2 DIFFERENCE EQUATIONS AND OCCURRENCE OF CHAOS IN MODELLING OF PHENOMENA IN THE ENVIRONMENTAL WORLD

In Chapter 3, we shortly elaborated two issues: (1) the choice of the model and (2) continuous-time versus discrete-time in building the model. In this chapter we will return to these issues, setting them into the context of the interrelations between mathematics and environmental sciences through some examples and our experience.

How to replace given differential equations by appropriate difference equations in the modelling of phenomena in the environmental world? Dilemma whether we should use either continuous-time or discrete-time in building the environmental model includes a lot of questions. This dilemma is neither simple nor it could be resolved unambiguously. Maybe the text that follows, which is derived from the paper by Zeilberger (2001) and different scientific internet forums (http://mathoverflow.net), picturesquely describes the fundamentality of this dilemma.

Although small discrete systems are unproblematic to work with, continuum models are easier to deal with than large discrete systems. Whether or not nature is fundamentally discrete, the most useful models are often continuous because the discreteness can only occur in very small scales. Discreteness is useful to include in the model if it occurs in the situation we are interested in. Thus, this is to a large extent a question of scales of interest in the procedure of modelling (Ilmavirta, 2014). In many cases (maybe most), approximate solutions are actually not simpler than exact solutions. For example, when we want to find the planar curve of a given length which encloses the largest area (the isoperimetric problem), then the problem can be reduced to solving a system of ordinary differential equations. If we want to do that analytically we will get a circle; if we choose to do it discretely we will get a sequence of curves which give better and better approximations of a circle. Now the question arises: How is the latter simpler? This is a serious issue in physics: continuous models often have lots of symmetry that you lose when you discretize them (Siegel, 2014). Also, physicists permanently use lattice approximations. Lattice models will typically break part of the symmetry of the system, which is a disadvantage both from a theoretical point of view and from a practical point of view. For instance, it is not possible to make a finite lattice model rotationally invariant although most laws of physics are rotationally invariant (Henriques, 2014). The presence or absence of symmetries in a physical system is model significant since each symmetry or its absence has physical meaning. Hence, symmetries that either exist or if they are absent just because of "metaphysical preferences about notational syntax" (Jorg, 2014) are not desirable wanted. Otherwise, this would correspond to a different physical system without a physical reason for the difference. Finally, one interesting fact about discreteness in environmental modelling noted in a scientific forum: "Discrete models are probably useful if nature has genuinely discrete structure and we are interested in phenomena at the scale where discreteness is visible. But on larger scales a discrete model would contain something that we cannot measure and might not even be interested in. Something that cannot be measured and does not have a significant impact on the behavior of the system should be left out of the model. This is related to the observation that continuum models often work well for large discrete systems" (Ilmavirta, 2014). Note, that there are a number of standard ways to replace an ordinary differential equation with a difference equation in environmental modelling. The corresponding techniques for solving them as well as partial differential equations are extremely challenging and are the basis for a lot of current research in applied mathematics. At this point, we shall leave the subject because our intention is not to delve deeply into the above issues. Our intention is just to explicate our experience in research work partly touching these issues.

Let us now consider an approach via difference equation, relying on the work by der Vaart (1973), using as an example disintegration of the nuclei of the radioactive atoms. The probability that any given nucleus disintegrates during a finite time interval (possibly short) of length Δt is $\lambda \Delta t$, where λ is the decay constant. It expresses the proportionality between the size of the population of nuclei and the rate at which the population decreases due to radioactive decay. It is expected that among large number of nuclei, $N(t)$ the number of disintegrated ones during the time interval $(t, t + \Delta t)$ can be expressed as $N(t)\lambda \Delta t$. Thus, we have the equation

$$N(t + \Delta t) - N(t) = -N(t)\lambda \Delta t, \tag{19.1}$$

which can be written in the form

$$\frac{N(t + \Delta t) - N(t)}{\Delta t} = -N(t)\lambda, \tag{19.2a}$$

whose right-hand side does not depend on Δt. Note, that this is in the spirit of Feynman's reasoning (der Vaart, 1973) that (Eq. (19.2a)) may be *replaced* by

$$\frac{dN(t)}{d(t)} = \lim_{\Delta t \to 0} \frac{N(t + \Delta t) - N(t)}{\Delta t} = -N(t)\lambda, \tag{19.2b}$$

thus we get a genuine differential equation with the well-known solution having the form

$$N(t) = N_0 e^{-\lambda t}, \tag{19.3}$$

where N_0 is the number of nuclei at time zero. This is the solution obtained by Joos (1939) who derived the differential equation for radioactive decay. Having in mind two assumptions: (1) the probability that any given nucleus disintegrates during the time element Δt is $\lambda \Delta t$ (this fact he just elucidated by the fact that experience shows λ to be constant) and (2) among a large number, N, of nuclei it will be then found, according to statistics, that during dt the number of $N\lambda dt$ will be practically disintegrated (der Vaart, 1973). However, in derivation of the differential equation for radioactive decay, Joos (1939) treated differentials as infinitesimals that are *not too small*. Note, that in derivation Eq. (19.2b) we had in mind Feynman's motivation, i.e., that for introduction of the mathematical concept of derivative as a model for the physical concept of speed.

The question now arises whether a difference equation can be found which does have the 'same' solutions as the differential equation (19.3). In that sense we have a look at the function defined by Eq. (19.3), which solves differential Eq. (19.2b) which originally comes from difference Eq. (19.2a). First we consider whether $N(t) = N_0 e^{-\lambda t}$, constitutes a solution to difference Eq. (19.2a). After substitution of this solution in this difference equation, it becomes obvious that the answer is negative, since $-\lambda \Delta t \neq e^{-\lambda \Delta t} - 1$. We can conclude only that this solution is an approximate solution of the difference Eq. (19.2a). Let us solve difference equation

$$N(t + \Delta t) = N(t)(1 - \lambda \Delta t) \quad \text{for } t = 0, \Delta t, 2\Delta t, \dots; \tag{19.4}$$

that can be also written in the form

$$N(\tau \Delta t) = N_0(1 - \lambda \Delta t)^\tau \quad \text{for } \tau = 0, 1, 2, \dots \tag{19.5}$$

Solution of Eq. (19.5) obviously depends on the $\lambda \Delta t$. Thus, if $0 < \lambda \Delta t < 1$, then the qualitative behavior of that solution is physically correct since $N(\tau \Delta t)$ is a decreasing function of τ, with $\lim_{\tau \to \infty}$ equal to 0 (black line in Fig. 19.1). If $1 < \lambda \Delta t < 2$, then $N(\tau \Delta t)$ is positive for even τ, negative for odd τ, but the curve still has $\lim_{\tau \to \infty}$ equal to 0 (red line in Fig. 19.1). Finally, if $\lambda \Delta t < 2$, then $N(\tau \Delta t)$ shows oscillations between positive and negative values (blue line in Fig. 19.1), while amplitude increasing when $\tau \to \infty$. From this example we see that depending on the size of

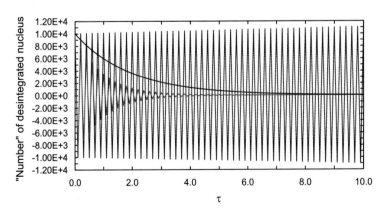

FIGURE 19.1

Solution of difference Eq. (19.4) as a function of dimensionless time (τ) for different values of $\lambda \Delta t$: (0.5)—*black line*; (1.9)—*red line* and (2.001)—*blue line*.

the time step Δt relative to the parameter λ, the solutions of the difference Eq. (19.2a) and differential Eq. (19.2b) may be qualitatively different. Let us suppose hypothetically that in this example the value of the parameter λ is unknown. Then the exact solutions of Eq. (19.2a) which are not different from the solutions of Eq. (19.2b) are those solutions having no negative values. This is a real assumption since the negative values of the solutions are physically impossible in the context in which Eqs. (19.2a) and (19.2b) are derived, where $N(t)$ is the number of nondisintegrated nuclei. However, in the situation when the value of the parameter λ is unknown (often for either original research or development work), we usually have a huge problem since then there is no guidance for the choice of the time step Δt.

Now, let us set ourselves the next task. Whether we can find a difference equation which has the "same" solutions as the differential Eq. (19.3). It seems that this task does not look difficult. Namely, if we use analytic expression for these solutions we get

$$N(t) = N_0 e^{-\lambda t}; \quad N(t + \Delta t) = N_0 e^{-\lambda(t+\Delta t)}; \quad N(t + \Delta t) = N(t)e^{-\lambda\Delta t}. \tag{19.6}$$

The third equation in (19.6) is the equation we are looking for. Certainly, its precise form could not be found without integrating the differential Eq. (19.2b). There is no problem that we convince ourselves that the difference Eqs. (19.2a) and (19.6) lead to the same differential equation, if one determines $N(t)$ from them according to the first equality in (19.2b), i.e., when $\Delta t \rightarrow 0$. Both of them (19.2a) and (19.6) represent the same physical property, i.e.,

$$\begin{aligned} N(t + \Delta t) - N(t) &= -\lambda N(t) \\ N(t + \Delta t) &= \left(e^{-\lambda\Delta t} - 1\right)N(t). \end{aligned} \tag{19.7}$$

Both the equations indicate that a certain number of nuclei present at Δt disintegrate in the interval t and $t + \Delta t$ and that the number is independent of the number of nuclei present at t but depending on Δt. Note that number of nuclei is indeed proportional to Δt. In the above reasoning with respect to the phenomenon of radioactive decay we know, although with a hindsight, that the function described by (19.3) is a good description of the observations. In contrast to that if the scientists want to study this phenomenon computationally then they might have chosen the conceivable difference Eq. (19.2a) that mathematically represents their assumptions. Following this approach they can expect to meet a "wrong" step length ($\Delta t > \lambda^{-1}$), in which case they would have found physically unrealistic curves corresponding to $\Delta t > \lambda^{-1}$. This is not the reason to conclude that their assumptions are wrong. In fact, unrealistic curves are artifacts of the computational simulation, so the best way to solve this equation is to analytically integrate Eq. (19.2a), in particular, when the phenomena under consideration are such that is more realistic to maintain the step length Δt away from zero (der Vaart, 1973).

Phenomenon of chaos in autonomous dynamical systems in environmental modelling. Since the issue (1) i.e., how to replace given differential equations by appropriate difference equations, has been already considered in Section 19.1, we will just add one comment arising from our environmental modelling experience. It seems more natural *to build the model* as a discrete difference equation from the start, without going through the painful, doubly approximate process of first, during the modelling stage, finding a differential equation to approximate a basically discrete situation, and then, for numerical computing purposes, approximating that differential equation by a difference scheme (der Vaart, 1973; Mihailović and Mimic, 2012). The issue (1) closely touches the issue (2). Namely, for some phenomena, described by equation(s), we already know the corresponding laws that can be deduced from symmetry conditions. However, in many other cases, we must determine equations from the general theoretical ideas and experimental data. Therefore, we can ask ourselves whether we can guarantee that these are the right equations. Even though we "know" equation(s), but we still are not sure about the values of parameters of these equations, since there are many generalizations we derived or designed for a very wide interval (Kreinovich and Kunin, 2003). Kolmogorov (1968) was the first who started in 1960s to analyze this issue. He pointed out two main reasons why a mathematically correct solution to the corresponding system of differential or difference equation cannot be physically possible: (1) there is difference in understanding the term "random" in mathematics and physics (also in environmental sciences) and (2) solutions of the corresponding systems of differential equations which lead to some numbers may be mathematically correct, but they are physically meaningless (Kolmogorov, 1968; Kreinovich, 2003). Finally, irregularities and chaotic fluctuations in solution of difference equation, describing autonomous dynamical systems (issue (3)), can come from two main reasons: (1) numerical, i.e., because we try to choose appropriate difference equation whose solution is "good" approximation to the solution of the given partial differential equation and (2) physical, i.e., occurrence of chaotic fluctuations in the considered system because, for example, the system cannot oppose an enormous amount of radiation, suddenly entering the system (Mihailović and Mimić, 2012).

The theory of dynamical systems is a well-developed and successful mathematical framework to describe time-varying phenomena. Its wide area of applications includes complex systems in different spatial and time scales. In particular, this broad scope of applications has provided a significant impact on the theory of dynamical systems itself and is one of the main reasons for its popularity over the last decades (Kloeden and Pötzsche, 2011). As a general principle, before abstract mathematical tools can be applied to real-world phenomena from the above areas, one needs corresponding models. At the conceptional level, in developing such models, one distinguishes an actual dynamical system from its surrounding environment. The system is given in terms of physical or internal feedback laws yielding an evolutionary equation. The parameters in this equation describe the current state of the environment that may change with time in autonomous dynamical systems. Hence we are led to autonomous difference equation of the form $x_{n+1} = f(x_n) n \in \mathbf{Z}^+$.

A difference equation of the form

$$x_{n+1} = f(x_n),$$ (19.8)

where $f : \mathbf{R}^d \to \mathbf{R}^d$, is called a first-order *autonomous difference* equation on the state space $\mathbf{R}^d$. There is no loss of generality in the restriction to first-order difference Eq. (19.8), since higher-order difference equations can be reformulated as Eq. (19.8) by the use of an appropriate higher dimensional state space. Successive iteration of an autonomous difference Eq. (19.8) generates the forward solution mapping $\pi : \mathbf{Z}^+ \times \mathbf{R}^d \to \mathbf{R}^d$ defined by $x_n = \pi(n, x_0) = f^n(x_0) = \underbrace{f \circ f \circ \cdots \circ f(x_0)}_{n \ times},$

which satisfies the *initial condition* $\pi(x_0) = x_0$ and the *semigroup property*

$$x_n = \pi(n, \pi(m, x_0)) = f^n(\pi(m, x_0)) = f^n \circ f^m = f^{n+m}(x_0) = \pi(m + n, x_0),$$ (19.9)

for all $m, n \in \mathbf{Z}^+, x_0 \in \mathbf{R}^d$. Here, and later, $\mathbf{Z}^+$ denotes the nonnegative integers.

For example, the difference Eq. (19.8) may represent a variable time-step discretization method for a differential equation $\dot{x} = f(x)$ that can be solved numerically by stepping either forward

$$x_{n+1} = x_n + \Delta\tau f(x_n)$$ (19.10)

or backward in time from the known initial condition

$$x_{n+1} = x_n + \frac{\Delta\tau}{1 - \Delta\tau\frac{\partial f}{\partial x}(x_n)} f(x_n),$$ (19.11)

where $\Delta\tau$ is the time step while n denotes the time iteration.

An example of occurrence of the chaos when solving either Eqs. (19.10) or (19.11) that represent time changes of the dimensionless environmental interface temperature response to the radiative forcing (Mihailović and Mimic, 2012) will be analyzed in more detail in Chapters 20 and 21. This equation, in which stepping forward in time is used, has a general form

$$\zeta_{n+1} = A_n\zeta_n - B_n\zeta_n^2$$ (19.12)

where A_n and B_n are the coefficients, while ζ_n is the dimensionless environmental interface temperature (Mihailović and Mimic, 2012). Irregularities in solution of difference Eq. (19.12) can come from two reasons: numerical and physical, as explained a few paragraphs above.

Fig. 19.2 shows the bifurcation diagram of Eq. (19.12). It is plotted with increments of 0.01 and 0.05 for A and B (taking the values 0.2, 0.4, 0.6, 0.8, and 1), respectively. From this figure are seen chaotic regions indicating chaotic fluctuations of ζ. However, inside the chaotic interval there exist open periodical "windows." It means that the dynamical system considered, i.e., temperature fluctuations on the environmental interface, is synchronized in some regions where the chaotic regime prevails.

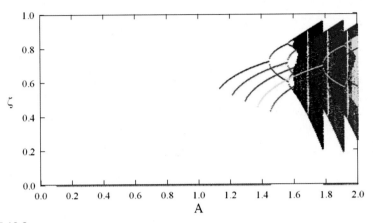

FIGURE 19.2

Bifurcation diagram of Eq. (19.12) as a function of coefficient $A \in (0,2)$ (the increment was 0.005) and five values of coefficient B (0.2, 0.4, 0.6, 0.8, 1 indicated by different colors going from the left to the right).

Reprinted with permission from Mihailović, D.T., Mimic, G., 2012. Kolmogorov complexity and chaotic phenomenon in computing the environmental interface temperature. Mod. Phys. Lett. B 26 (27), 1250175.

REFERENCES

der Vaart, H.R., 1973. A comparative investigation of certain difference equations and related differential equations: implications for model building. Bull. Math. Biol. 35, 195–211.

Henriques, A. (http://mathoverflow.net/users/5690/andr%c3%a9-henriques), 2014. Why Have Mathematicians Used Differential Equations to Model Nature Instead of Difference Equations, http://mathoverflow.net/q/182155.

Ilmavirta, J. (http://mathoverflow.net/users/55893/joonas-ilmavirta), 2014. Why Have Mathematicians Used Differential Equations to Model Nature Instead of Difference Equations, http://mathoverflow.net/q/182082.

Jorg, G. (http://mathoverflow.net/users/58777/guido-jorg), 2014. Why Have Mathematicians Used Differential Equations to Model Nature Instead of Difference Equations, http://mathoverflow.net/q/182155.

Joos, G., 1939. Lehrbuch der theoretischen Physik. Leipzig Akademische Verlagsgesellschaft. 3. Auflage.

Kloeden, P., Pötzsche, C., 2011. Nonautonomous Dynamical Systems in the Life Sciences. Springer Cham, Heidelberg.

Kolmogorov, A.N., 1968. Automata and life. In: Cybernetics Expected and Cybernetics Unexpected (Nauka, Moscow, 1968), p. 24 (in Russian).

Kreinovich, V., Kunin, I.A., 2003. Kolmogorov complexity and chaotic phenomena. Int. J. Eng. Sci. 41, 483–493.

Mihailović, D.T., 2012. Preface. In: Mihailović, D.T. (Ed.), Essays on Fundamental and Applied Environmental Topics. Nova Science Publishers, New York.

Mihailović, D.T., Mimic, G., 2012. Kolmogorov complexity and chaotic phenomenon in computing the environmental interface temperature. Mod. Phys. Lett. B 26 (27), 1250175.

Nobel Media AB, 2014. The Nobel Prize in Physics 1963. Nobelprize Org. http://www.nobelprize.org/nobel_prizes/physics/laureates/1963/.

Siegel, P. (http://mathoverflow.net/users/4362/paul-siegel), 2014. Why Have Mathematicians Used Differential Equations to Model Nature Instead of Difference Equations, http://mathoverflow.net/q/182138.

Wigner, E., 1960. Unreasonable effectiveness of mathematics in the natural sciences. Commun. Pure Appl. Math. 13, 1–4.

Zeilberger, D., 2001. "Real" analysis is a degenerate case of discrete analysis new progress in difference equations. In: Aulbach, B., Elaydi, S., Ladas, G. (Eds.), Proc. ICDEA 2001. Taylor and Frances, London.

Chaos in modelling the global climate system

20

20.1 CLIMATE PREDICTABILITY AND CLIMATE MODELS

A short survey on the predictability. Among the most interesting and fascinating phenomena that are predictable is the chaotic ocean/atmosphere/land system called weather and its longtime average, climate. While weather is not predictable beyond a few days, aspects of the climate may be predictable for years, decades, and perhaps longer (Keller, 1999). These two statements clearly summarize the current opinion and state in climate modelling community that deals with one of the aspects of the aforementioned subjects. However, the question of the weather and climate modelling and predictability has been initiated in the early sixties of the 20th century, when it was elaborated in the pioneering works by Lorenz (1960, 1963a, 1963b, 1964). He was the first person in the scientific world who explicitly pointed out the following points related to the nonlinear dynamics in atmospheric motion: (1) question of prediction and predictability, (2) importance of understanding the nonlinearity in modelling procedure, (3) demand for discovery of chaos, and (4) careful consideration of sensitivity of differential equations in modelling system on initial conditions (Mihailović et al., 2014). Subsequent three decades after appearance of these papers have been characterized by the strong interest for predictability of weather and climate on both theoretical and practical levels. The following topics have been set in the focus: (1) dynamics of error growth; (2) linear and nonlinear systems (normal modes, optimal modes, nonlinear geophysical systems, and scale selection in error growth); (3) predictability of systems with many scales; (4) limit of predictability; (5) weather predictability (growth of errors in General Circulation Models (GCMs) based on Lorenz's analysis); (6) predictability from analogs (targeted observations); (7) climate predictability (predictability of time-mean quantities, predictability of the second kind) and potential predictability; (8) seasonal mean predictability; and (9) El Niño—Southern Oscillation (ENSO) chaos, predictability of coupled models, and decadal modulation of predictability (Krishnamurthy, 1993; Lorenz, 1969, 1975, 1982; Hartmann et al., 1995; Shukla, 1981, 1985; Simmons and Hollingsworth, 2002; Lorenz and Kerry, 1998; Shukla et al., 2000; Zwiers and Kharin, 1998; Jin et al., 1994; Kirtman and Schopf, 1998). Because the focus of this chapter is complexity and predictability in climate modelling, we will not go deeper into any of these issues.

Developments in Environmental Modelling, Volume 29, ISSN 0167-8892, http://dx.doi.org/10.1016/B978-0-444-63918-9.00020-X **265**

Earth's atmosphere has evolved into a complex system in which life and climate are intricately interwoven. The interface between Earth and atmosphere as a "pulsating biophysical organism" is a complex system itself. Note that we use the term complex system in Rosen's sense as it was explicated in Chapter 1. Generally, predictability refers to the degree that a correct forecast of a system's state can be made either qualitatively or quantitatively. For example, while the second law of thermodynamics can tell us about the equilibrium that a system will evolve to and steady states in dissipative systems can sometimes be predicted, there exists no general rule to predict the time evolution of systems far from equilibrium, that is, chaotic systems, if they do not approach some kind of equilibrium. Their predictability usually deteriorates with time.

Lorenz (1984) discussed several issues in the predictability of weather systems. According to him, predictability is defined as the degree of accuracy with which it is possible to predict the state of weather system in the near and also the distant future (predictability in Lorenz's sense). In this paper it is assumed that weather predictions are made on the basis of imperfect knowledge of weather system's present and past states. This rather general statement is comprehensively elaborated by Hunt (1999). He described the fundamental assumptions and current methodologies of the two main kinds of environmental forecast (i.e., weather forecast). The first one is valid for a limited period of time into the future and over a limited space-time "target" and is largely determined by the initial and preceding state of the environment, such as the weather or pollution levels, up to the time when the forecast is issued and by its state at the edges of the region being considered. The second kind provides statistical information over long periods of time and/or over large space-time targets so that they only depend on the statistical averages of the initial and "edge" conditions. Environmental forecasts depend on the various ways in which models are constructed. These range from those based on the "reductionist" methodology (i.e., the combination of separate, scientifically based models for the relevant processes) to those based on statistical methodologies, using mixture of data and scientifically based empirical modelling. For example, limitations of the predictability in the world of atmospheric motions are concisely discussed in paper by James (2002). In this paper the predictability of a forced nonlinear system is numerically considered, proposed by Lorenz, as a compelling heuristic model of the midlatitude global circulation. In summary, as stated by Orell (2003) "Prediction problems have been described by Lorenz as falling into two categories. Problems that depend on the initial condition, such as short-to medium-range weather forecasting, are described as predictions of the first kind, while problems that depend on boundary rather than initial conditions, such as, in many cases, the longer-term climatology, are referred to as predictions of the second kind. Both kinds of prediction will be affected by error in the model equations used to approximate the true system" (Hunt, 1999; Collins, 2002; Orell, 2003).

The above insight of the predictability is underlined in the context of the "environmental predictability" (primarily linked to the climate change issues). We finish with the following question: *can we significantly improve the weather/climate predictions compared to the level they currently reached*? Since models that are in use give more or less different outputs and are based on somewhat different underlying assumptions, it is obvious that they can be improved. We can achieve that using two strategies. One could be based on common practice of analyzing obtained results and introducing local improvements and optimizations where necessary. The second strategy could rely on substantial changes of the very structure of the models. Such approach will rely on theoretical reconsiderations of the basics of modelling approach. As Hunt (1999) emphasized: "We concluded that philosophical studies of how scientific models develop and of the concept of determinism in science are helpful in considering these complex issues." What we should keep in mind is that there exists limitation of the modelling attempts on an epistemological level. To show that, we will use Gödel's incompleteness theorem about Number Theory (Gödel, 1931). Basically it says that no matter how one tries to formalize a particular part of mathematics, syntactic truth in the formalization does not coincide with the set of truths about numbers. In other words Gödel's theorem shows that formalizations are part of mathematics but not all of mathematics. There are many ways to look and "read" Gödel's theorem. One possible way is offered by Rosen (1985). According to him the first thing to bear in mind is that both Number Theory and any formalization of it are systems of entailment. It is the *relation* between them, or more specifically the extent to which these schemes of entailment can be brought into congruence, that is of primary interest. The establishment of such congruencies, through the positing of referents in one of them for elements of the other, is the essence of the *modelling relation*. In a precise sense, this theorem asserts that a formalization in which all entailment is syntactic entailment is too *impoverished in entailment* to be congruent to Number Theory, no matter how we try to establish such congruence. This kind of situation is termed *complexity* by Rosen (1977). Namely, in this light, Gödel's theorem says that Number Theory is more *complex* than any of its formalizations or, equivalently, that formalizations, governed by syntactic inference alone, are *simpler* than Number Theory. To reach Number Theory from its formalizations or, more generally, to reach a complex system from a simpler one requires some kind of limiting processes.

Let us return to the question we were asking ourselves after we had shortly considered climate modelling and predictability beyond the complexity. Our opinion is that there is a significant space for improvement of models and their capabilities to provide good forecasts. It can be done only if the modelling attempts are directed toward the following steps: from structures and states to processes and functions; from self-correcting to self-organizing systems; from hierarchical steering to participation; from conditions of equilibrium to dynamic balances of no equilibrium; from single trajectories to bundles of trajectories; from linear causality to circular causality; from predictability to relative chance; from order and stability to instability, chaos, and dynamics; from certainty and determination to a larger degree of risk, ambiguity, and uncertainty; from reductionism to emergentism; from being to becoming (Arshinov and Fuchs, 2003).

The current issues in modelling the global climate system. The target of global climate models is the Earth's climate system, consisting of the physical and chemical components of the atmosphere, ocean, land surface, and cryosphere. In climate simulations, the objective is to correctly simulate the spatial variation of climate conditions in some average sense. There exists a hierarchy of different climate models, ranging from simple energy balance models to the very complex global circulation models. These models attempt to account for as many processes as possible to simulate the detailed evolution of the atmosphere, ocean, cryosphere, and land system, at a horizontal resolution that is typically of order hundreds of kilometers. Climate model complexity is a result of the nonlinearity of the equations, high dimensionality, and the linking of multiple subsystems. However, the secret of understanding of climate model complexity lies in the nonlinear dynamics of the atmosphere and oceans, which is described by the Navier–Stokes equations whose solution is one of the most vexing problems in all of mathematics.

To better prepare the reader for the following section we will briefly go through the main current issues related to the modelling of the global climate system (Curry, 2011). (1) *Chaos.* Weather can be considered as being in the state of deterministic chaos, owing to its sensitivity to initial conditions. The source of the chaos is nonlinearities in the Navier–Stokes equations. Therefore, a consequence of sensitivity to initial conditions is that beyond a certain time (no more than seven days) the system will no longer be predictable. Climate models are also sensitive to initial conditions. However, in addition, in these models, coupling of a nonlinear, chaotic atmospheric model to a nonlinear, chaotic ocean model gives rise to something much more complex than the deterministic chaos of the weather model. Those coupled models give rise to bifurcation, instability, and chaos. The situation is further complicated because the coupled atmosphere/ocean system cannot be classified by the current theories of nonlinear dynamical systems, where definitions of chaos and attractor cannot be invoked in situations involving transient changes of parameter values (Stainforth et al., 2007a; Curry, 2011; Annan and Connolley, 2005). We will elaborate this issue in the Section 20.3 through examples originating from our modelling experience. (2) *Confidence in Climate Models.* The relevant issue is how well the climate model reproduces reality, that is, whether the model "works" and is fit for its intended purpose. In the absence of model verification or falsification, Stainforth et al. (2007b) describe the challenges of building confidence in predictions using current models and consider the implications for experimental design and the balance of resources in climate modelling research. (3) We are aware that our understanding of, and ability to simulate, the Earth's climate is rather limited. That fact causes the *climate model imperfection* (Stainforth et al., 2007b), which is divided into two types: uncertainty and inadequacy. The term "model uncertainty" means that we cannot reliably choose parameter values (or ensembles of parameter values), which will provide the most informative results. In addition, further complications arise from the choice of parameterization. Finally, model uncertainty is associated with uncertainty in model parameters, subgrid parameterizations, and also initial conditions. It is a well-known problem that numerical models of natural systems cannot be identical to the structure of those systems; that is, they cannot be isomorphic to the real system (Stainforth et al., 2007b). In other words, they are inadequate; that is, before we run any simulation of the future, we know in advance that models are unrealistic representations of many relevant aspects of the real-world system (Stainforth et al., 2007b; Beven, 2002; Smith, 2002). And, finally, atmospheric science has played a leading role in the development and use of computer simulation in scientific endeavors. Climate simulations of future states of weather and climate have important societal applications. Thus, we should have in mind this following statement by Heymann (2010): "Computer simulation in the atmospheric sciences has caused a host of epistemic problems, which scientists acknowledge and with which philosophers and historians are grappling with [sic]. But historically practice overruled the problems of epistemology. Atmospheric scientists found and created their proper audiences, which furnished them with legitimacy and authority. Whatever these scientists do, it does not only tell us something about science, it tells us something about the politics and culture within which they thrive...".

20.2 AN EXAMPLE OF THE REGIONAL CLIMATE MODEL APPLICATION

In this section we will present an example of the application of one the regional climate models (RCMs) with an overview of its outputs. The nested RCM technique consists of using initial conditions, time-dependent lateral meteorological conditions and surface boundary conditions to drive high-resolution RCMs. The driving data is derived from GCMs (or analyses of observations) and can include GHG and aerosol forcing. The basic strategy is, thus, to use the global model to simulate the response of the global circulation to large-scale forcings and the RCM to (1) account for sub-GCM grid scale forcings like complex topographical features and land cover inhomogeneity with the idea that it should be done in a physically based way and (2) enhance the simulation of atmospheric circulations and climatic variables at fine spatial scales. The nested regional modelling technique essentially originated from numerical weather prediction RCMs are now used in a wide range of climate applications, from palaeoclimate to anthropogenic climate change studies. They can provide high-resolution (up to 10−20 km or less) and multidecadal simulations and are capable of describing climate feedback mechanisms acting at the regional scale. A number of widely used limited area modelling systems have been adapted to, or developed for, climate application and recently for coupling atmospheric models with other climate process models, such as hydrology, ocean, sea-ice, chemistry/aerosol a time series and land-biosphere models (IPPC, 2007).

Here, we obtained the time series by dynamic downscaling of climate simulations conducted with the ECHAM5 GCM coupled with the Max Planck Institute Ocean Model (MPI-OM) (Jungclaus et al., 2006). The horizontal resolution of the

GCM was T63 (approximately 140 × 210 km for the mid-latitudes) with 31 vertical levels. The downscaling of the GCM climate simulations was performed with the coupled regional climate model EBU-POM (Djurdjevic and Rajković, 2012). The atmospheric part of the EBU-POM is the Eta/National Centers for Environmental Prediction (NCEP) model, and the oceanic part uses the POM. Eta/NCEP model is a grid point limited-area model. Numerical schemes in the model's dynamic core conserve energy and entropy together with other important basic and derived quantities (Janjić, 1977, 1984, 2001, 2003; Mesinger et al., 1988). The model's physics includes turbulence, convection, and large-scale precipitation (Janjić, 1990, 1994). For convection, the model uses the Betts-Miller-Janjić convective adjustment scheme (Janjić, 1994). The radiation and land-surface processes model used in the Eta model are described in Djurdjevic and Rajković, (2012). The ocean component, POM, is a primitive-equation grid-point model. This model has a free surface and a split time step, with a two-dimensional external mode and a three-dimensional internal mode with a complete thermodynamics scheme (Blumberg and Mellor, 1987). The two models are joined into a single model with an independently developed coupler that enables physically correct and efficient exchange of energy and mass fluxes at the interface between the atmosphere and the ocean (Djurdjevic and Rajković, 2008, 2012). Specifically, for this integration, the center of the atmospheric Eta model was at 41.5°N, 15°E, with ±19.9° boundaries in east–west direction, ±13.0° boundaries in north–south direction (Fig. 20.1), 0.25° horizontal resolution and 32 vertical levels (with the first level at 20 m and the top level at 10 hPa). The ocean model featured 0.2 × 0.2° of horizontal resolution and 21 vertical levels. The POM model was set over the Mediterranean Sea without the Black Sea; for other open seas, the sea surface temperature from the GCM was used as a bottom boundary condition. The coupling frequency was 6 min. To reduce systematic model error (model bias) in the key climate variables, the statistical method of bias correction (Piani et al., 2010) was applied to surface air temperatures and daily precipitation using modeled and observed daily time series of variables over the period 1961–90. The method assumes the construction of the correction functions based on the difference between the cumulative density function (cdf) of two data sets. The same correction functions are then applied for the whole scenario integration period up to 2100.

The projection of future climate at the regional and local scale is important for the development of local, national, and international policies to mitigate and adapt to the threat of climate change. Here, we show an example of the regional climate model projections of air temperature and precipitation, which are then used to establish shifts in the Köppen climate zones by comparing the results of downscaling with the EBU-POM model for the A1B and A2 scenarios over 2001–30 and 2071–2100 and the present climate simulations for the period 1961–90 (Mihailović et al., 2015).

The Köppen climate classification is a widely used climate classification system. Shifts of climate zones based on the Köppen climate classification were considered

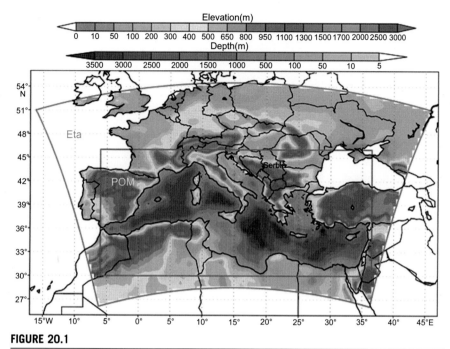

FIGURE 20.1

Model domains. Area bounded by outer line is the Eta Model Domain; the inner rectangle represents the domain boundary of the Princeton Ocean Model (POM).

Reprinted with permission from Kržič, A., Tošić, I., Djurdjević, V., Veljović, K., Rajković, B., 2011. Changes in climate indices for Serbia according to the SRES-A1B and SRES-A2 scenarios. Clim. Res. 49, 73–86.

on the basis of either global or regional climate model simulations (Hanf et al., 2012; Shi et al., 2012; Mahlstein et al., 2013; Mihailović et al., 2015). We derived the climate zones in Serbia, using climate simulations by the EBU-POM model (Fig. 20.2) for the period 1961−90 and according to the Köppen classification (Kottek et al., 2006), are as follows: *Cfwax″*, *Cfwbx″*, *Dfwbx″*, and *ET*. The dominant climate zone is *Cfwbx″*, where *C* = mild temperate/mesothermal climate; *f* = significant precipitation during all seasons; *w* = dry winters (in which the driest winter month average precipitation is less than one 10th the wettest summer month average precipitation); *b* = warmest month averaging below 22°C (but with at least 4 months averaging above 10°C); *a* = warmest month averaging above 22°C; *x″* = the second precipitation maximum occurs in autumn; *D* = continental/micro-thermal climate (a mean temperature above 10°C in the warmest months and a coldest month average below −3°C) and *ET* = polar and alpine climate with average temperatures below 10°C for all 12 months of the year (Kottek et al., 2006). Thus, the climate of Serbia can be described as moderate-continental. However, the climate of Serbia is influenced by the Alps, the Mediterranean Sea, the Genoa

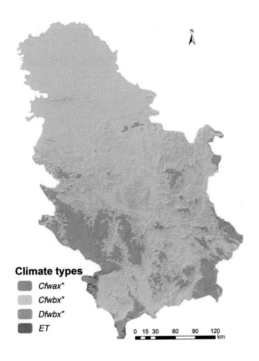

Climate types

- Cfwax"
- Cfwbx"
- Dfwbx"
- ET

0 15 30   60   90   120
km

FIGURE 20.2

Climate zones over Serbia according to the Köppen classification obtained from the simulations by EBU-POM model (Djurdjevic and Rajković, 2012) for the period 1961−90.

Reprinted with permission from Mihailović, D.T., Lalić, B., Drešković, N., Mimić, G., Đurđević, V., Jančić, M., 2015. Climate change effects on crop yields in Serbia and related shifts of Köppen climate zones under the SRES-A1B and SRES-A2. Int. J. Clim. 35, 3320−3334.

Gulf, the Pannonian basin, the Moravian valley, the Carpathian and Rhodope mountains, the hilly mountainous part with ravines and the highland plains, as well as the deep southward penetration of polar air masses, which leads to high spatial variability (Radinović, 1979). The comparison of Fig. 20.3a and 20.2 indicates that there are no changes in the *Dfwbx"* (18.2%) and *ET* (0.3%) climate zones for the period 2001−30 (the A1B scenario).

In contrast, there is an evident expansion of the *Cfwax"* climate type (10.4%), in particular in North Serbia (Vojvodina), i.e., the region with the highest crop production. As indicated in Fig. 20.3b, which depicts changes for the period 2071−2100, the *ET* climate type vanishes, and *Dfwbx"* is reduced on 1.1% of the total territory. The most evident change is seen in the expansion of the *Cfwax"* climate type (81.5%), which replaces the formerly dominant *Cfwbx"* type (17.4%). The future climate of Serbia will be warmer and drier according to the A1B scenario (Mihailović et al., 2015).

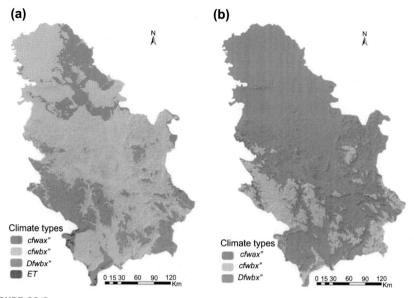

FIGURE 20.3

Köppen climate zones over Serbia for the period (a) 2001–30 and (b) 2071–2100 obtained from the EBU-POM model simulation under the A1B scenario.

Reprinted with permission from Mihailović, D.T., Lalić, B., Dresković, N., Mimić, G., Đurđević, V., Jančić, M., 2015. Climate change effects on crop yields in Serbia and related shifts of Köppen climate zones under the SRES-A1B and SRES-A2. Int. J. Clim. 35, 3320–3334.

As indicated in Fig. 20.4a, for the period 2001–30 (the A2 scenario), there is a decrease of 4.2 percentage points in the *Dfwbx"* climate type (14.0%), whereas the territory covered by *ET* (0.1%) climate type remains practically the same. As in the A1B scenario, there is an evident expansion of the *Cfwax"* climate type, but in this scenario, the expansion covers a wider area (17.6%). This is also more pronounced in North Serbia (Vojvodina). As indicated in Fig. 20.4b, which depicts changes for the period 2071–2100, the *ET* climate type vanishes, and *Dfwbx"* is reduced to 0.4% of the total territory. At this point, the expansion of *Cfwax"* climate type is 85.4%, replacing the dominance of *Cfwbx"* (14.2%). The future climate in Serbia will become warmer and drier, similar to the A1B scenario, but these changes are more pronounced in the A2 scenario (Mihailović et al., 2015).

It is interesting to see which is the level of impact of climate change on the thermal and moisture regimes of soils using the climate projections for soil temperature and moisture. Here, we will analyze the thermal and moisture regimes of Serbian RSGs (Reference Soil Groups) using regional climate simulation data based on the A1B scenario (Mihailović et al., 2016).

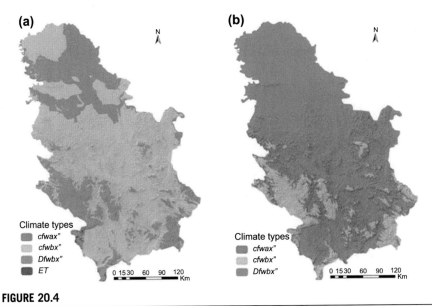

FIGURE 20.4

Köppen climate zones over Serbia for the period (a) 2001–30 and (b) 2071–2100 obtained from the EBU-POM model simulation under the A2 scenario.

Reprinted with permission from Mihailović, D.T., Lalić, B., Dreskovic, N., Mimić, G., Đurđević, V., Jančić, M., 2015. Climate change effects on crop yields in Serbia and related shifts of Köppen climate zones under the SRES-A1B and SRES-A2. Int. J. Clim. 35, 3320–3334.

The areas (Fig. 20.5a) containing the dominant RSGs in Serbia, according to the IUSS Working Group WRB classification (2014), are depicted in Fig. 20.5b, while a detailed description of soil types in Fig. 20.5b can be found in Mihailović et al. (2016).

Here, we consider the spatial distribution of the mean annual soil temperature and moisture. The mean annual soil temperature spatial distributions are due to the combined effects of land cover, RSG, and expected climate changes (Kang et al., 2000). The mean annual soil temperature in the top soil layers (0–40 cm) is given in Fig. 20.6 for different RSGs in Serbia. The mean annual soil temperature variations in this layer vary between approximately 1.9°C in southwestern Serbia to 2.4°C in northwestern and northern Serbia (Fig. 20.6a) (for the 2021–50 period). However, the changes obtained for the 2071–2100 period (Fig. 20.6b) vary between 2.8°C (southwest region) to 3.5°C (northwestern and northern regions). The areas with Chernozems, Eutric Cambisols, and Planosols (1, 5, and 6) display the highest soil temperatures for both periods (Fig. 20.6). As indicated in Mihailović (2016), the parts of the northern, northwestern, southern lowland, and western regions (altitude < 500 m, as illustrated in Fig. 20.5a) consist of nonirrigated arable land, permanently irrigated land, pastures, complex cultivation patterns, land principally

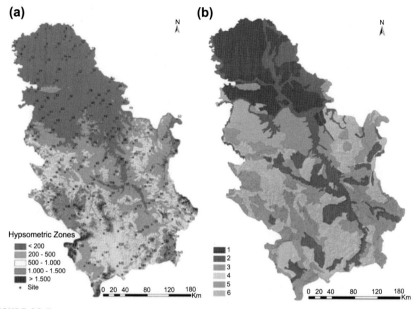

FIGURE 20.5

(a) The 150 sites in Serbia used in the example, with altitudes classified into five intervals, and (b) the dominant soil types in Serbia according to the IUSS Working Group WRB classification (2014). (1) Haplic Chernozems and Gleyic Chernozems (34 sites); (2) Vertisols, Gleysols, and Fluvisols (18 sites); (3) Umbrisols and Rendzic Leptosols (16); (4) Dystric Cambisols (52); (5) Eutric Cambisols (19); and (6) Planosols and Luvisols (11). The numbers in parentheses after the RSGs represent the numbers of sites with regional simulation data for each RSG.

Reprinted with permission from Mihailović, D.T., Dresković, N., Arsenić, I., Ćirić, V., Djurdjević, V., Mimić, G., Pap, I., Balaž, I., 2016 Impact of climate change on the thermal and moisture regimes of RSGs in Serbia: an analysis with data from regional climate simulations under SRES-A1B. Sci. Total Environ. 571, 398-409.

occupied by agriculture, natural vegetation, and sparse forest land cover classes. As previously stated, these soils possess lower water contents and larger thermal diffusivities, resulting in higher mean annual soil temperatures.

The areas with Umbrisols and Dystric Cambisols (3 and 4) are predominantly located beneath natural vegetation (broad-leaved forests, coniferous mixed forests, or pastures) and located in mostly mountainous regions (altitude > 500 m). Vertisols and Umbrisols (2 and 3) are dark soils with high water-retention capacities. They retain water throughout the year and possess smaller thermal diffusivities, resulting in smaller soil temperature variations. Note that the soil temperature changes strongly influence soil respiration, which impacts the CO_2 release soil organic matter decomposition (Conant et al., 2008), microbial processes (Rankinen et al., 2004), soil processes (Paul et al., 2004), agricultural production, biodiversity, and ecosystem functioning.

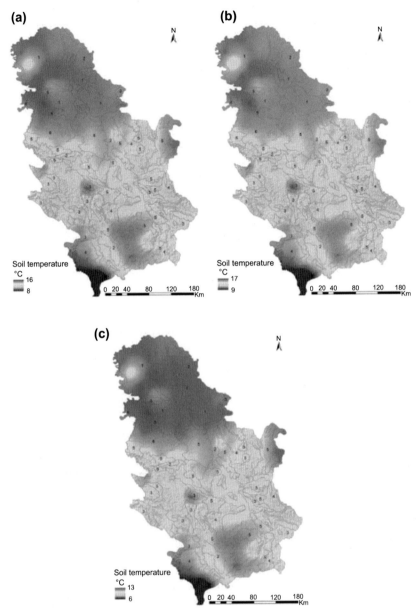

FIGURE 20.6

The mean annual soil temperature in the top soil layer (0—40 cm) of different RSGs in Serbia for 2021—50 (a) and for 2071—2100 (b) periods under the A1B scenario, against the 1961—90 period (c). This and all maps in further text are obtained from data sets of 150 sites using the GIS technique as in (Ninyerola et al., 2000).

Reprinted with permission from Mihailović, D.T., Drešković, N., Arsenić, I., Ćirić, V., Djurdjević, V., Mimić, G., Pap, I., Balaž, I., 2016 Impact of climate change on the thermal and moisture regimes of RSGs in Serbia: an analysis with data from regional climate simulations under SRES-A1B. Sci. Total Environ. 571, 398-409.

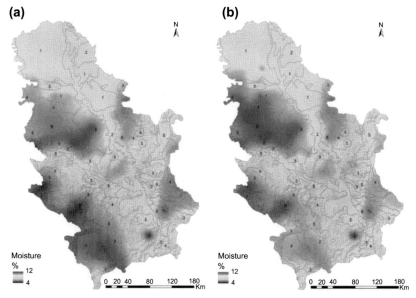

FIGURE 20.7

Relative mean annual soil moisture decrease in the top soil layers (0—40 cm) of different RSGs in Serbia for the 2021—50 (a) and 2071—2100 (b) periods under the A1B scenario, as compared to the 1961—90 period.

Reprinted with permission from Mihailović, D.T., Dresković, N., Arsenić, I., Ćirić, V., Djurdjević, V., Mimić, G., Pap, I., Balaž, I., 2016 Impact of climate change on the thermal and moisture regimes of RSGs in Serbia: an analysis with data from regional climate simulations under SRES-A1B. Sci. Total Environ. 571, 398-409.

The relative mean annual soil moisture variations across Serbia fluctuate between approximately 4.0% in parts of the southwestern, central, and northeastern regions to 9.6% in parts of the western, central, and eastern regions (Fig. 20.7a) (2021—50 period). The relative changes obtained for the 2071—2100 period (Fig. 20.7b) are slightly higher, varying between 5.4% and 11.9%. Fig. 20.7a and b illustrate that areas with Chernozems, Eutric Cambisols, and Planosols (1, 5, and 6) display the highest relative soil moisture changes during both periods (2021—50 and 2071—2100) (Mihailović et al., 2016). This soil water loss is likely due to thermal diffusivity differences based on RSG. Namely, these soils possess larger bulk densities, which result in larger thermal diffusivities (Nofziger, 2000), more intensive heating and substantial soil drying. Lower relative changes are expected in the areas with Vertisols, Umbrisols, and Dystric Cambisols (2, 3, and 4).

20.3 OCCURRENCE OF CHAOS AT ENVIRONMENTAL INTERFACES IN CLIMATE MODELS

The discovery of deterministic chaos attracted the attention of environmental modelers (Mihailović and Mimic, 2012). Many of models are very simple, yet even this simplicity can generate chaotic behavior, which resulted in publication of numerous papers and books, including those about chaos in environmental problems (Popova et al., 1997). Note that it is not instantaneously understandable how much an investigation of chaos says us about the real world. However, there is a situation when we have quite a clear case about the usefulness of considering the chaos in the real world, i.e., when it occurs unexpectedly in parts of the parameter space in some models. Then a full understanding of the dynamics of oscillating systems allows one, if necessary, to stabilize the periodic oscillations, thereby avoiding chaos (Ruelle, 1994). It is well known that the existence of chaotic behavior in sets of ordinary differential equations or difference ones is very sensitive to changes of parameters. Chaos often exists only in a very narrow range of parameters space and depends on the magnitude of their changes in the case of periodically forced systems (Moon, 1987). Therefore, the question whether chaotic behavior could be expected when we model the real environmental interfaces is: Surely, but intentionally fixing the parameters that describe the environmental interface, in order to avoid chaotic behavior is not the best way in designing the model usually because the basic variable (representing some important environmental quantity) changes with time under the influence of some field which is nonlinearly coupled to the variable. As we said, the mathematical formalism corresponding to this situation is either a difference or differential nonlinear equation with the parameters mentioned above actually describing this field. From this point of view the chaos can arise even after the change of some of the parameters. Also, we can consider the situation when the basic quantity is fixed, and the problem lies in the parameter space. Now, the change of one or more of the system parameters in time in a nonlinear manner, in fact causes the chaotic behavior of the parameter. This parameter is related to the original variable by a feedback and we again face the chaotic behavior of the system.

Here, we consider the issue addressed above in the sense whether we can find either domain or domains where physically meaningful solutions exist. We do that by considering the stability of physical solution of Eq. (3.1). Coefficients A_n and B_n in this equation change periodically during a day (see Section 3.1). Therefore, in the further analysis their changes can be given in a form $A_n = A\sin(\pi\varsigma_n)$ and $B_n = B\sin(\pi\varsigma_n)$ (Mihailović and Mimic, 2012). Stability, in mathematics, is a condition in which a slight disturbance in a system does not produce too disrupting an effect on that system. In terms of the solution of a differential equation, a function $f(\zeta)$ is said to be stable if any other solution of the equation that starts out sufficiently close to it when $\zeta = 0$, remains close to it for succeeding values of ζ. If the difference between the solutions approaches zero as ζ increases, the solution is called asymptotically stable. If a solution does not have either of these properties, it is called unstable. Stability of solutions is important in physical problems because if slight deviations from the mathematical model caused by unavoidable errors in measurement do not have a correspondingly slight effect on the solution, the mathematical equations describing the problem will not accurately predict the future outcome. We consider the stability of physical solution of Eq. (3.1) in sense of the Lyapunov exponent. Generally, the stability of a time-dependent system $\partial\zeta/\partial t = f(t)$ at the equilibrium ζ_0 with $\partial\zeta_0/\partial t = 0 = f(t_0)$ can be evaluated by assuming a small perturbation and linearizing $\partial(\zeta_0 + \zeta)/\partial t = f(\zeta_0) + (\partial f/\partial\zeta)_{x_0}\zeta'$ or $\partial\zeta'/\partial t = \lambda\zeta'$, where $\lambda = (\partial f/\partial\zeta)(\zeta_0)$ (Lyapunov exponent) with exponential solution $\zeta' = \zeta'(t=0)e^{\lambda t}$. The system is stable if $\lambda < 0$, and unstable if $\lambda > 0$. Eq. (3.1) is typically autonomous. However, since the ranges of coefficients A_n and B_n are known, we will analyze the behavior of the Lyapunov exponent for the corresponding autonomous equation which uniformly changes in intervals of A_n and B_n.

We calculate the Lyapunov exponent for the trajectory starting at ζ_0

$$\lambda = \lim_{n\to\infty} \frac{1}{n}\sum_{i=0}^{n-1} \ln|f'(\zeta_i)|. \tag{20.1}$$

The sign of λ is characteristic of the attractor type. For stable fixed point (steady state) and (limit) cycles, λ is negative; for chaotic attractors, λ is positive. For the map given by Eq. (3.1) we have, by definition

$$\lambda = \lim_{n\to\infty} \frac{1}{n}\sum_{i=0}^{n-1} \ln\left|(A\zeta_i - B\zeta_i^2)\pi\cos(\pi\zeta_i) + (A - 2B\zeta_i)\sin(\pi\zeta_i)\right|. \tag{20.2}$$

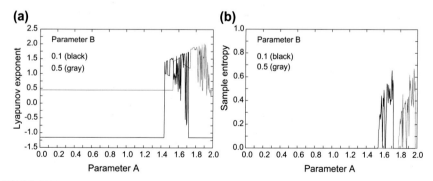

FIGURE 20.8

Lyapunov exponent (a) and sample entropy (b) for Eq. (3.1) given as a function of different values of coefficient $A \in (0,2)$ and of two values of coefficient B. The increment for A was 0.005.

Reprinted with permission from Mihailović, D.T., Mimic, G., 2012 Kolmogorov complexity and chaotic phenomenon in computing the environmental interface temperature. Mod. Phys. Lett. B 26, 1250175.

We also computed the sample entropy (SampEn) (see Section 9.1). The calculated Lyapunov exponent of Eq. (3.1) (SampEn), as a function of the coefficients A and B are depicted in Fig. 20.8. Each point in the above graphs was obtained by iterating 2000 times from the initial condition to eliminate transient behavior and then averaging over another 500 iterations. Initial condition: $\zeta = 0.25$. Looking at Fig. 20.8a we can see that for $B = 0.01$ the solution of Eq. (3.1) is stable ($\lambda < 0$) for intervals $A \in [0,1.44]$ and $A \in [1.72,2.0]$. Between them, the solution is unstable with chaotic fluctuations ($\lambda > 0$) but with sporadic windows of stability ($\lambda > 0$) occurring in irregular intervals. However, for $B = 0.5$ solution of Eq. (3.1) is always unstable $\lambda > 0$. Fig. 3.1 shows regions of stable and unstable solutions of Eq. (3.1) determined by the values of the Lyapunov exponent as a function the coefficients $A \in (0,2)$ and $B \in (0,0.5)$. The increment for changing values of A and B was 0.005. Calculated values of (SampEn) are shown in Fig. 20.8b ($r < 0.05$ and $m = 5$). From this figure it is seen that (SampEn) follows the shape of the Lyapunov exponent, i.e., it is equal to zero for $\lambda < 0$ or when the values of λ are close to zero. Otherwise, it has a value up to 0.6.

REFERENCES

Annan, J., Connolley, W., 2005. Chaos and Climate. http://www.realclimate.org/index.php/archives/2005/11/chaos-and-climate/.

Arshinov, V., Fuchs, C., 2003. Preface. In: Arshinov, V., Fuchs, C. (Eds.), Causality, Emergence, Self-organisation. NIAPriroda, Moscow, Russia, pp. 1−18.

Beven, K., 2002. Towards a coherent philosophy for modelling the environment. Proc. R. Soc. A 458, 2465−2484.

Blumberg, A., Mellor, G., 1987. Description of a three dimensional coastal ocean circulation model. In: Heaps, N. (Ed.), Three-Dimensional Coastal Ocean Models, vol. 4. American Geophysical Union, Washington, DC.

Collins, M., 2002. Climate predictability on interannual to decadal time scales: the initial value problem. Clim. Dyn. 19, 671−692.

Conant, R.T., Steinweg, J.M., Haddix, M.L., Paul, E.A., Plante, A.F., Six, J., 2008. Experimental warming shows that decomposition temperature sensitivity increases with soil organic matter recalcitrance. Ecology 89, 2384−2391.

Curry, J., 2011. Reasoning about climate uncertainty. Clim. Change 108, 723−732.

Djurdjević, V., Rajković, B., 2008. Verification of a coupled atmosphere−ocean model using satellite observations over the Adriatic Sea. Ann. Geophys. 26, 1935−1954.

Djurdjevic, V., Rajković, B., 2012. Development of the EBU-POM coupled regional climate model and results from climate change experiments. In: Mihailovic, D.T., Lalic, B. (Eds.), Advances in Environmental Modeling and Measurements. Nova Science Publishers Inc, New York, NY, pp. 23−32.

Gödel, K., 1931. On formally undecidable propositions of Principia Mathematica and related systems I (Translated by Martin Hirzel). Monatsh. Math. 38, 173−198.

Hanf, F., Korper, J., Spangehl, T., Cubasch, U., 2012. Shifts of climate zones in multi-model climate change experiments using the Koppen climate classification. Meteorol. Z 21, 111−123.

Hartmann, D.L., Buizza, R., Palmer, T.N., 1995. Singular vectors: the effect of spatial scale on linear growth of disturbances. J. Atmos. Sci. 52, 3885–3894.

Heymann, M., 2010. The evolution of climate ideas and knowledge. Wiley Interdiscip. Rev. Clim. Change 1, 581–597.

Hunt, J., 1999. Environmental forecasting and turbulence modeling. Phys. D 133, 270–295.

IPPC, 2007. Regional Climate Models (RCMs). https://www.ipcc.ch/ipccreports/tar/wg1/380.htm.

IUSS Working Group WRB, 2014. World Reference Base for Soil Resources 2014. International Soil Classification System for Naming Soils and Creating Legends for Soil Maps. World Soil Resources Reports No. 106. FAO, Rome.

James, N., 2002. Models of the predictability of a simple nonlinear dynamical system. Atmos. Sci. Lett. 3, 42–51.

Janjić, Z.I., 1977. Pressure gradient force and advection scheme used for forecasting with steep and small scale topography. Contrib. Atmos. Phys. 50, 186–199.

Janjić, Z.I., 1984. Nonlinear advection schemes and energy cascade on semi-staggered grids. Mon. Weather Rev. 112, 1234–1245.

Janjić, Z.I., 1990. The step-mountain coordinate: physical package. Mon. Weather Rev. 118, 1429–1443.

Janjić, Z.I., 1994. The step-mountain eta coordinate model: further developments of the convection, viscous sublayer and turbulence closure schemes. Mon. Weather Rev. 122, 927–945.

Janjić, Z.I., 2001. Nonsingular Implementation of the Mellor-Yamada Level 2.5 Scheme in the NCEP Meso Model. NOAA/NWS/NCEP. Office Note No. 437, 61 pp.

Janjić, Z.I., 2003. A nonhydrostatic model based on a new approach. Meteorol. Atmos. Phys. 82, 271–285.

Jin, F.-F., Neelin, J.D., Ghil, M., 1994. El Niño on the devil's staircase: annual subharmonic steps to chaos. Science 264, 70–72.

Jungclaus, J.H., Botzet, M., Haak, H., Keenlyside, N., Luo, J.J., Latif, M., Marotzke, J., Mikolajewicz, U., Roeckner, E., 2006. Ocean circulation and tropical variability in the coupled model ECHAM5/MPI-OM. J. Clim. 19, 3952–3972.

Kang, S., Kim, S., Oh, S., Lee, D., 2000. Predicting spatial and temporal patterns of soil temperature based on topography, surface cover and air temperature. For. Ecol. Manage. 136, 173–184.

Keller, C.F., 1999. Climate, modeling, and predictability. Phys. D 133, 296–308.

Kirtman, B.P., Schopf, P.S., 1998. Decadal variability in ENSO predictability and prediction. J. Clim. 11, 2804–2822.

Kottek, M., Grieser, J., Beck, C., Rudolf, B., Rubel, F., 2006. World map of the Koppen-Geiger climate classification updated. Meteorol. Z 15, 259–263.

Krishnamurthy, V., 1993. A predictability study of Lorenz's 28-variable model as a dynamical system. J. Atmos. Sci. 50, 2215–2229.

Kržič, A., Tošić, I., Djurdjević, V., Veljović, K., Rajković, B., 2011. Changes in climate indices for Serbia according to the SRES-A1B and SRES-A2 scenarios. Clim. Res. 49, 73–86.

Lorenz, E.N., 1960. The statistical prediction of solutions of dynamic equations. In: Proceedings of the International Symposium on Numerical Weather Prediction, pp. 629–635, Tokyo, Japan.

Lorenz, E.N., 1963a. The predictability of hydrodynamic flows. Trans. N.Y. Acad. Sci. Ser. 2 25, 409–423.

Lorenz, E.N., 1963b. Deterministic nonperiodic flow. J. Atmos. Sci. 20, 130−141.

Lorenz, E.N., 1964. The problem of deducing the climate from the governing equations. Tellus 16, 1−11.

Lorenz, E.N., 1969. The predictability of a flow which contains many scales of motion. Tellus 21, 289−307.

Lorenz, E.N., 1975. Climate predictability. In: The Physical Basis of Climate Modeling. GARP Publication Series, vol. 16. WMO, Geneva, Switzerland, pp. 132−136.

Lorenz, E.N., 1982. Atmospheric predictability experiments with a large numerical model. Tellus 34, 505−513.

Lorenz, E.N., 1984. Some aspects of atmospheric predictability. In: Burridge, D.M., Killn, E. (Eds.), Problems and Prospects in Long and Medium Range Weather Forecasting. Springer, Berlin, Germany, pp. 1−20.

Lorenz, E.N., Kerry, E.A., 1998. Optimal sites for supplementary weather observations: simulation with a small model. J. Atmos. Sci. 55, 399−414.

Mahlstein, I., Daniel, J.S., Solomon, S., 2013. Peace of shifts in climate regions increases with global temperature. Nat. Clim. Change 3, 739−743.

Mesinger, F., Janjić, Z.I., Nicković, S., Gavrilov, D., Deaven, D.G., 1988. The step-mountain coordinate: model description and performance for cases of Alpine lee cyclogenesis and for a case of an Appalachian redevelopment. Mon. Weather Rev. 116, 1493−1518.

Mihailović, D.T., Drešković, N., Arsenić, I., Ćirić, V., Djurdjević, V., Mimić, G., Pap, I., Balaž, I., 2016. Impact of climate change on the thermal and moisture regimes of RSGs in Serbia: an analysis with data from regional climate simulations under SRES-A1B. Sci. Total Environ. 571, 398−409.

Mihailović, D.T., Lalić, B., Drešković, N., Mimić, G., Đurđević, V., Jančić, M., 2015. Climate change effects on crop yields in Serbia and related shifts of Köppen climate zones under the SRES-A1B and SRES-A2. Int. J. Clim. 35, 3320−3334.

Mihailović, D.T., Mimic, G., 2012. Kolmogorov complexity and chaotic phenomenon in computing the environmental interface temperature. Mod. Phys. Lett. B 26, 1250175.

Mihailović, D.T., Mimić, G., Arsenić, I., 2014. Climate predictions: the chaos and complexity in climate models. Adv. Meteorol. 14. http://dx.doi.org/10.1155/2014/878249. Article ID 878249.

Moon, F.C., 1987. Chaotic Vibrations: An Introduction for Applied Scientists and Engineers. Wiley & Sons, New York, 309 pp.

Ninyerola, M., Pons, X., Roure, M.J., 2000. A methodological approach of climatological modeling of air temperature and precipitation through GIS techniques. Int. J. Clim. 20, 1823−1841.

Nofziger, D.L., 2000. Soil Temperature Variations With Time and Depth. Department of Plant and Soil Sciences, Oklahoma State University, Stillwater. Available from: http://soilphysics.okstate.edu/software/SoilTemperature/document.pdf.

Orell, D., 2003. Model error and predictability over different time scales in the Lorenz '96 systems. J. Atmos. Sci. 60, 2219−2228.

Paul, K.I., Polglase, P.J., Smethurst, P.J., O'Connell, A.M., Carlyle, C.J., Khanna, P.K., 2004. Soil temperature under forests: a simple model for predicting soil temperature under a range of forest types. Agr. For. Meteorol. 121, 167−182.

Piani, C., Haerter, J.O., Coppola, E., 2010. Statistical bias correction for daily precipitation in regional climate models over Europe. Theor. Appl. Climatol. 99, 187−192.

Popova, E.E., Fasham, M.J.R., Osipov, A.V., Ryabchenko, V.A., 1997. Chaotic behavior of an ocean ecosystem model under seasonal external forcing. J. Plankton Res. 19, 1495−1515.

Radinović, D., 1979. Weather and Climate of Yugoslavia. Civil Engineering Book, Belgrade, 283 pp.

Rankinen, K., Karvonen, T., Butterfield, D., 2004. A simple model for predicting soil temperature in snow-covered and seasonally frozen soil: model description and testing. Hydrol. Earth Syst. Sci. 8, 706–716.

Rosen, R., 1977. Complexity as a system property. Int. J. Gen. Syst. 3, 227–232.

Rosen, R., 1985. Anticipatory Systems: Philosophical, Mathematical and Methodological Foundations. Pergamon Press, New York, NY, USA.

Ruelle, D., 1994. Where can one hope to profitably apply the ideas of chaos? Phys. Today 47, 24–30.

Shi, Y., Gao, X.J., Wu, J., 2012. Projected changes in Koppen climate types in the 21st century over China. Atmos. Ocean. Sci. Lett. 5, 495–498.

Shukla, J., 1981. Dynamical predictability of monthly means. J. Atmos. Sci. 38, 2547–2572.

Shukla, J., 1985. Predictability. Adv. Geophys. 28, 87–122.

Shukla, J., Paolino, D.A., Straus, D.M., et al., 2000. Dynamical seasonal predictions with the COLA atmospheric model. Q. J. R. Meteorol. Soc. 126, 2265–2291.

Simmons, A.J., Hollingsworth, A., 2002. Some aspects of the improvement in skill of numerical weather prediction. Q. J. R. Meteorol. Soc. 128, 647–678.

Smith, L.A., 2002. What might we learn from climate forecasts? Proc. Natl. Acad. Sci. U.S.A. 99, 2487–2492.

Stainforth, D.A., Allen, M.R., Tredger, E.R., Smith, L.A., 2007b. Confidence, uncertainty and decision-support relevance in climate predictions. Philos. Trans. R. Soc. A 365, 2145–2161.

Stainforth, D.A., Downing, T.E., Washington, R., Lopez, A., New, M., 2007a. Issues in the interpretation of climate model ensembles to inform decisions. Philos. Trans. R. Soc. A 365, 2163–2177.

Zwiers, F.W., Kharin, V.V., 1998. Intercomparison of interannual variability and potential predictability: an AMIP diagnostic subproject. Clim. Dyn. 14, 517–528.

Chaos in exchange of vertical turbulent energy fluxes over environmental interfaces in climate models

21

21.1 CHAOS IN COMPUTING THE ENVIRONMENTAL INTERFACE TEMPERATURE

The environmental interface energy balance is usually defined with respect to an active layer of infinitesimal small thickness (Fig. 21.1). In this case the storage of energy in the layer can be neglected and the energy balance equation takes the form:

$$G{\downarrow}(1 - \alpha) + L{\downarrow} - L{\uparrow} + H + E + S = 0 \tag{21.1}$$

or, summarizing the radiation fluxes

$$R + H + E + S = 0, \tag{21.2}$$

where R is the net radiation, $G{\downarrow}$ is the global radiation, α is the albedo, $L{\downarrow}$ is the atmospheric counterradiation, $L{\uparrow}$ is the terrestrial emission, H is the sensible heat flux, E is the latent heat flux, and S is the soil heat flux. R and S are available energy terms while H and E represent turbulent fluxes. Regarding the sign convention, we use the following one: fluxes are considered positive when directed toward the surface (energy sources) and negative when directed away from the surface (energy sinks). Exceptions are $L{\uparrow}$ and a $R{\uparrow}$ (outgoing radiation fluxes), for which a minus sign is explicitly used in the energy balance equation. Note that this equation is mentioned as an example of diffusive coupling in the interaction of two environmental interfaces on the Earth's surface in Section 8.2.

In other situations, however, the active layer has a measurable thickness (Fig. 21.2). In this case the rate of change of energy stored in the layer, $C_g \partial T_g / \partial t$, must be included in the equation

$$C_g \frac{\partial T_g}{\partial t} = R + H + E + S, \tag{21.3}$$

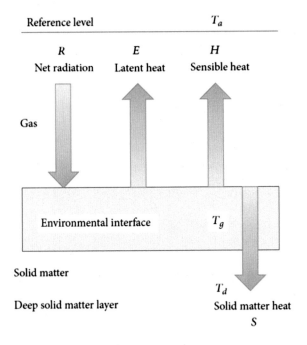

FIGURE 21.1

Energy balance equation terms at the environmental interface (Mihailović et al., 2014).

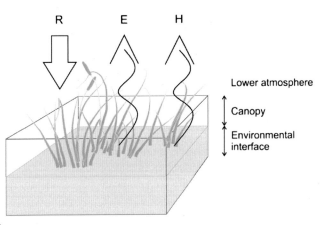

FIGURE 21.2

Energy balance equation terms at the environmental interface when the active layer has a measurable thickness.

where T_g is the environmental interface temperature. In many instances we also need to take the lateral fluxes (advection) into account. This situation is encountered for instance with vegetation or snow.

In atmospheric models of all scales (e.g., climate, regional, mesoscale, and small-scale models) the environmental interface temperature or shortly surface temperature is computed from either the energy balance equation at the atmosphere—surface interface in diagnostic form or the balance equation of a thin soil layer in prognostic form. In the diagnostic case, the soil heat flux parameterization is done very crudely. One possibility is to consider it as a constant part of the net radiation while the second is that the heat capacity of the earth is supposed to be zero with the ground heat flux also zero. On the other hand, Mahrer and Pielke (1977) have calculated the soil heat flux using a full treatment of soil heat diffusion in a multilevel soil model. Bhumralkar (1975) studied the application of procedures for calculating the surface temperature in the context of a general circulation model. He showed that the foregoing assumption of zero soil heat capacity results in an excessive diurnal range of temperature at the soil surface. He also showed that the heat flux into the soil could be represented by the sum of a temperature-derivative term and the difference between surface and deep soil temperature (Mihailović et al., 1999). Blackadar (1976) also introduced such an expression with slightly modified coefficients for use in a mesoscale model. Basically, he has established one of the most effective procedures for calculating the surface temperature using a prognostic equation based on the "force-restore" method. This "force-restore" method and its later use (Lin, 1980; Stull, 1988) and generalization (Dickinson, 1988) are still powerful tools in calculating the surface temperature whose variations in a diurnal range are less extreme than when the assumption of zero heat capacity is made.

An example of "force-restore" method application by solving Eq. (21.3) by the land surface scheme LAPS (Mihailović et al., 2010) is given in Fig. 21.3. This figure shows the calculated and observed temporal variations of soil surface temperature, T_g, beneath the soybean canopy at the experimental site in Marchfeld (Austria), while Fig. 21.3b depicts a comparison between the calculated and observed diurnal variations of soil surface temperature beneath the soybean canopy at Rimski Šančevi (Serbia). In both cases the soil surface temperature was measured using platinum resistance thermometers set in the top layer. Since soil temperature in the surface layer fluctuates much more than in the deeper soil layers, numerical models usually give values that can differ from the observed soil surface temperatures. This model feature is particularly pronounced under a sparse canopy. Although both soybean fields had a significant fraction of bare soil (soybean fractional cover was 0.65 and 0.60, respectively), the simulated values of the soil surface temperature compared well with the measurements for both datasets, suggesting that the model is well able to simulate accurately diurnal variation of the top soil temperature beneath the crop (Mihailović and Eitzinger, 2007).

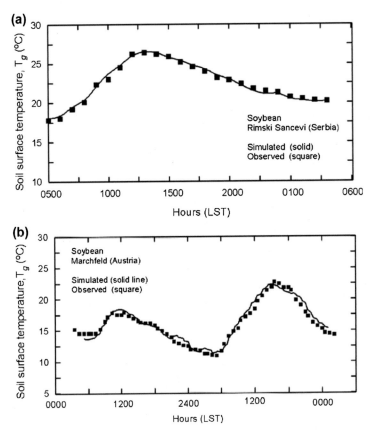

FIGURE 21.3

Two-day variation of the soil surface temperature, T_g simulated by the model proposed and observed beneath a soybean canopy for (a) 6–7 July 1995 (Marchfeld, Austria) and (b) 6 July 1982 (Rimski Šančevi, Serbia).

Reprinted with permission from Mihailović, D.T., Eitzinger, J., 2007. Modelling temperatures of crop environment. Ecol. Model. 202, 465–475.

An environmental interface that is peculiarly interesting in the calculation of environmental interface temperature in geophysical models is the rocky surface. This surface is often the dominant type of ground on the interface between the celestial objects and space or the atmosphere if it exists. That is the reason why, in the atmospheric and other numerical models, the calculation of rock surface temperature should be made with considerable attention. Calculation of the surface temperature, using the energy balance equation at the interface, is more complicated for rocks than for other solid materials, due to their particular thermal and physical properties (Arsenić et al., 2000).

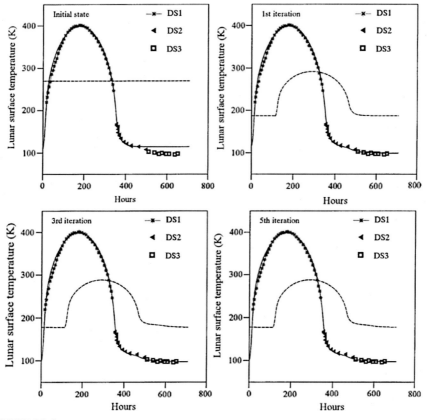

FIGURE 21.4

The calculated diurnal variation of the lunar surface temperature (*solid line*) and the deep ground temperature (*dashed line*) reached through five iterations, compared to observations (Arsenić et al., 2000). For verification of the method suggested, three available data sets with the Earth-based observations concerning the lunar surface temperature obtained in the infrared spectral area were used. They are indicated by DS1 (Stimpson and Lucas, 1972), DS2 (Jones et al., 1975), and DS3 (Jones et al., 1975).

Reprinted with permission from Arsenic, I., Mihailovic, D.T., Kapor, D.V., Kallos, G., Lalic, B., Papadopoulos, A., 2000. Calculating the surface temperature of the solid underlying surface by modified "Force Restore" method. Theor. Appl. Climatol. 67, 109–113.

To solve this problem, Arsenić et al. (2000) have modified the "force-restore" method into a self-consistent procedure for simultaneous determination of both surface and the deep ground temperature. Their modification is verified by calculating the lunar surface temperature (Fig. 21.4). There is a physical reason for this choice since the Moon has no exchange of heat by the latent and sensible heat fluxes due to absence of the atmosphere. Thus, on the right-hand side of the force-restore Eq. (21.3) remain only two terms, one due to radiation and other due to the ground heat flux whose imbalance, determined by the errors in estimation of the deep ground temperature, will be more emphasized than in the presence of the atmosphere.

When calculating the environmental interface temperature in environmental numerical models, we often encounter surfaces that have such thermal characteristics that the coefficients in the energy balance Eq. (21.3) may vary significantly. Therefore, it causes an unexpected behavior of the parameter that can lead to the appearance of the chaos (see Section 20.3). One example for environmental interface is depicted in Fig. 21.2. For this interface, visible radiation provides almost all of the received energy. Some of the radiant energy is reflected back to the space. The interface also radiates some of the energy received from the Sun. The quantity of the radiant energy remaining on the environmental interface is the net radiation, which drives physical processes important to our further considerations. Since all of the energy transfer processes occur in the finite time interval, the energy balance equation at any environmental interface can be written in terms of finite differences of ground and air temperatures as seen from Eq. (8.7). Under some conditions, this equation can be transformed into the logistic Eq. (8.8) in which x is the dimensionless temperature (Mihailović, 2010), i.e.,

$$x_{n+1} = rx_n(1 - x_n), \qquad (21.4)$$

where $r = 1+\tau$, $\tau = \Delta t/\Delta t_p$ is the dimensionless time, Δt is the time step, and Δt_p is the scaling time of energy exchange at the environmental interface. With this time we indicate the time which is needed for establishing the balance between all kinds of energy (by radiation, convection, and conduction as seen from Eq. (21.3)) reaching the environmental interface (expressed through quantity $\Sigma = C_R - C_L - C_H - C_D$, where the right-hand side coefficients are defined in Section 8.2) and thermal capacity of that interface expressed through the environmental interface soil heat capacity per unit area C_g. Thus, the scaling time Δt_p can be written as

$$\Delta t_p = \frac{C_g}{\Sigma}. \qquad (21.5)$$

Fig. 21.5 depicts the scaling time Δt_p as a function of energy exchange ε at the environmental interface. We have calculated the scaling time Δt_p curves for two different surfaces, one for the lunar surface and the second one for the soil Vertisols using thermal properties from Arsenić et al. (2000) and Mihailović et al. (1992), respectively. For both surfaces the soil heat capacity per unit area C_g was calculated as $C_g = c_t C_v \sqrt{k_t}$, where $c_t = 78.7926$ s$^{1/2}$, C_v is the volumetric heat capacity of the environmental interface (in this case soil Vertisols and lunar ground) while k_t is its thermal diffusivity (Zhang and Anthes, 1982; Mihailović, 1991). The range of Σ values was taken to be as it is commonly used in environmental models. From Fig. 21.5, it is seen that because of thermal properties, Δt_p has higher values for the lunar ground than for the soil Vertisols.

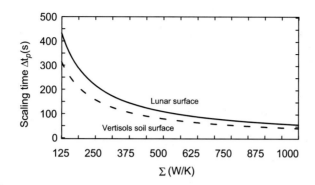

FIGURE 21.5

The scaling time Δt_p in dependence on energy exchange Σ at the environmental interface. Σ includes coefficients expressing all kinds of energy reaching that interface.

Now it raises the question whether we can find either domain or domains where physically meaningful solutions of Eq. (21.4) exist. We do that by analyzing Eq. (21.4). It can be written in the form

$$x_{n+1} = \left(1 + \frac{\Delta t}{\Delta t_p}\right) x_n (1 - x_n). \tag{21.6}$$

In this equation the solutions are not in the chaotic region if the following condition is satisfied

$$1 + \frac{\Delta t}{\Delta t_p} < 3.57, \tag{21.7}$$

since when $1 + \Delta t / \Delta t_p$ is increased beyond the accumulation point (3.56994...), chaos onsets. The condition (Eq. (21.7)) can be written as

$$\frac{\Delta t}{\Delta t_p} < 2.57 \tag{21.8}$$

or, after using relation (Eq. (21.5)) in the form

$$\Delta t < 2.57 \frac{C_g}{\Sigma}. \tag{21.9}$$

Since $\Sigma = C_R - C_L - C_H - C_D$ then, finally we get

$$\Delta t < 2.57 \frac{C_g}{C_R - C_L - C_H - C_D}. \tag{21.10}$$

The last relation can be understood as a criterion for choice of the time step used in environmental models for numerical solving of the energy balance equation to avoid situations when the environmental interface cannot oppose an enormous amount of energy (including radiation, convection, and conduction), suddenly entering system. Note that this criterion depends only on energy reaching

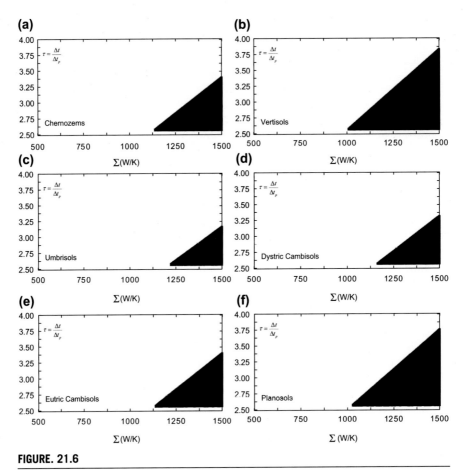

FIGURE. 21.6

The dimensionless time $\tau = \Delta t / \Delta t_p$ in dependence on energy exchange Σ at the environmental interface. The black triangles indicate the regions in which Eq. (21.4) exhibits chaotic behavior for different soil types used in environmental modelling.

the environmental interface and its thermal property. One way of looking at the above inequality is to consider Δt_p as the relaxation time often accoutered in the theory of nonequilibrium processes. In this case, inequality (Eq. (21.10)) means that by choosing time step Δt to be smaller than the relaxation time, we can observe then the energy exchange at a smaller time scale thus avoiding the trap of the chaos.

Fig. 21.6 depicts the dimensionless time $\tau = \Delta t / \Delta t_p$ as a function of energy exchange Σ at the environmental interface for different soil types used in environmental modelling. In $\tau = \Delta t / \Delta t_p$, during the simulations, the time step Δt was ranged in the interval between 20 and 100 s. The black triangles indicate the regions in which Eq. (21.4) exhibits chaotic behavior, i.e., when the environmental interface cannot respond to an enormous amount of energy reaching that one.

21.2 A DYNAMIC ANALYSIS OF SOLUTIONS FOR THE ENVIRONMENTAL INTERFACE AND DEEPER SOIL LAYER TEMPERATURES REPRESENTED BY THE COUPLED DIFFERENCE EQUATIONS

Now we consider an environmental interface for which difference equations for calculating the environmental interface temperature and deeper soil layer temperature are represented by the coupled difference equations, i.e., maps. First equation has its background in the energy balance equation while the second one in the prognostic equation for deeper soil layer temperature (Mihailović et al., 1999). In Mihailović and Mimić (2012), it is shown that the ground surface is treated as a complex system in which chaotic fluctuations occur while we compute its temperature. This system, as a dynamic system, is very sensitive to initial conditions and arbitrarily small perturbation of the current trajectory that may lead to its unpredictable behavior. In this paper the lower boundary condition, i.e., the deeper soil layer temperature was constant, but it can also vary in time making with the energy balance equation a coupled system of equations. That system, often used in environmental models, is of interest to be analyzed by the methods of nonlinear dynamics (Mimić et al., 2013). Following idea by Mimić et al. (2013), our analysis will include (1) examination of period one fixed point and (2) bifurcation analysis. Focusing part of analysis is calculation of the Lyapunov exponent for a specific range of values of system parameters and discussion about the domain of stability for this coupled system. To end, we calculate Kolmogorov complexity of time series generated from the coupled system.

Using Eq. (21.3) and prognostic equation for the deeper soil layer temperature T_d (Mihailović et al., 1999)

$$\frac{\partial T_d}{\partial t} = \frac{1}{\tau_d}(T_g - T_d), \tag{21.11}$$

where $\tau_d = 86{,}400$ s Mimić et al. (2013) have shown that this system of partial differential equations, under some conditions, can be written in the form of two coupled difference equations

$$z_{n+1} = Az_n - Bz_n^2 + Cy_n \tag{21.12a}$$

$$y_{n+1} = Dz_n + (1-D)y_n, \tag{21.12b}$$

where $z = (T_g - T_a)/T_0$ is the dimensionless environmental interface temperature, $y = (T_d - T_a)/T_0$ is the dimensionless deeper soil layer temperature, T_a is the temperature at some reference level, $T_0 = 288K$, $A = 1 + \frac{\Delta t}{C_g}(C_R - C_H - C_L bd - C_D)$, $B = C_L dT_0 \frac{b^2 \Delta t}{2C_g}$, $C = \Delta t \frac{C_D}{C_g}$, and $D = \frac{\Delta t}{\tau}$, while the meaning and values of the constants b and d are available in Mimić et al. (2013) and Section 8.2. Introducing the replacement $z_n = Ax_n/B$, where x is modified dimensionless environmental interface temperature and following Mimić et al. (2013) can write

$$x_{n+1} = Ax_n(1 - x_n) + \frac{CB}{A}y_n \tag{21.13a}$$

$$y_{n+1} = \frac{DA}{B}x_n + (1 - D)y_n. \tag{21.13b}$$

Analysis of values of parameters A, B, C, and D, based on a large number of energy flux outputs from the land surface scheme runs, indicates that their values are ranged in the following intervals: (1) $A \in [0,4]$ and (2) B, C, and D are ranged in the interval $[0,1]$. Thus, A is the logistic parameter, which from now on will be denoted by r. All other groups of parameters in the system (Eqs. (21.13a) and (21.13b)) have the values in the same interval $[0, 1]$. Let us underline that under some circumstances those parameters can be equal. Correspondingly, we replaced all of them by introducing the coupling parameter ε. Finally, the system (Eqs. (21.13a) and (21.13b)) can be written in the form of coupled maps, i.e.,

$$x_{n+1} = rx_n(1 - x_n) + \varepsilon y_n \tag{21.14a}$$

$$y_{n+1} = \varepsilon(x_n + y_n). \tag{21.14b}$$

We now examine the effect of coupling two nonlinear maps given by Eqs. (21.14a) and (21.14b), with the logistic parameter $r \in [0,4]$ and the coupling parameter $\varepsilon \in [0,1]$. This map displays a wide range of behavior as the parameters r and ε change. We consider a system of difference equations of the form $X_{n+1} = F(X_n)$ with the notation $F(X_n) = (rx_n(1-x_n) + \varepsilon y_n, \varepsilon(x_n + y_n))$, where $X_n = (x_n, y_n)$ is a vector representing the dimensionless environmental interface temperature and the deeper soil layer temperature, respectively. We look for the fixed point of mapping given by (Eqs. (21.14a) and (21.14b)) using the criterion $X = F(X)$. Thus, we get $(0, 0)$ and $((r+\varepsilon^2/(1-\varepsilon)-1)/r, \varepsilon/(1-\varepsilon)[(r+\varepsilon^2/(1-\varepsilon)-1)/r])$ as two fixed points. Now, for the fixed point $(0, 0)$ we have two eigenvalues $\lambda_{1,2} = (r + \varepsilon \pm \sqrt{r^2 - 2r\varepsilon + 5\varepsilon^2})/2$. Using the one with the plus sign, which has higher absolute value, and the criterion that a fixed point is attractive if $|\lambda| < 1$ and it is repulsive if $|\lambda| > 1$, we localize regions in the (ε, r) plane, which tell us for what pair of parameter values the fixed point $(0,0)$ would be either attractive or repulsive. Applying the same procedure for the other fixed point given by $((r+\varepsilon^2/(1-\varepsilon)-1)/r, \varepsilon/(1-\varepsilon)[(r+\varepsilon^2/(1-\varepsilon)-1)/r])$ and with the eigenvalues $\lambda_{3,4} = \frac{1}{2(\varepsilon-1)}(-2+\varepsilon+3\varepsilon^2+r-\varepsilon r) \pm \sqrt{(2-\varepsilon-3\varepsilon^2-r+\varepsilon r)^2}$ $-4(\varepsilon-1)(-2\varepsilon+3\varepsilon^2+\varepsilon^3+\varepsilon r-\varepsilon^2 r)$, we get exactly the same regions of attraction and repulsion in the (ε, r) plane as depicted in Fig. 21.7.

Bifurcation diagrams for x and y are given in Fig. 21.8 as a function of the logistic parameter r and for this coupled maps were plotted with r ranging from 0 to 4 and for $\varepsilon = 0.1$. For each value of r, we used the final point of the previous r value and then 500 iterations are plotted. We noticed that maximum values of y strongly depend on

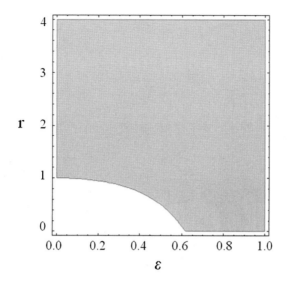

FIGURE 21.7

Graphical interpretation of fixed points for the coupled maps (Eqs. (21.14a) and (21.14b)) as a function of logistic parameter r and coupling parameter ε. Both fixed points are in the following regions: (1) attractive (white) and (2) repulsive (light blue (gray in print versions)).

Reprinted with permission from Mimić, G., Mihailović, D.T., Budinčević, M., 2013. Chaos in computing the environmental interface temperature: nonlinear dynamic and complexity analysis of solutions. Mod. Phys. Lett. B. 27, 1350190–1350200, 10.1142/S021798491350190X.

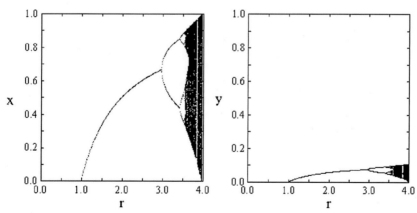

FIGURE 21.8

Bifurcation diagrams for the coupled maps (Eqs. (21.14a) and (21.14b)) with r ranging from 0 to 4 and $\varepsilon = 0.1$. Initial conditions were $x_0 = 0.2$ and $y_0 = 0.4$.

Reprinted with permission from Mimić, G., Mihailović, D.T., Budinčević, M., 2013. Chaos in computing the environmental interface temperature: nonlinear dynamic and complexity analysis of solutions. Mod. Phys. Lett. B. 27, 1350190–1350200, 10.1142/S021798491350190X.

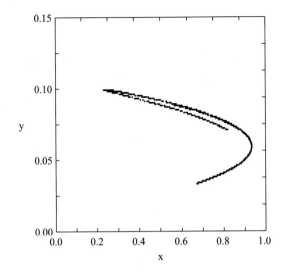

FIGURE 21.9

Phase diagram of the map (15)–(16) for $r = 3.7$, $\varepsilon = 0.1$, and initial point $x_0 = 0.2$, $y_0 = 0.4$.

Reprinted with permission from Mimić, G., Mihailović, D.T., Budinčević, M., 2013. Chaos in computing the environmental interface temperature: nonlinear dynamic and complexity analysis of solutions. Mod. Phys. Lett. B. 27, 1350190–1350200, 10.1142/S021798491350190X.

the coupling parameter ε. Let us note that value of the coupling parameter is small. Thus, the bifurcation diagram of x is close the logistic map. Bifurcations start after the parameter r reaches value 3 and chaotic regime exists after $r = 3.5$ on both the diagrams.

We have also plotted the phase diagram for x and y, which is depicted in Fig. 21.9. This plot was obtained by iterating x (from 0 to 1) and y (from 0 to 0.15). In those calculations, 1000 iterations were applied for the initial state $(x_0 = 0.2, y_0 = 0.4)$ after 200 steps of stabilization of the (x,y) pair. From this figure, it is seen that this plot is similar to the Henon's attractor (Henon, 1976). It was expected, because for $D = 1$ dynamical system given with Eqs. (21.14a) and (21.14b) is similar to the Henon map.

Irregularities in the solution of the system (Eqs. (21.14a) and (21.14b)) can come from two reasons. They are (1) numerical, i.e., because we try to choose appropriate difference equation whose solution is "good" approximation to the solution of the given differential equation and (2) physical, i.e., occurrence of chaotic fluctuations in the considered system because the environmental interface cannot oppose an enormous radiative forcing, suddenly reaching the interface (Fig. 21.10).

Let us note that the assumption T_g, $T_d \geq T_a$ is violated in dependence on atmospheric conditions. However, there exist conditions for which this criterion is satisfied since the ground surface ("skin") temperature can be even for 10°C higher than

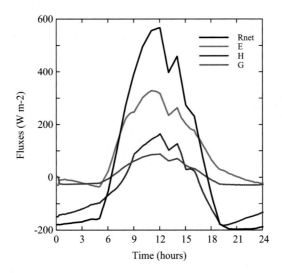

FIGURE 21.10

Diurnal cycle of energy balance equation components. Second peak in net radiation term can be noticed in all other fluxes. Simulation was done using LAPS land surface scheme (Mihailović, 1996).

Reprinted with permission from Mimić, G., Mihailović, D.T., Budinčević, M., 2013. Chaos in computing the environmental interface temperature: nonlinear dynamic and complexity analysis of solutions. Mod. Phys. Lett. B. 27, 1350190–1350200, 10.1142/S021798491350190X.

air temperature at 2 m height (Stull, 1988). We wanted to point out that in these kinds of situations, there is a possibility for occurrence of chaotic phenomena, which can cause uncertainties in calculations of the ground surface temperature. This is because of drawback of currently designed environmental models to calculate the ground surface temperature under these conditions.

Looking at the system (Eqs. (21.13a) and (21.13b)), we have to keep in mind some a priori mathematical limitations. As we take the range of x, y between 0 and 1, it is seen from Eq. (21.14b) that parameter ε has to be less or equal to 0.5. Further, from Eq. (21.14a), where the maximum value for x is 0.5 and for y is 1, we get a new condition that $r/4 + \varepsilon \leq 1$. Since Eq. (21.14a) has the form of a logistic map, we know that chaos is present in a case when the logistic parameter r is in interval $(3, 4)$ so parameter of coupling ε should be very small, i.e., beyond 0.1. Thus, selection of initial conditions is also very important in the evolution of the system. Therefore, it raises the question whether we can find either domain or domains where physically meaningful solutions exist. We do that by considering the stability of physical solutions, using the Lyapunov exponent, which is a measure of convergence or divergence of near trajectories in phase space. The sign of the Lyapunov exponent is characteristic of the attractor type: for a stable fixed point it is negative, and for a chaotic attractor it is positive.

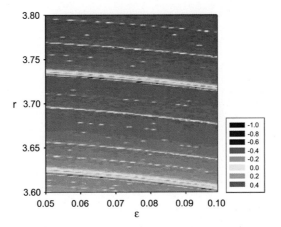

FIGURE 21.11

Lyapunov exponent of the coupled system (Eqs. (21.13a) and (21.13b)), which shows presence of straight regions of stability in highly developed chaos.

Reprinted with permission from Mimić, G., Mihailović, D.T., Budinčević, M., 2013. Chaos in computing the environmental interface temperature: nonlinear dynamic and complexity analysis of solutions. Mod. Phys. Lett. B. 27, 1350190–1350200, 10.1142/S021798491350190X.

We analyze the behavior of the Lyapunov exponent for the corresponding autonomous equation which uniformly changes in intervals of r and ε. For our system $X_{n+1}{=}F(X_n)$, $X_n = \begin{pmatrix} x_n \\ y_n \end{pmatrix}$ and for any initial point (x_0,y_0) from attracting region, characterization of asymptotic behavior of the orbit is given by the largest Lyapunov exponent (Furstenberg and Kesten, 1960) as in Eqs. 7.5 and 7.6 in Section 7.3. In our case the Jacobian matrix ξ_s is given by

$$\xi_s = \begin{bmatrix} r(1-2x_s) & \varepsilon \\ \varepsilon & \varepsilon \end{bmatrix}. \tag{21.15}$$

Calculating the Lyapunov exponent for the coupled system Eqs. (21.14a) and (21.14b) with values of parameters $\varepsilon \in (0.05,0.1)$ and $r \in (3.6,3.8)$ we got the results depicted in Fig. 21.11. It is shown that the Lyapunov exponent mostly has positive values which approve presence of chaos in this system, but there are still some strait regions where the Lyapunov exponent is negative and where the solutions of the coupled system are stable, i.e., domains of stability.

For nonlinear time series analysis, we use Kolmogorov complexity with an idea to calculate Kolmogorov complexity of time series produced for chaotic states of coupled system with a range of parameters $\varepsilon \in (0.05,0.1)$ and $r \in (3.6,3.8)$.

It is seen from Fig. 21.12 that the complexity of the system strongly depends on parameter r. Higher value of Kolmogorov complexity implies highly developed chaos. Although, there are still regions, colored with purple, which are related to domains of stability with a stable solution and no chaotic behavior of system.

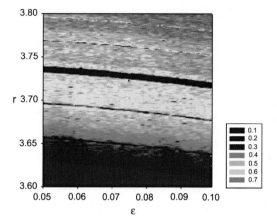

FIGURE 21.12

Kolmogorov complexity for the coupled system of Eqs. (21.14a) and (21.14b) as a function of parameters r and ε.

Reprinted with permission from Mimić, G., Mihailović, D.T., Budinčević, M., 2013. Chaos in computing the environmental interface temperature: nonlinear dynamic and complexity analysis of solutions. Mod. Phys. Lett. B. 27, 1350190–1350200, 10.1142/S021798491350190X.

REFERENCES

Arsenic, I., Mihailovic, D.T., Kapor, D.V., Kallos, G., Lalic, B., Papadopoulos, A., 2000. Calculating the surface temperature of the solid underlying surface by modified "Force Restore" method. Theor. Appl. Climatol. 67, 109–113.

Bhumralkar, C.M., 1975. Numerical experiments on the computation of ground surface temperature in an atmospheric general circulation model. J. Appl. Meteorol. 14, 1246–1258.

Blackadar, A.K., 1976. Modeling the nocturnal boundary layer. In: Proceedings of the Third Symposium on Atmospheric Turbulence, Diffusion and Air Quality, Am. Meteorol. Soc., pp. 46–49.

Dickinson, R.E., 1988. The force-restore model for surface temperatures and its generalizations. J. Clim. 1, 1086–1097.

Furstenberg, H., Kesten, H., 1960. Products of Random Matrices. Ann. Math. Statist. 31, 457–469.

Hénon, M., 1976. A two-dimensional mapping with a strange attractor. Comm. Math. Phys. 50, 69–77.

Jones, W.P., Watkins, J.R., Calvert, T.A., 1975. Temperatures and thermo physical properties of the lunar outermost layer. Moon 13, 475–495.

Lin, J.D., 1980. On the force-restore methods for the prediction of ground surface temperature. J. Geophys. Res. 85, 3251–3254.

Mahrer, Y., Pielke, R.A., 1977. The effects of topography on the sea and land breezes in a two-dimensional numerical model. Mon. Weather Res. 105, 1151–1162.

Mihailović, D.T., 1991. A model for the prediction of the soil temperature and the soil moisture content in three layers. Z. Meteorol. 41, 29–33.

Mihailovic, D.T., 1996. Description of a land-air parameterization scheme (LAPS). Glob. Planet. Change 13, 207–215.

Mihailović, D.T., 2010. Climate modeling beyond the complexity: challenges in model building. In: Alexandrov, V., Gajdusek, M.F., Knight, C.G., Yotova, A. (Eds.), Global Environmental Change: Challenges to Science and Society in Southeastern Europe. Selected Papers Presented in the International Conference Held 19–21 May 2008, Sofia (Bulgaria). Springer, Dordrecht, Heidelberg, London, New York.

Mihailović, D.T., de Bruin, H.A.R., Jeftić, M., van Dijken, A., 1992. A study of the sensitivity of land surface parameterizations to the inclusion of different fractional covers and soil textures. J. Appl. Meteorol. 31, 1477–1487.

Mihailović, D.T., Eitzinger, J., 2007. Modelling temperatures of crop environment. Ecol. Model. 202, 465–475.

Mihailovic, D.T., Kallos, G., Arsenic, I., Lalic, B., Rajkovic, B., Papadopoulos, A., 1999. Sensitivity of soil surface temperature in a force-restore equation to heat fluxes and deep soil temperature. Int. J. Climatol. 19, 1617–1632.

Mihailović, D.T., Lazić, J., Leśny, J., Olejnik, J., Lalić, B., Kapor, D.V., Ćirišan, A., 2010. A new design of the LAPS land surface scheme for use over and through heterogeneous and non-heterogeneous surfaces: numerical simulations and tests. Theor. Appl. Climatol. 100, 299–323.

Mihailović, D.T., Mimic, G., 2012. Kolmogorov complexity and chaotic phenomenon in computing the environmental interface temperature. Mod. Phys. Lett. B 26, 1250175. http://dx.doi.org/10.1142/S0217984912501758.

Mihailović, D.T., Mimić, G., Arsenić, I., 2014. Climate predictions: the chaos and complexity in climate models. Adv. Meteorol. http://dx.doi.org/10.1155/2014/878249.

Mimić, G., Mihailović, D.T., Budinčević, M., 2013. Chaos in computing the environmental interface temperature: nonlinear dynamic and complexity analysis of solutions. Mod. Phys. Lett. B 27, 1350190–1350200. http://dx.doi.org/10.1142/S021798491350190X.

Stimpson, L.D., Lucas, J.W., 1972. Lunar thermal aspects from surveyor data. In: Lucas, J.W. (Ed.), Thermal Characteristic of the Moon, vol. 28. The MIT Press, Cambridge Massachusetts, and London, England, pp. 121–149.

Stull, R.B., 1988. An Introduction to Boundary Layer Meteorology. Kluwer Academic, Dordrecht, 660 pp.

Zhang, D., Anthes, R.A., 1982. A high resolution model of the planetary boundary layer sensitivity tests and comparisons with SESAME-79 data. J. Appl. Meteorol. 21, 1594–1609.

Synchronization and stability of the horizontal energy exchange between environmental interfaces in climate models

22.1 SYNCHRONIZATION IN HORIZONTAL ENERGY EXCHANGE BETWEEN ENVIRONMENTAL INTERFACES

There are three major sets of processes that must be considered when constructing a climate model: (1) radiative (the transfer of radiation through the climate system, e.g., absorption and reflection); (2) dynamic (the horizontal and vertical transfer of energy, e.g., advection, diffusion and convection); and (3) surface process (inclusion of processes involving land/ocean/ice and the effects of albedo, emissivity, and surface—atmosphere energy exchanges). If the nonlinearities in these processes are treated improperly, then while designing the model, the complexity and thus its reliability will not be retained in the highest degree. In Section 21.1 we have considered surface—atmosphere energy exchanges with emphasis on the possible occurrence of the chaos in solving the energy balance equation for calculating the environmental interface temperature in climate models. Here, following Mihailović et al. (2012) we analyze the horizontal energy exchange between environmental interfaces which is described by the dynamics of driven coupled oscillators (Mihailović et al., 2012). To study their behavior, when changes are introduced in the coupling parameter, the logistic parameter, and the horizontal energy exchange intensity (parameter of exchange, in further text), we considered the dynamics of two maps serving the diffusive coupling (Mihailović et al., 2012).

The horizontal exchange of energy between environmental interfaces is a diffusion-like process. The dynamics of energy exchange behavior on environmental interface is typically expressed as a logistic map $\Phi(x) = rx(1-x)$, where x is the dimensionless temperature of environmental interface and r is a logistic parameter representing vertical turbulent energy flux intensities over an environmental interface (Mihailović et al., 2012, 2014a; Mimić et al., 2013). However, we use an alternative form of this map, which includes a parameter p that represents the total turbulent energy exchange between a single environmental interface and the surrounding environment (see Section 10.3), i.e.,

$$\Phi_{r,p}(x) = rx^{p}(1 - x^{p}).$$ (22.1)

The dynamics of map (Eq. (22.1)) is governed by two parameters, p and r, which express the intrinsic property of the environmental interfaces and the influence of the environment, respectively. Since these and many other processes on environmental interface are defined as diffusion-like, it is interesting to see: (1) how these processes can be better represented in climate models by introducing parameter of exchange p in the diffusive coupling associated with the horizontal energy exchange (Fig. 22.1); and (2) how the horizontal energy exchange intensity dynamics are affected by the changes of parameters that represent influence of the environment, environmental interface coupling, and horizontal energy exchange intensity. In considering these problems (Mihailović et al., 2014a), we have included observational heterarchy, which is considered in Chapters 12 and 13.

The time development of the environmental surface dynamics $x_{i,n}$, for two interfaces, is expressed as

$$x_{i,n+1} = (1 - c)\Phi_{r,p}(x_{i,n}) + f\big(\Phi_{r,p}(x_{j,n})\big),$$ (22.2)

where n is the time iteration, $i, j = 1,2$, $x_{i,n} \in [0,1]$, $c \in [0.0, 1.0]$ the coupling parameter as a measure of diffusion of the energy exchange between environmental interfaces, f the map representing the horizontal energy exchange between environmental interfaces, $\Phi_{r,p}$ is one of maps in the pair $(\Psi_{r,p}, \Phi_{r,p})$ whose composition is

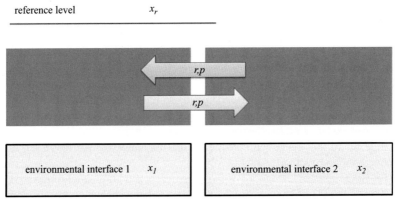

FIGURE 22.1

Schematic diagram of horizontal energy exchange between two environmental interfaces. Parameters p and r express intrinsic property of the environmental interfaces and the influence of the environment, respectively

Reprinted with permission from Mihailović, D.T., Mimic, G., Arsenic, I., April 2014a. Climate predictions: the chaos and complexity in climate models. Adv. Meteorol. Article ID 878249, 1–14.

preserved by a prefunctor $\langle F \rangle$. Here, we apply the framework of an observational heterarchy to the two environmental interface systems. If Intent and Extent are denoted by $\Phi_{r,p}$ and $\Psi_{r,p}$, respectively, the time development of the concentration is expressed as $x_{i,n+1} = (1-c)\Phi(x_{i,n}) + \Psi(x_{j,n})$ (Gunji and Kamiura, 2004). In this expression, if $\Psi_{r,p}(\mathbf{X}) = f(\Phi_{r,p}(x))$ then it can be reduced to Eq. (22.2).

We perform our analysis following the procedure described in Gunji and Kamiura (2004) and Chapter 13. First, we address the synchronization of the coupling for two environmental interfaces given by Eqs. (22.1) and (22.2), and then we will show that changes in the above-mentioned parameters can modify the dynamics and enhance robust behavior in a multienvironmental interface system. Synchronization is a collective phenomenon in various multicomponent physical as well as the climate systems (Pikovsky et al., 2001; Arenas et al., 2008; Chen et al., 2003), where the exchange of information (coupling) among the components can be either global or local. Here, we consider that chaotic systems are synchronized only when the largest Lyapunov exponent of the driven system is negative (Zhou and Lai, 1998). We calculated this exponent using Eqs. (7.5) and (7.6).

Fig. 22.2 depicts the normalized frequency of synchronization F_p ($\lambda < 0$) for a system of two passively coupled environmental interfaces (Eqs. (22.1) and (22.2)), as a function of coupling parameter c, averaged over all values of the parameter of exchange p and logistic parameter r. The value of the normalized frequency of synchronization F_p is calculated as in expression (13.5) where $\sum N_n(\lambda < 0)$ and $\sum N_p(\lambda > 0)$ are the numbers of negative and positive values of the Lyapunov exponent λ, respectively. These numbers were calculated for fixed values of c, with p and r changing in the intervals (0,1) and (1, 4), respectively, with a step of 0.05. From this figure it is seen that for $c < 0.8$, F_p is nearly constant. After that value it starts to decline, indicating a decrease of number of states which are synchronized.

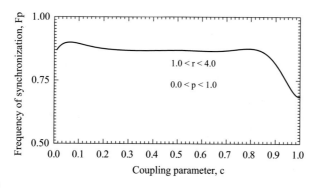

FIGURE 22.2

Normalized frequency of synchronization, $F_p(\lambda < 0)$ for a system of two environmental interfaces coupled as a function of coupling parameter c. An averaging was done over all values of logistic parameter r and parameter of exchange p.

Now we deal with simulations of active coupling in a multienvironmental interface system. Here, we estimate whether a coupled map system described above can achieve synchronization under influence of changes in parameters. The dynamics of two-environmental interface system is expressed via a system of Eqs. (22.1) and (22.2). We note that the dynamical system defined by these equations is called the passive coupling. To see how perturbation enhances robust behavior in the framework of observational heterarchy we considered a multienvironmental interface system represented by closed contour of coupled environmental interfaces exchanging the energy horizontally. Then the system of coupled difference equations for N environmental interfaces exchanging the energy can be written in the form of matrix equation (Mihailović et al., 2014a, 2014b). Simulations with the passive coupling, defined by Eqs. (22.1) and (22.2), were performed following Mihailović et al. (2014b). The results of simulations are shown in Fig. 22.2. In this figure Lyapunov exponent λ is plotted against coupling parameter c for passive coupling, for different values of the parameter of exchange p and the logistic parameter r. Simulations were performed with the closed contour of $N = 100$ interfaces. The Lyapunov exponent was calculated using Eqs. (7.5) and (7.6).

In calculating λ, for each c from 0.0 to 1.0 with step 0.005, 10^4 iterations were applied for an initial state, and then the first 10^3 steps were abandoned. To see how the passive coupling modifies the synchronization of horizontal energy exchange between environmental interfaces, we used a randomly chosen p and a logistic r parameter with the values of 4.0, 3.82, and 3.6, respectively (Fig. 22.3). Fig. 22.3a depicts that in close to chaotic regime ($r = 3.6$), regardless of the value p, the Lyapunov exponent is practically always negative ($\lambda < 0$) and therefore the process of the horizontal energy exchange in a multienvironmental interface system is always synchronized. Otherwise, as seen from Fig. 22.3c, in the chaotic regime ($r = 4.0$) the Lyapunov exponent is always positive ($\lambda > 0$), regardless of the value p, and therefore the process of the horizontal energy exchange in a multienvironmental interface system is always unsynchronized. Finally, in the region when $r = 3.82$ (Fig. 22.3b) there is continuity of regions with negative and positive values of the Lyapunov exponent.

22.2 STABILITY OF HORIZONTAL ENERGY EXCHANGE BETWEEN ENVIRONMENTAL INTERFACES

In Section 10.3 we have introduced a dynamical system approach (Eq. 10.22) that provides more realistic results in modelling of energy exchange over the heterogeneous grid-box than the flux aggregation methods that suffer from Schmidt's paradox, which is an effect occurring in the subgrid scale parameterization. This dynamical system approach can be also successfully applied in the modelling of horizontal energy exchange between the either small or large scale of heterogeneous environmental interfaces. In that approach the horizontal energy exchange is taken

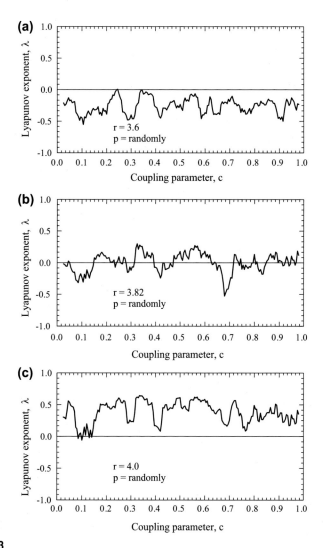

FIGURE 22.3

Diagram of Lyapunov exponent λ against coupling parameter c for the so-called passive coupling (Gunji and Kamiura, 2004) for different values of parameter of exchange p and logistic parameter r. p is randomly chosen, while r takes values 4.0, 3.82, and 3.6 respectively. Simulations were performed with the closed contour of $N = 100$ environmental interfaces.

into account and it is represented by a matrix of coupling parameters. Since it is, in general, very difficult to specify the quantities in that matrix, we derived a sufficient condition for the asymptotic stability that can be applied for any coupling matrix. We have proved two theorems that consider the flux aggregation effect over a heterogeneous grid-box (Mihailović et al., 2015).

As mentioned in Section 10.3, the requirement of such a model $\sum_{j \in N} c_{i,j} = 1$, for all $i \in N$ implies that the matrix C has to be nonnegative stochastic matrix, see Berman and Plemmons (1994). Since the class of nonnegative stochastic matrices plays a fundamental role in probability theory, theory of Markov processes, and in many different applications of matrix theory, a lot of its important properties have been discovered over the years. Here we will use the well-known fact that the spectral radius of such a matrix is always one, i.e., $\rho(C) = 1$. First step in the understanding of the dynamical behavior of (Eq. 10.22) is to examine its equilibrium states, and then to determine the stability of the evolution process around these states. Therefore, we begin with the analysis of the existence of equilibrium points. We will follow with the analysis by Mihailović et al. (2015).

An equivalent condition for state vector $\tilde{x}$ to be an equilibrium point of (Eq. 10.22) is that $C\Phi(\tilde{x}) = \tilde{x}$, which obviously holds for $\tilde{x} = 0 = [0 \ 0 \ ... \ 0]^T$. Therefore, 0 is a trivial equilibrium point. In the following, we are interested in the existence and stability of nontrivial equilibrium points of (Eq. 10.22).

First, we analyze the properties of logistic functions $\Phi_{ri,pi}$, for $i \in N$. As seen in Fig. 10.6, logistic functions $\Phi_{r,p}$ have exactly one nontrivial fixed-point for every choice of parameters r and p.

To prove that the nontrivial equilibrium point of the coupled system of EIs exists, we will analyze Jacobian of the right-hand side of (Eq. 10.22) and identify the subspace of $[0,1]^n$ where $C\Phi$ is a contraction, and then use the Banach fixed-point theorem. Here, the abbreviation EI is used as an acronym for environmental interfaces as in Section 10.3.

Given a state $x \in (0,1]^n$, Jacobian of the right-hand side of (Eq. 10.22) at x is

$$\nabla C\Phi(x) = C\nabla\Phi(x) = CF(x),$$

where $F(x) = \mathrm{diag}(\phi'_{r_1,p_1}(x_1)\phi'_{r_2,p_2}(x_2)...\phi'_{r_n,p_n}(x_n))$ and

$$\phi'_{r,p}(x) = rp\frac{1 - 2x^p}{x^{1-p}}.$$

We start with the following lemma which identifies a set where the modified logistic function $\phi_{r,p}$ is a contraction.

Lemma 1 Given $p \in (0,1]$ and $r \in [1,4]$,

1. then there exists a unique solution $\alpha(r,p)$ of the problem

$$2x^p + \frac{1}{rp}x^{1-p} = 1, x \in \left(0, \left(\frac{1}{2}\right)^{\frac{1}{p}}\right),$$

2. if $r \geq \frac{1}{p}$, there exists a unique solution $\beta(r,p)$ of the problem

$$2x^p - \frac{1}{rp}x^{1-p} = 1, x \in \left(\left(\frac{1}{2}\right)^{\frac{1}{p}}, 1\right), and$$

3. if $r \geq \left(\frac{1-p}{2(1-2p)}\right)^{\frac{1-p}{p}} \frac{1-2p}{p^2}$, there exists a unique solution $\beta(r,p)$ of the problem

$$2x^p - \frac{1}{rp}x^{1-p} = 1, x \in \left(\left(\frac{1}{2}\right)^{\frac{1}{p}}, \left(\frac{1-p}{2(1-2p)}\right)^{\frac{1}{p}}\right),$$

and there exists a unique solution $\gamma(r,p)$ of the problem

$$2x^p - \frac{1}{rp}x^{1-p} = 1, x \in \left(\left(\frac{1-p}{2(1-2p)}\right)^{\frac{1}{p}}, 1\right).$$

Moreover, denoting

$$f(p) := \begin{cases} \left(\frac{1-p}{2(1-2p)}\right)^{\frac{1-p}{p}} \frac{1-2p}{p^2} & ,p < \frac{1}{3}, \\ \frac{1}{p} & ,p \geq \frac{1}{3}, \end{cases}$$

it follows that $|\phi_{r,p'}(x)| < 1$ if and only if either of the following cases hold:

- $x \in (\alpha(r,p),1)$ and $r < f(p)$, or
- $x \in (\alpha(r,p),\beta(r,p)) \cup (\gamma(r,p),1]$, $r \geq f(p)$, and $p < \frac{1}{3}$, or
- $x \in (\alpha(r,p),\beta(r,p))$, $r \geq f(p)$, and $p \geq \frac{1}{3}$.

Proof. To prove this lemma we analyze the inequality $rp\frac{|1-2x^p|}{x^{1-p}} < 1$, for $x \in (0,1]$. First, denote

$$\varphi^+(x) := rp\frac{1 - 2x^p}{x^{1-p}} - 1, \text{ and } \varphi^-(x) := rp\frac{2x^p - 1}{x^{1-p}} - 1, \qquad (22.3)$$

and observe that $\varphi^+(x)$ and $\varphi^-(x)$ are continuous functions for $x \in (0,1]$.

Now, since $\varphi^+(\varepsilon) > 0$, for sufficiently small $\varepsilon > 0$, and $\varphi^+\left(\left(\frac{1}{2}\right)^p\right) < 0$, there exists $\alpha(r,p)$ such that $\varphi^+(\alpha(r,p)) = 0$. Since it is easy to see that φ^+ always has exactly one zero on the interval $\left(0, \left(\frac{1}{2}\right)^{\frac{1}{p}}\right]$, value $\alpha(r,p)$ is uniquely defined and it can be efficiently computed by Newton's method. Therefore, we have obtained that

$$\varphi^+(x) < 0 \text{ for all } x \in \left(\alpha(r,p), \left(\frac{1}{2}\right)^{\frac{1}{p}}\right]. \qquad (22.4)$$

Next, we analyze φ^- on the interval $\left(\left(\frac{1}{2}\right)^{\frac{1}{p}}, 1\right]$ and compute its derivative:

$$[\varphi^-(x)]' = rp\frac{1 - p - 2(1 - 2p)x^p}{x^{2-p}}.$$

Observe that if $p \geq \frac{1}{3}$, $[\varphi^-(x)]' > 0$ for all $x \in \left(\left(\frac{1}{2}\right)^{\frac{1}{p}}, 1\right]$, implying that φ^- is an increasing function on the given interval. On the other hand, if $p < \frac{1}{3}$, φ^- has a local maximum in $x_{st} = \left(\frac{1-p}{2(1-2p)}\right)^{\frac{1}{p}}$. In the following, the sign of the φ^- in x_{st} will be

of special interest, so, observe that $\varphi^-(x_{st}) < 0$ if $r < \left(\frac{1-p}{2(1-2p)}\right)^{\frac{1-p}{p}} \frac{1-2p}{p^2}$, and

$\varphi^-(x_{st}) \geq 0$, otherwise.

Therefore, we distinguish the following three cases:

- First, if $r < f(p)$, then we have that the maximal value (obtained in either $x = x_{st}$ or in $x = 1$) of φ^- on the interval is negative, and, thus, we obtain that

$$\varphi^-(x) < 0 \text{ for all } x \in \left(\left(\frac{1}{2}\right)^{\frac{1}{p}}, 1\right]. \tag{22.5}$$

- Second, if $r \geq f(p)$ and $p \geq \frac{1}{3}$, we have that the function φ^- is increasing on the interval with $\varphi^-(1) > 0$. So, the equation

$$2x^p - \frac{1}{rp}x^{1-p} = 1$$

has a unique solution $\beta(r,p)$ such that $\left(\frac{1}{2}\right)^{\frac{1}{p}} < \beta(r,p) < 1$, which can be efficiently computed using Newton's method. Furthermore, we obtain

$$\varphi^-(x) < 0 \text{ for all } x \in \left(\left(\frac{1}{2}\right)^{\frac{1}{p}}, \beta(r,p)\right). \tag{22.6}$$

- Finally, if $r \geq f(p)$ and $p < \frac{1}{3}$, then $\varphi^-(x_{st}) \geq 0$ and, therefore, there exist values $\beta(r,p) < \gamma(r,p)$ such that $\varphi^-(x) \geq 0$ for all $x \in (\beta(r,p),\gamma(r,p))$. These two values can be obtained solving the equation $\varphi^-(x) = 0$, i.e., the equation

$$2x^p - \frac{1}{rp}x^{1-p} = 1$$

has exactly two solutions $\beta(r,p)$ and $\gamma(r,p)$ such that

$$\left(\frac{1}{2}\right)^{\frac{1}{p}} < \beta(r,p) < \left(\frac{1-p}{2(1-2p)}\right)^{\frac{1}{p}} < \gamma(r,p) \leq 1,$$

which can, again, be computed using Newton's method. So, in this case, we obtain that

$$\varphi^-(x) < 0 \text{ for all } x \in \left(\left(\frac{1}{2}\right)^{\frac{1}{p}}, \beta(r,p)\right) \cup (\gamma(r,p), 1). \tag{22.7}$$

Now, collecting Eqs. (22.4)–(22.7), the proof is completed.

In the following, observe that if the spectral radius of the Jacobian $\nabla C\Phi$ is strictly less than one on a certain subspace S of $(0,1]^n$, then the map $C\Phi$ will be a contraction on S. On the other hand, due to the constraints of the model we have $\rho(C) = ||C||_\infty = 1$ and will use infinity norm of $\nabla C\Phi(\mathbf{x})$ instead of its spectral radius to obtain reasonably good results. Namely, in the following theorem we identify subspace $S = S_1 \times S_2 \times \ldots \times S_n$ of $(0,1]^n$ such that for $\mathbf{x} \in S$ we have that $||\nabla C\Phi(\mathbf{x})||_\infty < 1$.

Before we proceed, to simplify the notation, for $i \in N$, denote $\alpha_i := \alpha(r_i, p_i)$, $\beta_i := \beta(r_i, p_i)$, and $\gamma_i := \gamma(r_i, p_i)$. Moreover, let us recall the standard definition: an equilibrium state $\hat{\mathbf{x}}$ of a (discrete) dynamical system (Eq. 10.22) is called locally asymptotically stable if all the eigenvalues of the Jacobian matrix of (Eq. 10.22) computed in $\hat{\mathbf{x}}$ are inside an open unit disc in the complex plane.

Theorem 1. Given a system of coupled EIs whose evolution is described by (Eq. 10.22), denote the following three set of indices

$$N_1 := \{i \in N : r_i < f(p_i)\},$$

$$N_2 := \left\{i \in N : r_i \geq f(p_i) \text{ and } p_i < \frac{1}{3}\right\}, \qquad (22.8)$$

$$N_3 := \left\{i \in N : r_i \geq f(p_i) \text{ and } p_i \geq \frac{1}{3}\right\}$$

and for $i \in N$ define sets S_i by

$$S_i := \begin{cases} (\alpha_i, 1], & i \in N_1, \\ (\alpha_i, \beta_i) \cup (\gamma_i, 1) & i \in N_2, \\ (\alpha_i, \beta_i) & i \in N_3. \end{cases} \qquad (22.9)$$

If discrete dynamical system (Eq. 10.22) has an equilibrium point $\hat{\mathbf{x}}$ such that $\hat{\mathbf{x}} \in S = S_1 \times S_2 \times \ldots \times S_n$, then $\hat{\mathbf{x}}$ is an asymptotically stable equilibrium state of (Eq. 10.22).

Proof. Having an equilibrium point $\hat{\mathbf{x}}$ such that $\hat{\mathbf{x}} \in S$, previous lemma implies that for every $i \in N$

$$||F(\mathbf{x})||_\infty = \max_{i \in N}\left(r_i p_i \frac{|1 - 2x_i^{p_i}|}{x_i^{1-p_i}}\right) < 1.$$

On the other hand, $\rho(\nabla C\Phi(\mathbf{x})) = \rho(CF(\mathbf{x})) \leq ||CF(\mathbf{x})||_\infty$ and, consequently,

$$\rho(\nabla C\Phi(\mathbf{x})) \leq ||C||_\infty ||F(\mathbf{x})||_\infty = \max_{i \in N}\left|\phi_{r_i, p_i}'(x_i)\right| < 1.$$

Therefore according to Theorem in Appendix B in Mihailović et al. (2014b), $\hat{\mathbf{x}}$ is asymptotically stable equilibrium of (Eq. 10.22).

According to the previous theorem, the set S can be considered as a space of states which are the potential asymptotically stable equilibrium points.

To better visualize the subspace S where map $C\Phi$ is a contraction, we have plotted Figs. 22.4 and 22.5. Fig. 22.5 illustrates different forms of S_i depending of

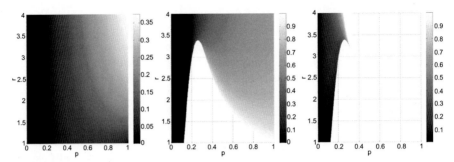

FIGURE 22.4

From left to right, plots of the values of $\alpha(r,p)$, $\beta(r,p)$, and $\gamma(r,p)$ for $r \in [1,4]$ and $p \in (0,1]$, respectively

Reprinted with permission from Mihailović, D.T., Kostić, V., Mimić, G., Cvetković, L.j. 2015. Stability analysis of turbulent heat exchange over the heterogeneous environmental interface in climate models. App. Math. Comp. 265, 79–90.

parameter values r_i and p_i, while in Fig. 22.4 we have computed values of α_i, β_i, and γ_i as surfaces over the region $(r_i, p_i) \in [1,4] \times [0,1]$. Therefore, gaps between these surfaces for each fixed pair (r_i, p_i) represent set S_i (following their form given in Fig. 22.5).

Moreover, if we are able to show that there exists a closed subset $A \subseteq S$ invariant under the map $C\Phi$, i.e., $C\Phi : A \to A$, then, due to the Banach fixed-point theorem, map $C\Phi$ has a unique fixed point $\widetilde{\mathbf{x}} \in A$ such that for the states of (Eq. 10.22)

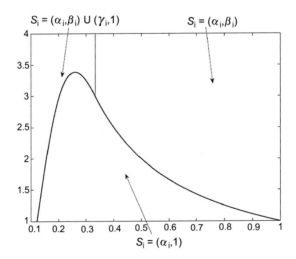

FIGURE 22.5

Plot of the function $f(p)$ for $p_i \in [0,1]$, and different forms of $p_i \in [0,1]$ depending on value of $r_i \in [1,4]$.

Reprinted with permission from Mihailović, D.T., Kostić, V., Mimić, G., Cvetković, L.j. 2015. Stability analysis of turbulent heat exchange over the heterogeneous environmental interface in climate models. App. Math. Comp. 265, 79–90.

$$\lim_{k \to \infty} \mathbf{x}^{(k)} = \mathbf{x},$$

for any $\mathbf{x}^{(0)} \in A \subseteq S$. In other words, such an equilibrium point is the asymptotically stable one.

Here, we note that for $i \in N_2 \cup N_3$ we have small lengths of the stability intervals, and, therefore, if we have that $N_2 \cup N_3 \neq \emptyset$, it is hard to construct invariant subset A such that $C\Phi : A \to A$ in order to obtain the existence of nontrivial asymptotically stable equilibrium state.

On the other hand, for the values of the logistic and the affinity parameters such that $N_2 \cup N_3 = \emptyset$, we now show existence of the nontrivial asymptotically stable equilibrium of the system of arbitrarily coupled EIs.

Theorem 2. Given a system of coupled EIs whose evolution is described by (Eq. 10.22), denote $\tau = \max_{i \in N_1} \frac{r_i}{4}$. If $N = N_1$ and

$$\max_{i \in N} \alpha_i < \min_{i \in N} r_i \tau^{p_i} (1 - \tau^{p_i}), \tag{22.10}$$

then, for arbitrary coupling matrix C, there exists a unique nontrivial asymptotically stable equilibrium state of the dynamical system given by (Eq. 10.22).

Proof. Let us denote $\alpha = \max_{i \in N} \alpha_i$. Then, according to Eq. (22.10), there exists $\varepsilon > 0$ such that $\alpha + \varepsilon < \min_{i \in N} r_i \tau^{p_i} (1 - \tau^{p_i})$. First we show that for arbitrary $i \in N$, $\phi_{ri,pi} : [\alpha + \varepsilon, \tau] \to [\alpha + \varepsilon, \tau]$. Since the maximum of $\phi_{ri,pi}$ equals $\frac{r_i}{4}$, it follows that $\phi_{ri,pi}(x_i) \leq \tau$, for all $x_i \in [0,1]$. So, it remains to check only the lower bound of $\phi_{ri,pi}(x_i)$ for $x_i \in [\alpha + \varepsilon, \tau]$. But, due to the choice of ε, we have that for $\alpha + \varepsilon \leq \phi_{ri,pi}(\tau)$, which, together with monotone properties of the function $\phi_{ri,pi}$ (see Fig. 22.4), implies that $\phi_{ri,pi}(x_i) \geq \alpha + \varepsilon$ for all $x_i \in [\alpha + \varepsilon, \tau]$.

Therefore, denoting $A := [\alpha + \varepsilon, \tau]^n$, we have obtained that $\Phi(\mathbf{x}) \in A$, for all $\mathbf{x} \in A$. In other words, for $\mathbf{x} \in A$, we have that $(\alpha + \varepsilon)\mathbf{1} \leq \Phi(\mathbf{x}) \leq \tau \mathbf{1}$, where $\mathbf{1} := [1\ 1 \ldots 1]^T$, and $\leq$ is understood component-wise. But then, since C is the stochastic matrix, $C\mathbf{1} = \mathbf{1}$ and C, being nonnegative, implies that $(\alpha + \varepsilon)\mathbf{1} = (\alpha + \varepsilon)C\mathbf{1} \leq C\Phi(\mathbf{x}) \leq \tau C\mathbf{1} = \tau \mathbf{1}$. Thus, we have obtained that $C\Phi : A \to A$.

Since $N = N_1$, according to Eq. (22.9), $S = (\alpha_1, 1) \times (\alpha_2, 1) \times \ldots \times (\alpha_n, 1)$ so it follows that $A \subseteq S$, which implies that $C\Phi$ is a contraction on A. Therefore, we can apply Banach fixed-point theorem to $C\Phi : A \to A$, and the statement of the theorem follows.

Here, we note that the last theorem gives only sufficient condition for the existence of a nontrivial asymptotic stable equilibrium point of (Eq. 10.22) where the horizontal energy exchange given by matrix C can be arbitrary. Additionally, the spectral radius is replaced by the infinity norm and the restrictive assumption $N = N_1$ is used. Consequently, we have obtained a relatively demanding condition on parameters r_i and p_i, $i \in N$.

Here, we will illustrate, by application of the above theorems, how Schmidt's paradox can be overcome. It will be done by considering the aggregation effect over a heterogeneous grid-box, through a numerical example following Mihailović et al. (2015). Namely, we will use a heterogeneous grid-box over the Prospect Park,

New York, USA, shown in Fig. 10.5. This grid-box consists of three patches: urban part (surrounding buildings), vegetative part (mixture of trees and grass), and water surface (lake). Here parameter p, that represents the total turbulent energy exchanges between a single EI and the surrounding environment, is expressed through its cover percentage in the grid-box, i.e., p_1 of the grid-box is covered by concrete, p_2 by grass, and p_3 by water. The values of the parameter p are $p_1 = 0.28$, $p_2 = 0.46$, and $p_3 = 0.26$, see Fig. 10.5. Each of these EI surfaces has different response to the forcing by the solar radiation. Regarding the coupling parameter C we can say that its quantification is very difficult. In this example we first consider vertical energy exchange for each patch and then we will show how horizontal energy exchange can have a stabilizing role in the dynamics of the energy exchange in and over the heterogeneous grid-box. If we use logistic map to describe the process of energy exchange (Mimić et al., 2013), then the response of the EI surface is represented with logistic parameter r

$$x_i^{n+1} = r_i x_i^n \left(1 - x_i^n\right), i = 1, 2, 3 \tag{22.11}$$

Now we calculate dimensionless temperature expressing the energy exchange for each patch, starting from some initial state. As it is known, concrete is a good absorber of solar radiation and also is emitting the long wave radiation intensively, so r_1 has high value, e.g. $r_1 = 3.33$. In the case of water, there is a strong flux of latent heat above, so we could set $r_3 = 3.10$. Situation over the grass is less turbulent and a suitable value for r_2 would be below 3, e.g. $r_2 = 2.15$.

We will start with the set of initial conditions $x_1^0 = 0.8$, $x_2^0 = 0.75$, and $x_3^0 = 0.7$ which corresponds to an interesting case of initially high energy exchange. After 10^4 iterations, we have the bifurcation of the logistic map over the patches for concrete (1) and water (3), as indicated in Fig. 10.5 to the following values of dimensionless temperature: $x_1^{f_1} = 0.8296$ and $x_1^{f_2} = 0.4707$, and $x_3^{f_1} = 0.7646$ and $x_3^{f_2} = 0.5580$, respectively. Dimensionless temperature of the grass stabilizes numerically at $x_2^f = 0.5349$. Here, superscript f indicates the final state(s) of the iterations in Eq. (22.11). The iterative procedure in these calculations is illustrated at the left panels of Fig. 22.6.

Now, if we want to derive, from this data, representative value of the dimensionless temperature over the whole grid-box, we have to make an averaging of their values. Thus, we get the four different representative values:

$$x_{a_1} := p_1 x_1^{f_1} + p_2 x_2^f + p_3 x_3^{f_1} = 0.6771, \quad x_{a_2} := p_1 x_1^{f_1} + p_2 x_2^f + p_3 x_3^{f_2} = 0.6234,$$

$$x_{a_3} := p_1 x_1^{f_2} + p_2 x_2^f + p_3 x_3^{f_1} = 0.5766 \text{ and } x_{a_4} := p_1 x_1^{f_2} + p_2 x_2^f + p_3 x_3^{f_2} = 0.5229,$$

and, consequently the aggregated logistic parameters $r_{ai} := (1 - x_{ai})^{-1}$, $i = 1, 2, 3, 4$, are:

$$r_{a_1} = 3.0973, \quad r_{a_2} = 2.6555, \quad r_{a_3} = 2.3621, \text{ and } r_{a_4} = 2.0961. \tag{22.12}$$

These values would indicate that there exists a bifurcation in this heterogeneous grid-box due to bifurcation in the patches consisting of concrete and water leading to

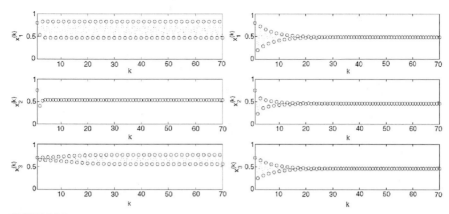

FIGURE 22.6

Evolution of the decoupled logistic Eq. (22.11), left panels, and coupled logistic Eq. (10.22), right panels, for the grid-box described in Fig. 3.15, whose coupling matrix C is given by Eq. (22.13). In plots, k is the number of iterations.

Reprinted with permission from Mihailović, D.T., Kostić, V., Mimić, G., Cvetković, L.j. 2015. Stability analysis of turbulent heat exchange over the heterogeneous environmental interface in climate models. App. Math. Comp.
265, 79–90.

the case when the occurrence of the Schmidt's paradox in numerical treatment of sensible heat fluxes is expected.

Here, we note that the Schmidt's paradox occurs because of neglecting the horizontal energy exchange between different environmental interfaces when either flux aggregation method is used or the methods of parameter aggregation and flux aggregation are combined. However, these methods are still in use in climate modelling community. A reason for that lies in the fact that horizontal energy exchange inside a heterogeneous grid-box is very difficult to describe or at least parameterize. Therefore, this paradox has a source in the subgrid scale surface flux parameterization in atmospheric as well as in climate models. To investigate this phenomenon, we have introduced the dynamical system approach, where the horizontal energy exchange is represented by the matrix C of coupling parameters.

As we have illustrated, in the dynamical systems approach, the (logistic) parameter aggregation has no mathematical sense whenever the bifurcation ($r > 3$) or chaos ($r \geq 3.45$) happens in one of the patches. Therefore, in our model, the asymptotic stability is the proper and realistic indicator that corresponds to the notion of a physical indicator (the overall r in this example) of the nature of energy exchange over the heterogeneous grid-box. Since it is difficult to specify the elements of the matrix C we have derived the sufficient conditions for the asymptotic stability that can be applied for any coupling matrices C.

In this example we obtain that $N = N_1 = \{1,2,3\}$, and

$$S_1 = (0.0518, 1], \quad S_2 = (0.1032, 1], \quad \text{and} \quad S_3 = (0.0425, 1].$$

Furthermore, since $\tau = 0.8325$ and

$$\max_{i=1,2,3} \alpha_i = 0.1032 \quad < 0.1376 = \min_{i=1,2,3} r_i \tau^{p_i}(1 - \tau^{p_i}),$$

we can apply Theorem 1 and conclude that the whole grid-box stabilizes in the unique equilibrium state.

Therefore, we can conclude that, independently of the rates of the horizontal exchange, this heterogeneous grid-box exhibits stable energy exchange despite expected bifurcations in isolated patches.

Now, let us consider an idealized example when the rate of energetic influence of the grass EI to the concrete EI is estimated as 20%, while rates of influence of the concrete and the water to the grass is estimated to be 10% each, and the influence of the water EI to the grass EI by 10%, i.e.,

$$C = \begin{bmatrix} 0.8 & 0.2 & 0 \\ 0.1 & 0.8 & 0.1 \\ 0 & 0.1 & 0.9 \end{bmatrix}. \tag{22.13}$$

The evolution of dimensionless temperatures for the corresponding coupled system (Eq. 10.22) is given at the right panels of Fig. (22.6). As we can see, in this case, the system stabilizes in the first 50 iterations. In our numerical experiments the rate of convergence (corresponding to the stabilization time scale) highly depends on the structure of the coupling matrix. It is noted that for matrices that exhibit strict diagonal dominance property (Kostić, 2014), the stabilization occurs faster. To prove this theoretically would be an interesting challenge, since it could explain, at least in a certain sense, the role of the horizontal exchange in the heterogeneous grid-boxes. Therefore, Schmidt's paradox related to numerical treatment to real fluxes can be avoided using a new approach of discrete dynamical system of coupled logistic equations. Moreover, this explains that the horizontal energy exchange can have a stabilizing role in the dynamics of the energy exchange in and over the heterogeneous grid-box.

In the next example, we consider the situation discussed in Raddatza et al. (2013), i.e., when the sensible heat flux comes from an unconsolidated sea-ice surface. To that end, consider the idealized example of the heterogeneous $2 \times 2 \text{ km}^2$ grid-box consisting of two environmental interfaces. The first EI is the consolidated ice that covers 75% of the grid-box, while the second EI is the area of open water (covering 25%). From the reason of the simplicity, but without loosing generality, we will define the two-component system (ice and water EIs). Note that one possible model could be a model with many water openings making a higher resolution system. This would be especially interesting in estimating the influence of the horizontal exchange rates in the case of functionally different water openings.

Let us suppose that the logistic parameters for ice are $r_1 = 2.6$ stable exchange with small fluxes), while for water is $r_2 = 3.5$ (larger fluxes leading to higher sensitivity of the system's behavior in terms of chaotic logistic behavior). Now, we are interested in estimating the stability of such heterogeneous grid-box. Note that, based on Raddatza et al. (2013), one would not expect to consider such system as a stable one, since water openings can produce the sensible heat fluxes contributing to an instability over them.

As expected, after computing the stability sets of Theorem 2, in this case, we obtain

$$S_1 = (0.2224, 0.6585) \text{ and } S_2 = (0.0404, 0.2119) \cup (0.5123, 1].$$

Therefore, Theorem 2 could not be applied to conclude the stable behavior independently of the horizontal exchange rates inside the grid-box. But, to discuss this situation more thoroughly, we start with the initial dimensionless temperatures for ice and water $x_1^0 = 0.3$ and $x_2^0 = 0.9$, respectively. Then, in Figs. 22.7 and 22.8 we depict the behavior of the uncoupled EIs (left panels) and the coupled ones (right panels). Namely, the right panels of Fig. 22.7 show the chaotic behavior of air over the water openings generated by their heat fluxes (left lower panel) that can destabilize the ice flux (left upper panel). On the other hand, we see in the right panels of Fig. 22.7 that the system is not stable. Namely, the ice EI component (upper panel) and the water EI component (lower panel) both exhibit bifurcation. This solution corresponds to the situation when both EIs' influence rates to one another are estimated by 20%. Contrary to that, assuming that the water openings are relatively small so that their energy is mostly communicated to ice ($c_{21} = 50\%$), we obtain a completely different picture, as shown in Fig. 22.8. In this situation, the ice EI could overcome the flux disturbances and maintain the stable behavior of the grid-box.

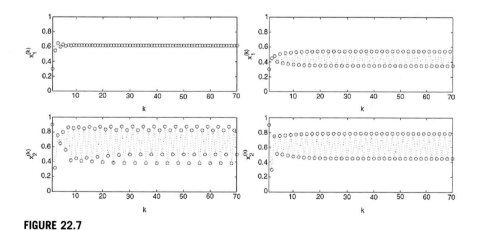

FIGURE 22.7

Evolution of the decoupled logistic Eq. (22.11), left panels, and coupled logistic Eq. (10.22) right panels, for the ice-and-water grid-box example, where the coupling matrix parameters are $c_{11} = c_{22} = 0.8$.

Reprinted with permission from Mihailović, D.T., Kostić, V., Mimić, G., Cvetković, Lj. 2015. Stability analysis of turbulent heat exchange over the heterogeneous environmental interface in climate models. App. Math. Comp. 265, 79–90.

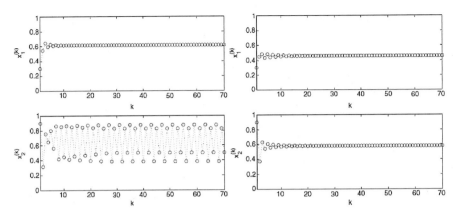

FIGURE 22.8

Evolution of the decoupled logistic Eq. (22.11), left panels, and coupled logistic Eq. (10.22) right panels, for the ice-and-water grid-box example, where the coupling matrix parameters are $c_{11} = 0.8$ $c_{22} = 0.5$.

Reprinted with permission from Mihailović, D.T., Kostić, V., Mimić, G., Cvetković, L.j. 2015. Stability analysis of turbulent heat exchange over the heterogeneous environmental interface in climate models. App. Math. Comp.

265, 79–90.

REFERENCES

Arenas, A., Diaz-Guilera, A., Kurths, J., Moreno, Y., Zhou, C., 2008. Synchronization in complex networks. Phys. Rep. 469, 93–153.

Berman, A., Plemmons, R., 1994. Nonnegative Matrices in the Mathematical Sciences. SIAM Publications, Philadelphia.

Chen, Y., Rangarajan, G., Ding, M., 2003. General stability analysis of synchronized dynamics in coupled systems. Phys. Rev. E 67, 026209.

Gunji, Y.-P., Kamiura, M., 2004. Observational heterarchy enhancing active coupling. Physica D 198, 74–105.

Kostić, V., 2014. On general principles of eigenvalue localizations via diagonal dominance. Adv. Comput. Math. 41, 55–75.

Mihailović, D.T., Budinčević, M., Perišić, D., Balaž, I., 2012. Maps serving the combined coupling for use in environmental models and their behaviour in the presence of dynamical noise. Chaos Solitons Fractals 45, 156–165.

Mihailović, D.T., Mimic, G., Arsenic, I., April 2014a. Climate predictions: the chaos and complexity in climate models. Adv. Meteorol. 1–14. Article ID 878249.

Mihailović, D.T., Kostić, V., Balaž, I., Cvetković, L., 2014b. Complexity and asymptotic stability in the process of biochemical substance exchange in a coupled ring of cells. Chaos Solitons Fractals 65, 30–43.

Mihailović, D.T., Kostić, V., Mimić, G., Cvetković, L.j., 2015. Stability analysis of turbulent heat exchange over the heterogeneous environmental interface in climate models. App. Math. Comp. 265, 79–90.

Mimić, G., Mihailović, D.T., Budinčević, M., 2013. Chaos in computing the environmental interface temperature: nonlinear dynamic and complexity analysis of solutions. Mod. Phys. Lett. B 27. Article ID 1350190.

Pikovsky, A., Rosenblum, M., Kurths, J., 2001. Synchronization: A Universal Concept in Nonlinear Sciences. Cambridge University Press, Cambridge.

Raddatza, R.L., Galleya, R.J., Candlisha, L.M., Asplina, M.G., Barber, D.G., 2013. Integral profile estimates of sensible heat flux from an unconsolidated sea-ice surface. Atmosphere Ocean 51, 135−144.

Zhou, C., Lai, C.-H., 1998. Synchronization with positive Lyapunov exponents. Phys. Rev. E 58, 5188−5191.

PART

Synchronization and stability of the biochemical substance exchange between cells VII

Environmental interfaces and their stability in biological systems

23

23.1 BUILDING BLOCKS OF ENVIRONMENTAL INTERFACES

Iconic example of environmental interfaces in the biological world are membranes. They compartmentalize living systems into cells and organelles allowing them to evolve different modes of functioning. Membrane-associated proteins act as the main intermediaries for receiving and filtering external signals. They determine a set of small molecules that are allowed to enter the cells or they can block or facilitate extraction of some molecules. All this defines how cells respond to changes of environmental conditions by altering gene expression and dynamics of metabolic pathways. In the biological world, there are myriad differences between different mechanisms and corresponding functions of membrane-related signal transits, but one of the fundamental abilities that allow very existence of signal transit is perception by receptors. Physiology of signal perception and transduction is a wide field that covers types of receptors, mechanisms of their actions, biochemical processes, and topology of their distribution (Mayne et al., 2016; Schenk and Snaar-Jaglaska, 1999; Soyer et al., 2006). While these research fields offer essential information on each specific case, they cannot be readily converted to a broader understanding of the role of interfaces in biological systems. Therefore, we will offer a short overview of several important issues that should not be neglected in both theoretical and modelling treatment of environmental interfaces.

We will start with a rather straightforward remark that each perceptive entity can observe or register only those changes in its vicinity which activate its functional operations. In other words, receptors imprint the forms of their actions on the structures with which they interact. In that manner, each identification is a functional identification, i.e., potential action. More generally, the externality of any such entity is prestructured in accordance with its functional purposes and objectives, through a process of assimilating external changes within the operative pattern of that entity. Therefore, it is incorrect to talk about objective structures (different molecules, signals, etc.) in the vicinity of any living system. The mere fact that something can be perceived means that it becomes connected to the needs of the living system. And conversely, since each organism constructs its environment as a field of interests, only functionally relevant changes can appear in it. As an inherent consequence of such assimilation of reality, the external medium of living systems is transformed into a functionally treatable environment where external, arbitrary events can be conveyed into the internal pattern of reactions. In this way, the system displaces itself from direct contact with the environment and can afford to be indifferent to a subset of external changes, thereby opening possibilities for developing different functional strategies.

We will focus our attention to protein receptors as rudiments of perceptivity that are able to detect and successively assimilate segments of external changes by their own functioning. Usually applied scheme, where the folded protein, with a static spatial structure, passes through a cyclic series of conformational changes after interaction with other structures (small molecules, DNA, or other proteins), beginning and ending with an elementary stable state, is quite inaccurate and oversimplified. During the last few decades, several important aspects of protein's functioning have been discovered, which has necessarily led to a revision of the previous picture.

First is protein's structural instability. In proteins we can distinguish two dynamically very different regions: liquid-like and solid-like. The second group is represented by domains (α helices and β sheets) as the fundamental, unchangeable structural units of proteins, while the first consists of residues and represents a matrix surrounding solid-like fragments (Hinsen et al., 2002; Kneller and Smith, 1994; Kurzynski, 1998). Domain movements are slow with high amplitudes, while movements in liquid-like regions have much higher frequencies and smaller amplitudes (Hinsen, 2000). Owing to the existence of "energy walls" that stem from local deformations, these low-energy movements are restricted only to the local region and are not influenced by movements within other regions. Functionally speaking, their continuous movements "lubricate" large-scale structural changes by providing low-energy pathways between conformational states (Thune and Badger, 1995).

Second one is the protein—solvent integration. Movements of protein domains and its overall functionality in general can only be achieved if the protein is hydrated (Bellissent-Funel, 2000). During hydration, a sort of micelle is formed around the protein where H_2O molecules are immobilized and oriented in a complex multilayered structure. Within such micelles, the translational and rotational degrees of freedom of isolated water molecules are transformed into vibrational modes of the protein—water complex (Smith et al., 2002). Along with hydration, an essential influence for achieving the functionality of the protein is the high viscosity of intracellular solvent, thus making biological reactions indescribable by the transition state theory (Sumi, 2001). More precisely, since all fluctuations and molecular translocations are slowed down, activation energy decrease and solvent fluctuations are able to produce conformational fluctuations in proteins. Such motions of segments, often mutually independent, indicate that the fluidity of a protein's structure has an important role in providing the energy for enzymatic transformations. As emphasized by Ferdinand (1976), that energy can only be provided from the translational energy of solute molecules colliding with the enzyme—substrate complex, since the substrate itself has become tied down in the active site and cannot provide translational energy for participating in the reaction. Today we have several models which describe that process of vector translation of collision-obtained energy to particular domains within the enzyme, but common to all of them is the perspective originating in Lumry's work (Lumry, 1971, 1980). According to him, the enzyme is a structure which transforms the free energy of a medium's fluid movements into chemical processes. Which of the existent models is closest to the real situation is not at issue here. What is of essential importance, and what should be again emphasized is the fact that all of them view the enzyme as a transducer of energy which is able to transform thermal fluctuations of environment into vector movements and use them for catalytic transformations.

Third important factor is the so-called static disorder. Until recently, it was considered that each population of genetically identical proteins consists of identical copies with identical functional properties. However, it has been shown that the final configuration of the protein is not completely a result of settling into the ideal optimum of energetic states, but the conformational variations of folded proteins are a reflection of the medium's current state in which folding takes place, as a combination of funneling and independent formation of domains (Dill et al., 1993; Finkelstein and Shakhnovic, 1989; Karplus and Shakhnovich, 1992; Onuchic et al., 1997; Veitshans et al., 1997). Therefore, each population of proteins is composed of a great number of conformationally different units (conformers) each having its own specific energetic minimum, and—more importantly—they each display broad and asymmetrical distribution of activity within a population, a so-called static disorder (Xie and Lu, 1999; Xie, 2002).

At the end of this short overview, we can return to the initial question: what is unique in protein functioning which distances them from the machine-like paradigm? The first reason for insisting on such a distinction is the fact that the protein and its vicinity operate as an integrated unit where the actual function or functional state of the protein is very closely connected to the state of the surrounding medium. Second, the spatial organization of a protein is in constant move, causing continuous alterations of its functioning. This mode of action widely exceeds the explanatory capacity of the usual models applied for autonomous program-driven machines.

Dealing with such structures raises many difficulties if someone tries to simply extrapolate the problem into a common input/output framework. First, regulation is usually perceived as dependent on specific devices for specific types of stimuli. Here the situation is fundamentally different. Proteins possess active sites which can be considered equivalent to functional receptors; their activation initiates a defined sequence of configuration changes which is also in accordance with common thinking about regulation. But on the other hand, their spatial fluctuations (with all the functional consequences) do not have any connection to their perceptivity. In other words, we cannot straightforwardly talk about specific parameters of the medium which influence a protein's dynamics, or about a separate causal factor. The protein physically immobilizes part of its environment, constructing a functional unity with it; but its dynamics does not include receptors as mediators. Which external changes will induce a protein's fluctuation—and how—is not predefined by a scope of receptors or by any processing of signals; it is fully dependent on a continuity of actual context. Here we cannot talk about intracellular space as an indifferent scheme which can be arbitrarily filled without functional consequences (as it is a case with "objective" space) because through such fusion of movement/state, the entire intrasystemic medium (nonperceptive and nonschematized) is transformed into a regulative factor which can (re)route functional processes. Proteins thus avoid the hazard of overlooking changes but, by possessing active sites, they are not mere passive fluctuating structures. Being such transitional forms, they can functionally structuralize their environment but at the same time are not completely connected to such structuralization; they partially remain in an unmediated contact with the surrounding medium which provides energy for a segment of the protein's activity by which, on the other hand, they withstand the medium's fluctuations. Therefore, generation and regulation of processes in living systems cannot be treated in full accordance with the classic paradigm of an input/output model with localized well-defined regulators. Proteins are the first step in transforming meaningless external variations into a schematized and functionally treatable construction.

23.2 EMERGENCE OF FUNCTIONALITY

All interfaces in the biological world are characterized by an inherent relation between perceptivity and assimilation. Accordingly, if something is extracted from the continuity of external changes through perceptivity, it is necessarily associated with operative patterns. Also, since we are dealing with material systems, the necessity of action/reaction is inherent to them and there is no possibility for infinite recursions, as is the case with formal systems. Therefore, association escapes its solipsistic character and has bimodal consequences: it becomes an act of prescribing the forms of operations to others, and on the other hand it is a process of functional self-labeling. Only such assimilation can be considered functionally meaningful because passive association becomes liable to interpretations by other elements of the system, which form a set of possible transformations for that system. Therefore, when analyzing living systems, it is not completely accurate to apply only the framework of causality—where elements are causes of systemic conditions—but is also necessary to deal with the element's interpretations of systemic states. It is very important to note this difference, since interpretations can only happen in a situation where the system is able to impute meanings and where subsequent reactions are based on the structure of that imputation. In that way, arbitrary 'distance', in the form of mutual interpretations along biological interfaces, weakens the requirements of physical and chemical relations, rendering different strategies of organization possible. It has several fundamental consequences.

First, the division established by perceptivity is division into perceptive realities and those beyond perception. And for simple noncognitive living systems, what is beyond perception is nonexistent. Inserting such distance relieves the system of the pressure of continuous environment changes, since only some aspects or indirect consequences of environmental changes can be registered. In that way, the system as an integrated whole achieves more space for building its own processes since it is no longer under pressure to respond immediately to each environmental change (the existence of a physical barrier is a necessary precondition). In this way, what is beyond immediate perception is not a systemic, but merely local category, where rules cannot be uniformly transferred among different perceptive domains. Therefore, within a system, significant parts of processes are enclosed in their own domains, beyond the supervision of other parts. Since interactions are localized within boundary regions and are dependent on the pattern of mutual representation of subsystems, it is an excellent basis for developing a wide spectrum of ad hoc, local solutions. Moreover, not only are changes beyond the scope of perceptivity invisible, but fluctuations within particular perceptive boundaries are also erased. Everything that is inside perceptively defined operative units, all differences, varieties, internal dynamics are erased and are unobservable for elementary perceptive elements—i.e., proteins. In this way, the internal systemic environment becomes the subject of radical transformations.

Second, by establishing perceptive boundaries for every aspect of interactions, external changes are no longer simply contiguous values obtainable by homogeneous succession within a scale of changes. On the contrary, if they are perceived as changes on different sides of the boundary, following the chain of transformations (which is a consequence of their appearance), diametrically opposite results can be obtained, no matter how close (according to absolute measures) they can be. As an inherent consequence of such configuration, a more or less undistinguishable continuum of changes is transformed into a set of separated operative unities with defined focuses of transformations. In this way the external world is transformed into an assembly of operational absolutes where each absolute is connected to a predefined reaction which is purposeful within the system. Despite being a basis for systemic organization, this situation also makes intrasystemic environment a highly conflicting place. What is usually presented as a harmonious flow of metabolic pathways is actually achieved only by the use of power; i.e., by using the possibility of influencing behavior and transformations of others by combination of transformations (i.e., physical transformations of shape or structure), transpositions (displacement of other elements into different contexts, which is also a process of assigning different meanings to elements), and assimilations.

Since the actual modalities of each of these categories can be very organism-specific, further evaluation will be focused on the general properties of conflict resolution. In metabolic space we can distinguish few phases: (1) entrance into observable space, (2) establishment of local power constellation and reconstruction of boundaries, and (3) normative encirclement and transformation of the object. First phase is actually a first step which leads to generation of conflict. Here, some change of environment becomes perceptively visible either by physical approaching toward some functional element or subsystem (e.g., translocation of molecules in cytoplasm), or by increased frequency of occurrence of some set of changes (e.g., increased level of damaged proteins can lead to activation of cell's heat shock response). After that, establishment of local power constellation is a necessary consequence because a subsystem whose perceptive threshold has been traversed, by the mere "recognition" of some external change, enters into indirect interaction with subsystem(s) which already perceived that change. "Already perceived" does not mean only a struggle for assimilation of some change with same functional context, but what is more important (because first situation is resolved only at the level of local conflicts with no influence on organizational rearrangement), it can also be a situation where different subsystems try to assimilate same external change with their own different operative patterns. In the second case, resolution of conflicts overgrown significance of local competitions because it means activation of new operative patterns, changes in perceptive focuses, and treatment of environment, thus inducing activation of global organizational rearrangements. Although local power constellation is always the main determinant of conflict resolving, if systems amplify expression of certain groups of proteins or regulators (activating number of positive or negative feedbacks), global balance of interactions (which is essentially stochastic) will necessary be forced toward desirable state. Last phase, which chronologically is last but its possibility of appearance is precondition for all previous phases, is normative encirclement of elements (objects) and their transformation. Only at this stage we can talk about implementation of power in the sense of its definition because just here subsystems effectuate control over actions (transformations) of external elements. In other words, only when assimilation is conjoined with transformation of assimilated elements in accordance with norms of "observer" we can talk about functional assimilation and whole process of conflict raising and resolving gain sense.

In short, in a system which is basically stochastic, establishment of control, and arrangement of particular processes into organized metabolic pathways is achieved only through continual conflict resolving between subsystems with their own tendencies of assimilations which are established and changed (*in establishment and in changing*) during these interactions. In contrast from usual representation of metabolism as a predefined set of algorithmic processes, flow of transformations in living systems is generated during that process itself and decisions about the next step is always realized only for the particular case. Although assimilation of some nutrient by the cell usually results in a well-known catabolic pathway, it is not an issue here. In a living system, during a process, at each temporal moment t we can define only a limited set of possible states at the moment $t + 1$. After actualization of some of these states, a new set of states will be generated as a function of actual system configuration. Owing to such pattern of process generation, determinants of the next stage are not only usual control mechanisms as a feedback, attachment of inhibitor/activator, and so on. In addition, a very important role is reserved for factors described here: reproductive cycles and assimilative conflicts. These aspects can be completely omitted from analysis if we apply algorithm-like reasoning as only valid paradigm.

Although perceptive assimilation plays a significant role in organizing a system, it is far from the final stage in building a functional autonomous system. Besides direct physical and chemical interactions, there are also upper-level organizational differentiations which are based only on segments of material dynamics. If we mark relations in some material process as relations[1] (first-level relations), then relations based on segments of that processes are relations[2] (second-level relations), relations among relations[2] are relations[3] (third-level relations) and so on. It is very important to note that in living systems there is no exclusive primacy of such linear stratification of hierarchies, but establishment of direct branching to higher or lower levels is very common (e.g. global regulators as H-NS or Hsp in bacteria). It is obvious that such a network of relations cannot be established without the existence of material elements, but from the mere fact that it is based only on segments of material dynamics, a certain distance entails. In other words, upper-level organization is always partially indifferent to material fluctuations and its structure can withstand different perturbations without being changed by them. As long as processes are performed in the usual manner and with the usual dynamics, organization remains unaffected by material changes and has the status of an a priori given controller for them. It does not imply view of organization as a set of nonmaterial principles which regulate material structure. A reason why "organization" transcends in time material structures and to some extent influences its shaping and behavior is a fact that it is not based on monitoring actual configuration of certain proteins or their precise actual relations, but on a flow of processes which are only a segment of totality of material dynamics. Also, it should be emphasized that construction of each organizational level does not mean mere incorporation of previous levels into subsequent ones but is based on a gradual restructuration of relations within new contexts.

Since the process of such organizational differentiations is formed in accordance with actual context, instead of strict determination by formal, centralized rules, there is a wide spectrum of possible formation matrices. It certainly does not imply unlimited freedom of differentiations, because within relatively closed and stable systems there is always a finite field of structural possibilities determined by the system. However, owing to the continuous reconstruction of a living system, only the formation of functionally meaningful subsystems can be established during a prolonged period of time because it is in just such a case that they build and became embedded in a network of positive feedbacks which maintain them during and through intrasystemic reproductive cycles. All other subsystems can be freely formed, but the temporal intervals of their existence are negligible because they are not accompanied by maintenance mechanisms. In any case, through internal differentiations, the system reconstitutes itself into a number of subsystems, where system and environment distinction repeat separately for each microdomain.

When stable functional subsystems are established (or more precisely: are *in* establishment), a wide spectrum of new possibilities for further organizational building is formed. First, the formation of subsystems, partially provided by reflexive reproduction, restrains the horizons of that same reproduction, stabilizing it through the formation of different domains of protocommunication. In other words, differentiation into subsystems generates relatively stable, partially separated groups of mutual coordination (protocommunication), which are nonunderstandable for others. Therefore they are forced to generate "public" output signals, i.e., communication channels visible to other subsystems. The subject matter of protocommunication (but not of physical/chemical interactions, which are unaffected) within a particular subsystem narrows greatly and is restricted only to a small, functionally meaningful domain of total possibilities. Therefore, during assimilation of functional elements, the otherwise open spectrum of possible transformations is reduced and gradual structuration of processes takes place. It is not completely correct to define it as a binding of some elements to a particular subsystem (e.g., x protein is a member of xy metabolic pathway) because its association is always only actual, determined by actual context and never predefined (however, the pattern of association is usual, but this is not the issue here). Therefore, subsystemic protocommunicative space can be best described as a discontinuous classification where relations are established through continuous overlapping and impositions. Establishing such an enclosure has a number of consequences for building up a living system's organization.

First, the set of elements which constitute a subsystem is treated by other subsystems as a unity. Since particular groups are functionally conjoined and are mutually identified as unities using defined output signals for interactions, local problems are allocated to separated levels of time, priorities, and functioning.

The same factor enables much more freedom for intrasubsystemic dynamics because the formation of input/output relations fixates external control only of output signals while the richness of internal dynamics remains invisible for a particular set of observers. It should be emphasized that each phase of internal dynamics can be observed or controlled at some instance, but it is not in conflict with this model because possibility of observation is always relative and therefore separated into different levels of subsystemic protocommunicative domains determined by actual pattern of subsystemic overlapping.

The next consequence is a proliferation of internal environments. In accordance with the perceptivity of constitutive elements and constructed models of interactions with other subsystems, each functional subsystem develops its own reductive pattern of its environment. In that manner, one more or less homogenous intrasystemic medium is transformed into a vast number of separate, local, meaningful environments. With the development of such differentiation, the internality of a system becomes more thoroughly encompassed by functional meanings, while the subsystems themselves are relieved from pressure for excess processing.

At the same time, enclosure in protocommunicative domains leads to the possibility of autonomously determining the intensity and normativity of reproductive cycles; in other words, it leads to the development of subsystemic self-referential autonomy. In this way, each subsystem encloses itself according to its own modes of differentiation, its own construction of reality, and its own functional matrix. Of course, they cannot develop absolute autonomy, but it is very important to emphasize that each subsystem's mode of functioning is generated as a result of self-organization based on local conditions. Through it, in an environment which might seem homogeneous, chaotic or full of noise to an external observer, subsystems generate a causative basis for their own activity (operations).

Finally, it is essential to bear in mind that direct interactions of two or more subsystems should not be analyzed in accordance with geometrical analogies (as touch, intersection, and so on). The main reason is a fact that during their direct interaction, boundaries of one subsystem can be transferred into operative space of other subsystem(s) thus enabling manipulations of each other's complexity. It does not imply violation of input/output model because scope of that model is development of protocommunication through exchange of signals, while accent here is on building of local power relations without participation of indirect mediation by signals. Also, manipulation of other's complexity does not mean opening of boundaries (because it is in principle impossible), but overlapping of competences. Then, subsystemic determinants, constraints imposed on process formations, contextual dependent fluctuations of identities, and modes of assimilations of environment are reconsidered by other subsystem(s). Final result can be only full success or failure of assimilation, but we should take a closer look at transient formation of new (sub-) subsystems, induced by such situations. Here, two or more subsystems partially penetrate into each other and for their own constructions use complexity of the other by interpreting their elements in accordance with own operative pattern. They do not exchange information/signals in a manner of emission and receiving of public available outputs. On the contrary they interact through allocation of different operative horizons to the same set of material structures. It is very important to emphasize that aspect of conflicts, because relations which will result in a certain final condition are constituted through it. It is clear that such situations cannot be preserved during prolonged period of time (and thereby cannot be established as a stable subsystemic construction with all consequences: successive construction of regulations, functional relations…) because the system is always under pressure to operationalize its states, but, and it should be clearly underlined, it is a specific mode of interactions for living systems.

Most of the mechanisms explained above are formed and function only at the local level. Until now this was only implied, but here it is explicitly postulated as one of the genuine properties of living systems which ultimately leads to the internal incoherence of their organization. Of course, it is not only spatial locality which determines the proliferation of separated operative frameworks. As has already been said, perceptivity is not indifferent toward its objects; assimilation is an inseparable unity composed of association-prescription-transformation. Therefore, unlike logical relations which are indifferent toward their objects (for example, the logical relation $A < B$ does not deform their values just because it compares them), perceptive relations by rule perform this kind of deformations (Piaget, 1973). In this manner, relations between two or more observed objects will always be relations of overestimation or underestimation, thus making a comparison impossible based on uniform logical rules. Instead, the context-based formation of local operative fields emerges as a main determinant of their functional composition. On the other hand, attending to the relations between systems (subsystems, functional elements) in general, we can see that by assimilating only a segment of the environment, each perceptive entity or functional subsystem places operative focus on that part of reality. Its construction of an operative pattern is under continuous pressure from the insecurity of such noninclusion. It is clear that local operations generated through these processes are not forced to be consistent with other operations generated at different spatial and temporal points within the same system. Certainly, organisms are highly integrated systems, due to long internal coevolution, where externally visible segments of subsystemic internal dynamics have developed in accordance with perception of them by observers, which themselves evolved in accordance with their perception of externality—but the basis for all functional strategies developed during that process is the actual (and inherent) condition of incompleteness and the insecurity of perceptive inclusion. Here we should bear in mind that complete transparency would lead to destruction of any operative focus for further construction of a functional system. Moreover, inability of full penetration into actions of others is a triggering factor for development of strategies of indirect communication. Also, only through continuous reproduction of conflict situations (i.e., nonpredefined process flow) can the system achieve sufficiently high fluidity in transformations from one operative pattern to another without needing to construct specific regulative pathways for each situation. In this way, at each temporal moment, local contextual fragments are determinants for the pattern and constitution of functional elements, routing flow of processing paths to itself and finally establishing systemic incoherency. Of course, the nonexistence of overall logical coherence does not mean that generation of systemic events without systemic causation is allowed, but only that mutual coordination of local events and straightforward generation of hierarchies is not necessary. Each attained level of local settlement is related only to a particular part of the system, i.e., it is genuinely incomplete, necessarily leading to new imbalances and initiating cycles of new reconstructions. Although seemingly a paradox, only through this kind of organization can a living system establish itself as an autonomous entity which is flexible enough to allow instantaneous establishment of a wide spectrum of autonomous, local structures of organization.

Finally, if the incoherency of living systems is their essential property, the question of their unification remains open. How they are encircled into a functional unity with clear demarcation between the external and internal environment? Without presuming to give a final answer, two basic mechanisms could be named. First, the external environment is constituted exclusively through perceptive assimilation, which means that the living system repeats the same types of endogenous differentiations and imprints its own organization onto the external environment. In this way environment is so arranged that it can be manipulated. Further, as a result of the specific integration of proteins and surrounding medium, the internal environment is not characterized by such exclusive primacy of schematization. Each event, therefore, at each level of organization—even beyond perceptive assimilation—can be effective, importing a significant moment of indeterminacy into the organization of living systems. Here, the determination of functional states does not overlap with perceptive boundaries, thus enabling the internal medium to merge into a united system of interactions with simultaneously different operative focuses (i.e., a united set of internal environments). The second mechanism is the formation of relatively closed system of mutual impositions. As has already been said, a route toward any intrasystemic conflict settlement is always relative to the actual context. However, the development of functional subsystems enables modification of local balances by shifting the temporal patterns of reproductive cycles. Local assignments of values thus become infused into a more or less stable configuration of functional estimations which determine the routes of tendencies in process generation.

23.3 FUNCTIONAL STABILITY

The usual view tends to reduce the functioning of living systems to a network of linear channels of signals flowing between well-defined convergence points. By changing several paradigms, the multiple refractions of these channels and points emerge at the forefront of the analysis. The construction and reconstruction of material structures, changes in observed identification of elements, assimilation of external changes into operative signals, their divergence into different possibilities of process development and convergence into local normative closures, continuous reorganization of operative horizons, all constitute a dynamics which is trapped in multileveled (material, temporal, and organizational) reconstructive cycles. Owing to such organizational patterns, maintaining such structures is the same as their existence: They cannot at any moment passively exist, from the mere fact that they are already formed. Since a living system is not a finished system, but rather a system in continuous self-construction, the maintenance of such structures is fully dependent on their functioning, which in turn lasts only because of the continuous formation of preservation mechanisms materialized by the temporalized reconstitution of elements. Therefore, functioning is not reduced merely to a set of internal transformations, but it is inseparably composed of elements and their transformations. By constantly producing itself (by reflexive reproduction) the local nonparametric state of the system (i.e., the context) emerges as a new regulatory force. Also, each deconstructed functional element (protein/subsystem) indicates not only the loss of one functional unit, but at the same time the loss of one object of interaction with the environment, along with the spatial and temporal patterns imposed by it, thus changing the balance of power between different subsystems. Therefore, when talking about living cells, we cannot talk about the mere allocation of material structures, but rather of internal spatial and temporal perceptions which undergo reconfigurations with each external change. In this sense, we can identify the permanent mutual dependence of dyadic modifications: changes in "objective" material composition (i.e., uptake of nutrients, catabolic or anabolic transformation), and changes in the perception of such "objective" changes, each of them regulating the other. On the other hand, the mere fact that each living system is composed of perceptive elements—and consequently by parametric differentiation—leads to the constitution of well-known regulatory models. There is, however, an important difference regarding the usual perspective. Strictly speaking, in contrast to the classical model which supposes the ultimate primacy of uniform and unambiguous parameters penetrating the whole system and constituting its eternal organization, in organisms we can speak of classical input/output regulations only in short, separate segments. Unambiguous parameters are certainly also achieved in living systems, and this should be clearly stressed, but this happens only in temporarily short segments which are crystallized within (and in accordance with) a much broader situation. The entire intrasystemic environment constitutes one comprehensive controlling form which is not centralized (but rather segmented) and is not linked with (pre)defined signaling pathways. Internality, through temporalization, determines the dynamics of reproductive cycles and, through contextualization, determines the actual modes of identifying elements, therefore being able to continuously change the status of parameters. In this way, it comprises all processes, which are then reduced from the eternal level to a transitory formation of nonambiguity, during continuous process of reconstruction.

There may be a tendency to identify such systems as unstable. Not wanting to open a discussion about the roots of such thinking or its correctness, we will emphasize only two factors. From the perspective of the system itself, it is much more appropriate to state the diametrically opposite situation, far from the constant threat of self-destruction. On the one hand, by developing an operative patterns in dealing with the environment, the system actually stabilizes it—because the receptive reduction of environment is not only its reduction in the sense of depletion, but at the same time is an enhancement of it, in the sense of being able to ascribe "nonobjective" relations to elements in the environment, thus laying a basis for further development of their functional treatment. And stabilization of the environment allows further construction of intrasystemic functional strategies. On the other hand, with the temporalization of elements, an increase in local perturbations does not need to be followed by expansion throughout the whole system, because it can always be compensated by new self-constructions. It is correct that such an organized system is not grounded on the steadiness of material structures or nonambiguity of rules, but we should ask if these are the only factors of systemic perseverance and functioning.

Another issue is how to model such systems. Usual approach is to consider functioning of living systems in terms of machine analogy where all aspects are well defined. Alternative approach would be to model them as systems where neither compartments nor signals are completely predefined. Issues that we will consider in this Part are stability of communication between cells under the influence of other environmental and intracellular processes.

REFERENCES

Bellissent-Funel, M.-C., 2000. Hydration in protein dynamics and function. J. Mol. Liq. 84, 39—52.

Dill, K.A., Fiebig, K.M., Chan, H.S., 1993. Cooperativity in protein folding kinetics. Proc. Natl. Acad. Sci. U.S.A. 90, 1942—1946.

Ferdinand, W., 1976. The Enzyme Molecule. Willey, New York.

Finkelstein, A.V., Shakhnovic, E.I., 1989. Theory of cooperative transitions in protein molecules. 2. Phase diagram for a protein molecule in solution. Biopolymers 28, 1681—1694.

Hinsen, K., 2000. Domain motions in proteins. J. Mol. Liq. 84, 53—63.

Hinsen, K., Petrescu, A.-J., Dellerue, S., Bellissent-Funel, M.-C., Kneller, G.R., 2002. Liquid like and solid like motions in proteins. J. Mol. Liq. 98—99, 381—398.

Karplus, M., Shakhnovich, E., 1992. Protein folding: theoretical studies of thermodynamics and dynamics. In: Creighton, T.E. (Ed.), Protein Folding. Freeman, New York, pp. 127—195.

Kneller, G.R., Smith, J.C., 1994. Liquid-like side-chain motions in myoglobin. J. Mol. Biol. 242, 181—185.

Kurzynski, M., 1998. A synthetic picture of intramolecular dynamics of proteins. Toward a contemporary statistical theory of biochemical processes. Prog. Biophys. Mol. Biol. 69, 23—82.

Lumry, R., 1971. Some fundamental problems in the physical chemistry of protein behavior. In: King, T., Klingemberg, M. (Eds.), Electron and Coupled Energy Transfers in Biological Systems. Dekker, New York, p. 1.

Lumry, R., 1980. Dynamical factors in protein-protein association. In: Nimai, N., Sugai, S. (Eds.), Dynamic Properties of Polyion Systems. Kodansha Publ Co., Tokyo and Elsevier, Amsterdam.

Mayne, C.G., Arcario, M.J., Mahinthichaichan, P., Baylon, J.L., Vermaas, J.V., Navidpour, L., Wen, P.C., Thangapandian, S., Tajkhorshid, E., 2016. The cellular membrane as a mediator for small molecule interaction with membrane proteins. Biochim. Biophys. Acta. http://dx.doi.org/10.1016/j.bbamem.2016.04.016.

Onuchic, J.N., Luthey-Schulten, Z., Wolynes, P.G., 1997. Theory of protein folding: the energy landscape perspective. Annu. Rev. Phys. Chem. 48, 545—600.

Piaget, J., 1973. Introduction Al'épistémologiegénétique (1 La penséemathématique: Stojanović, S., Serbian Trans.). IzdavačkaknjižarnicaZoranaStojanovića, Novi Sad.

Schenk, P.W., Snaar-Jagalska, B.E., 1999. Signal perception and transduction: the role of protein kinases. Biochim. Biophys. Acta 1449, 1—24.

Smith, J.C., Merzel, F., Verma, C.S., Fischer, S., 2002. Protein hydration water: structure and thermodynamics. J. Mol. Liq. 101, 27—33.

Soyer, O.S., Salathe, M., Bonhoeffer, S., 2006. Signal transduction networks: topology, response and biochemical processes. J. Theor. Biol. 238, 416—425.

Sumi, H., 2001. Solvent fluctuations and viscosity-dependent rates of solution reactions in a regime indescribable by the transition state theory. J. Mol. Liq. 90, 185—194.

Thune, T., Badger, J., 1995. Thermal diffuse x-ray scattering and its contribution to understanding protein dynamics. Prog. Biophys. Mol. Biol. 63, 251—276.

Veitshans, T., Klimov, D.K., Thirumalai, D., 1997. Protein folding kinetics: time scales, pathways, and energy landscapes in terms of sequence dependent properties. Fold. Des. 2, 1—22.

Xie, X.S., 2002. Single-molecule approach to dispersed kinetics and dynamic disorder: probing conformational fluctuation and enzymatic dynamics. J. Chem. Phys. 117, 11024—11032.

Xie, X.S., Lu, H.P., 1999. Single-molecule enzymology. J. Biol. Chem. 274, 15967—15970.

Synchronization of the biochemical substance exchange between cells

24.1 A MODEL REPRESENTING BIOCHEMICAL SUBSTANCE EXCHANGE BETWEEN CELLS: MODEL FORMALIZATION

Communication between cells is ubiquitous in the biological world. From single cell bacteria to complex eukaryotic organisms, cellular communication is a way for creating more complex structures through integration and coordination of functioning. Organisms evolved various ways for ensuring that transfer of signals can be performed timely and efficiently, both between organisms and within single organism. However, at the molecular level, basic scheme of signals exchange remains in the same form: signaling molecules should reach cellular receptor, which in turn activates regulatory response, modulating production of targeted molecular species. These species then either directly or indirectly influence production of arriving signals. In this general scenario, several points should be noted. Since communication is established by exchange of biochemical substances (substances in the further text) through surrounding environment, this process is heavily influenced by the state of environmental factors. In single cell organisms, environmental fluctuations are even more prominent since substances have to be released into the external environment, which is not included in the homeostasis created by the organism. Additionally, even in clonal population, and under heavily controlled environment, significant level of fluctuations of constituting parameters will remain, due to protein disorder (Dunker et al., 2001, 2002) and the so-called intrinsic noise (Elowitz et al., 2002; Swain et al., 2002). Finally, due to thermal and conformational fluctuations, biochemical processes are inherently random (Longo and Hasty, 2006). Although the biological "noise" is not strictly defined, in this section we will use it in the context of variations in the functioning of a biological system that results from the presence of random internal as well as the external fluctuations.

If we consider the described process of substances' exchange between cells that creates a complex system which should maintain its functionality under strong influence of both internal and external fluctuations, we are approaching the problem of robustness (Barkai and Shilo, 2007; Kitano, 2004, 2007). Although the robustness and stability are not the main focus of consideration, in Mihailović and Balaž (2011) we touched on the problem of how the system can avoid functional collapse by switching between several stable states. Some elaborated formal treatments of this problem are still in infancy. One of the main reasons for that is the fact that its focus is beyond already developed tools of dynamical systems theory. It indicates that a new, more general approach for describing such systems has to be developed.

In this section, our focus is only a segment of the problem, for the specific group of cases. Our question is: how the system, which is basically stochastic, and is inherently influenced by noise, can maintain its functioning? First, we give a short overview of general mechanism for substances exchange between two cells, representing cooperative communication process. Then, we identify the main parameters of the process and derive a system of two coupled logistic equations as an appropriate model of the given process. In Mihailović and Balaž (2011) we investigate synchronization of the model and its sensitivity to fluctuations of environmental parameters. It should be emphasized that our goal is not development of an accurate quantitative model of substances exchange between cells. Rather, we are interested in the formalization of the basic shape of the process, and creating the appropriate strategy, that allows further investigation of robustness and influence of noise induced by external fluctuations of environmental parameters.

Empirical background. Communication between cells is one of the main prerequisites for assembling them into the higher organized structures. It is ubiquitous in the living world, from bacteria where quorum sensing (Waters and Bassler, 2005) and colony formation (Stoodley et al., 2002) are efficient mechanisms for rapid switching between different phenotypes to sophisticated humoral control in vertebrates which ensures proper functioning of the organism as an integrated system. Despite great variety of specific mechanisms and even greater number of molecules included, the general scheme remains fairly universal (see for example, Purves et al., 2004) as is seen in Fig. 24.1.

Signaling molecules are ones which are deliberately extracted by the cell into extracellular environment, and which can affect behavior of other cells of the same or different type (species or phenotype) by means of active uptake and subsequent changes in genetic regulations. They can be excreted as either a side product of other metabolic processes, or as purposefully synthesized and transported from the cell. Once present in the extracellular environment, they can be transported to other cells that can be affected. Since active uptake is one of the milestones of the process, a very important factor is a current set of receptors and transporters in cellular membrane, during the communication process. At the same time they constitute the backbone of the whole process while simultaneously being a very important source of perturbations of the process due to protein disorder and intrinsic noise. As a result, the process of exchange is constantly under inherent fluctuations of the aforementioned parameters. Another important factor is surrounding environment which could interfere with the process of exchange. It includes distance between cells, mechanical and dynamical properties of the fluid which serves as a channel for

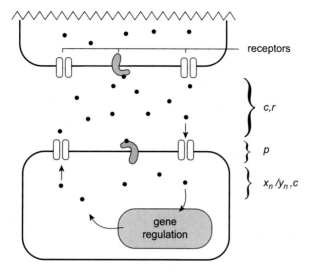

FIGURE 24.1

Schematic representation of cellular communication. Here, c represents concentration of signaling molecule in extracellular environment coupled with intensity of response they can provoke while r includes collective influence of environmental factors which can interfere with the process of communication. x_n/y_n represents concentration of signaling molecules in intracellular environment, while p denotes cellular affinity to uptake the substance.

Reprinted with permission from Balaz, I., Mihailovic, D.T., 2010. Modeling the intercellular exchange of signaling molecules depending on intra- and inter-cellular environmental parameters. Arch. Biol. Sci. Belgrade 62 (4), 947–956.

exchange, and various abiotic and biotic factors which influence physiology of the involved cells. Final requisite phase is induction of change in the receiving cell. As a result, metabolic state of the cell changes, which can be detected by measuring concentration of specific molecules. Therefore, their concentration inside the cell can serve as an indicator of dynamics of the whole process of communication. These signaling molecules can be either the same for all involved cells or they can be different, acting directly or indirectly on production of arriving signals. Additionally, the influence of affinities in functioning of living systems is also an important issue. It can be divided into the following aspects: (a1) affinity of genetic regulators toward arriving signals which determine intensity of cellular response and (a2) affinity for uptake of signaling molecules. First aspect is genetically determined and therefore species specific. Second aspect is more complex and is influenced by affinity of receptors to binding specific signaling molecule, number of active receptors, and their conformational fluctuations (protein disorder).

Model description. As it is obvious from the empirical description, we can infer the successfulness of the communication process by monitoring: (1) number of signaling molecules, both inside and outside of the cell and (2) their mutual influence. Concentration of signaling molecules in the extracellular environment is subject to various environmental influences and taken alone often can indicate more about state of the environment than about the communication itself. Therefore, we choose to follow concentration of signaling molecules inside the cell as the main indicator of the process. In that case, parameters of the system are (1) affinity by which cells perform uptake of signaling molecules (a2), that depends on number and state of appropriate receptors, (2) concentration of signaling molecules in extracellular environment within the radius of interaction, (3) intensity of cellular response (a1), and (4) influence of other environmental factors which can interfere with the process of communication. In this case we postulate that the third parameter can be taken collectively, within the one variable, indicating overall disposition of the environment to the communication process.

Since concentration of signaling molecules can be regarded as their population for fixed volume, and since we are focused on mutual influence of these populations, it points out to the use of the coupled logistic equations. In that case investigation of conditions under which two equations are synchronized and how this synchronization behaves under continuous noise can give some answers on the question of maintaining functionality in the system which is inherently influenced by noise and where elementary events are basically stochastic. Therefore, having in mind that cellular events are discrete (Barkai and Shilo, 2007), we consider a system of difference equations of the form

$$\mathbf{X}_{n+1} = \mathbf{F}(\mathbf{X}_n) \equiv \mathbf{L}(\mathbf{X}_n) + \mathbf{P}(\mathbf{X}_n), \tag{24.1}$$

with notation

$$\mathbf{L}(\mathbf{X}_n) = ((1-c)rx_n(1-x_n), (1-c)ry_n(1-y_n)), \quad \mathbf{P}(\mathbf{X}_n) = \left(cy_n^p, cx_n^{1-p}\right), \tag{24.2}$$

where $\mathbf{X}_n = (x_n, y_n)$ is a vector representing concentration of signaling molecules inside the cell, while $\mathbf{P}(\mathbf{X}_n)$ denotes simulative coupling influence of members of the system which is here restricted only to positive numbers in the interval $(0,1)$. The starting point $\mathbf{X}_0$ is determined so that $0 < x_0, y_0 < 1$. Parameter r is in this case the logistic parameter, which in logistic difference equation determines an overall disposition of the environment to the given population of signaling molecules and exchange processes. Affinity to uptake signaling molecules is indicated by p. Since fixed point is $\mathbf{F}(0) = 0$, in order to ensure that zero is not at the same time the point of attraction we defined $p \in (0,1)$ as an exponent. Finally, c represents coupling of two factors: concentration of signaling molecules in extracellular environment and intensity of response they can provoke. This form is taken because the effect of the same intracellular concentration of signaling molecules can vary greatly with variation of affinity of genetic regulators for that signal, which is further reflected on the ability to synchronize with other cells. Therefore, c influence both, rate of intracellular synthesis of signaling molecules, as well as synchronization of signaling processes between two cells so the parameter c is taken to be a part of both $\mathbf{L}(\mathbf{X}_n)$ and $\mathbf{P}(\mathbf{X}_n)$. However, relative ratio of these two influences depends on current empirical setting. For example, if for both cells $\mathbf{X}_n$ is strongly influenced by extracellular concentration of signals, while they can provoke relatively smaller responses, then the form of equation will be

$$x_{n+1} = (1 - c)rx_n(1 - x_n) + cy_n^p \tag{24.3a}$$

$$y_{n+1} = (1 - c)ry_n(1 - y_n) + cx_n^{1-p}, \tag{24.3b}$$

where $0 < c < 1$, $0 < p < 1$, and $r > 0$. Using the fact that for $0 \le x \le 1$ and $1 > p > 0$ we have $x \le x^p \le 1$, then it is possible to consider system Eqs. (24.3a) and (24.3b) in a simpler form. After its majorization and minorization, respectively, we reach the systems

$$x_{n+1} = (1 - c)rx_n(1 - x_n) + c \tag{24.4a}$$

$$y_{n+1} = (1 - c)ry_n(1 - y_n) + c \tag{24.4b}$$

and

$$x_{n+1} = (1 - c)rx_n(1 - x_n) + cy_n \tag{24.5a}$$

$$y_{n+1} = (1 - c)ry_n(1 - y_n) + cx_n. \tag{24.5b}$$

System (Eqs. 24.4a and 24.4b) is an uncoupled system of logistic difference equations defined on domain $D = (I \times I)$ where $I = (-\delta, 1 + \delta)$, and $\delta < 0$ is the smallest solution of the equation $x = (1 - c)rx(1 - x) + c$. In this system, all information about bifurcations and chaotic behavior we get by its comparison with the standard form $x_{n+1} = \rho x_n(1 - x_n)$ where $\rho = (r(1 - c) + 4c)/(1 - 2\delta)$. A comprehensive analysis of the system (Eqs. 24.5a and 24.5b) in more detail can be found in Mihailović et al. (2010).

At the end we summarize the above consideration. Modelling of cellular processes usually takes the form of explicit kinetic or stoichiometric models. Due to their specificity, they fail to treat some phenomena common for the whole class of different empirical cases. Analyzing the general scheme of communication between cells, we focused on persistence of the process under constant and significant presence of parameter fluctuations. Following discreteness of cellular processes, we developed a model based on coupled difference equations to further investigate stability of their synchronization. Additionally, we had in mind that the described class of problems is now considered under the notion of robustness. We expect that in the future more abstract mathematical tools will be developed to treat that problem. However, they should be incorporated with already existing formulations from dynamical systems theory to connect more abstract notion of functionality preserving with its underlying dynamics. We believe that the approach offered here could serve as one of the connecting links.

24.2 SYNCHRONIZATION OF THE BIOCHEMICAL SUBSTANCE EXCHANGE BETWEEN CELLS: EFFECT OF FLUCTUATIONS OF ENVIRONMENTAL PARAMETERS TO BEHAVIOR OF THE MODEL

A continuing interesting problem in nonlinear science is the interplay between chaos and perturbation. It comes from the fact that complex systems are often under fluctuations of different magnitude, and it is important to assess how the dynamics of deterministic chaotic system is affected by the noise. Some pioneering works on this problem comprise physics, biology, and biophysics (Schaffer et al., 1993; Billings and Schwartz, 2002; Schwartz et al., 2004; Liui and Ma, 2005). It is well known that chaos, in general, is robust under stochastic fluctuations, which can induce chaos for parameters just before a bifurcation to chaos. Namely, stochastic perturbations can induce chaotic dynamics where there is no naturally occurring chaos, far away from any bifurcation leading to chaos. This is possible due to the complex topology associated with two nearby unstable orbits (Schaffer et al., 1993). Intention of this section is to study the sensitivity of the proposed model representing biochemical substance exchange between cells (Balaž and Mihailović, 2010, 2011) to fluctuations of environment parameters of different order of magnitude using maximal Lyapunov exponent and cross-sample entropy.

The effort toward formalizing a model of process of substance exchange between cells is still in its infancy and much remains to be completed to build a mature theory. For a model to be useful, it must be able to predict characteristics and behavior of the system. This means that the model has to be framed to explicitly describe constraints that bind the system. That effort is still going up over stairways that have not a complete structure. One of the hard tasks in building those stairways is the issue related to stability and robustness. Exploring the difference between "stable" and "robust" is related to almost every aspect of what we instinctively find interesting about robustness, not only in natural, but also in engineering, and social systems. It is argued here that robustness is a measure of feature persistence in a system that compels us to focus on fluctuations, and often assemblages of perturbations, qualitatively different in nature from those addressed by stability theory. Moreover, to address feature persistence under these sorts of perturbations, we are naturally led to study issues including: (1) the coupling of dynamics with organizational architecture, (2) implicit assumptions of the environment, (3) the role of a system's evolutionary history in determining its current state and thereby its future states, (4) the sense in which robustness characterizes the fitness of the set of "strategic options" open to the system, and (5) the capability of the system to switch among multiple functionalities (Jen, 2003; Kitano, 2007). Defining any scientific term is a nontrivial issue, but in this section, the following definition will be used: "robustness" is a property that allows a system to maintain its functions against internal and external perturbations. It is important to realize that robustness is concerned with maintaining functions of a system rather than system states, which distinguishes robustness from stability or homeostasis (Kitano, 2004).

Following the model introduced by Balaž and Mihailović (2011), we investigate behavior of two cells, exchanging biochemical substances, using methods of nonlinear dynamics—calculating maximal Lyapunov exponent and cross-sample entropy. In the above paper, we derived the two-dimensional mapping given by Eqs. (24.3a) and (24.3b) that describes communicative interaction between two cells.

According to the assumption in model design, the dynamical behavior of the substance concentrations x_n and y_n depends on three factors: (1) its own concentration c within the radius of interaction in the surrounding environment (Balaž and Mihailović, 2011), (2) parameter r, and (3) affinity p for binding on cellular receptors. The first factor is determined by underlying feedback mechanism of intracellular regulations, while the second one represents level of the suitability of the environment to the communication between two cells (Mihailović et al., 2010).

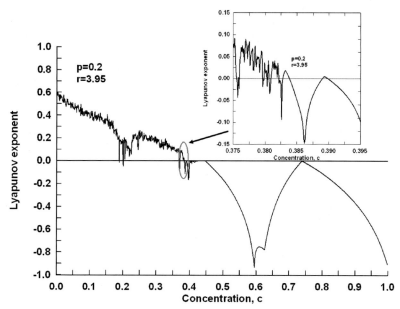

FIGURE 24.2

Lyapunov exponent of coupled maps with no fluctuations (Eq. 24.3) as a function of concentration c ranging from 0 to 1. Ellipsis indicates the region used for analyzing the effect of fluctuations.

Reprinted with permission from Mihailovic, D.T., Balaz, I., 2011. A model representing biochemical substances exchange between cells. Part II: effect of fluctuations of environment parameters to behavior of the model. J. App. Funct. Anal. 6, 77–84.

The third factor depends on protein disorder (Dunker et al., 2001, 2002) which is used to be constant in this model. The map displays a wide range of behavior as the parameters are varied including periodic, quasiperiodic, and chaotic motion. The variation of the Lyapunov exponent as a function of concentration c is depicted in Fig. 24.2 for $p = 0.2$ and $r = 3.95$. The part of this curve in elliptic area is chosen for analysis of the effect of fluctuations on synchronization of the system.

To characterize the asymptotic behavior of the orbits, we need to calculate the largest Lyapunov exponent λ, which is given for the initial point $\mathbf{X}_0$ in the attracting region by

$$\lambda = \lim_{n \to \infty} (\ln\|\mathbf{J}^n(\mathbf{X}_0)\|/n) \tag{24.6}$$

where $\mathbf{J}$ is the Jacobi matrix. With this exponent, we measure how rapidly two nearby orbits in an attracting region converge or diverge. In practice, using $\mathbf{J}^k(\mathbf{X}_k) = \mathbf{J}^k(\mathbf{X}_0) = \mathbf{J}(\mathbf{X}_{k-1})\ldots \mathbf{J}(\mathbf{X}_1)\mathbf{J}(\mathbf{X}_0)$, we compute the approximate value of λ by substituting in (Eq. 24.6) successive values from $\mathbf{X}_{n_0}$ to $\mathbf{X}_{n_1}$, for n_0, n_1 large enough to eliminate transient behaviors and provide good approximation. If $\mathbf{X}_0$ is part of a stable periodic orbit of period k, then $\|\mathbf{J}^k(\mathbf{X}_0)\| < 1$ and the exponent λ is negative, which characterizes the rate at which small perturbations from the fixed cycle decay, and we can call such a system synchronized.

Effect of fluctuations. Here we will investigate the behavior of the coupled maps given by Eqs. (24.3a) and (24.3b) in the presence of fluctuations of environmental parameters or other noise. As has been shown in the case of uncoupled nonlinear

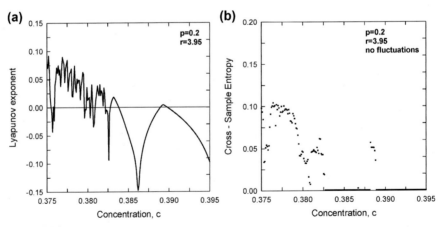

FIGURE 24.3

Lyapunov exponent (a) and cross-sample entropy (Cross-SampEn) (b) for coupled maps with no fluctuations (Eq. 24.3) as a function of concentration c ranging from 0.375 to 0.395.

Reprinted with permission from Mihailovic, D.T., Balaz, I., 2011 A model representing biochemical substances exchange between cells. Part II: effect of fluctuations of environment parameters to behavior of the model. J. App. Funct. Anal. 6, 77–84.

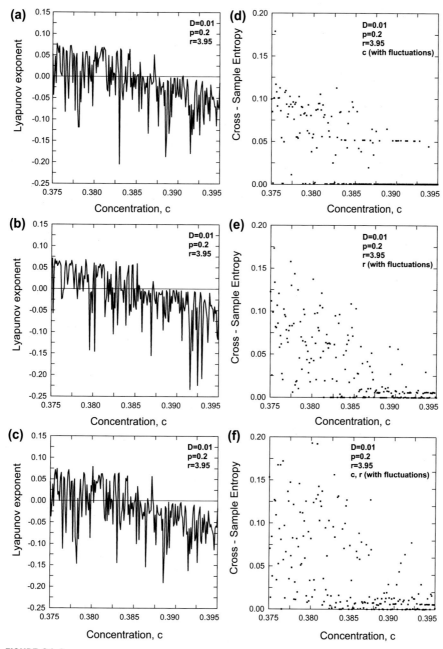

FIGURE 24.4

Lyapunov exponent (a)–(c) and Cross-SampEn (d)–(f) for coupled maps (Eq. 24.3) for fluctuation with amplitude **D** = 0.01 as a function of concentration c ranging from 0.375 to 0.395.

Reprinted with permission from Mihailovic, D.T., Balaz, I., 2011 A model representing biochemical substances exchange between cells. Part II: effect of fluctuations of environment parameters to behavior of the model. J. App. Funct. Anal. 6, 77–84.

oscillators, the addition of parametric fluctuations has a pronounced effect on the dynamics of such systems (Hogg and Huberman, 1984). In particular, the presence of noise introduces a gap in the bifurcation sequence of period-doubling systems and renormalizes the threshold of the appearance of chaotic behavior in the model considered. It is therefore of interest to investigate the effect of noise on the exchange substances between two cells. This is because cells are intrinsically noisy biochemical reactors: low reactant numbers can lead to significant statistical fluctuations in molecule numbers and reaction rates (Thattai and Oudenaarden, 2001).

The effect of fluctuations was modeled by adding uniformly distributed random numbers to the map of Eq. (24.1). Specifically, we considered the map

$$x_{n+1} = (1 - c)r\left(1 + \tau\delta_n^{(1)}\right)x_n(1 - x_n) + cy_n^{1-p} + \xi\mathbf{D}\delta_n^{(1)} \tag{24.7a}$$

$$y_{n+1} = (1 - c)r\left(1 + \tau\delta_n^{(2)}\right)y_n(1 - y_n) + cx_n^{1-p} + \xi\mathbf{D}\delta_n^{(2)}, \tag{24.7b}$$

where τ and ξ take value 0 or 1 while $\delta_n^{(1)}$ and $\delta_n^{(2)}$ are random numbers uniformly distributed in the interval $[-1, 1]$ and $\mathbf{D}$ is the amplitude of the fluctuations. In numerical simulations we used three kind of fluctuations: (1) $(\tau = 0, \xi = 1)$— fluctuations of c; (2) $(\tau = 1, \xi = 0)$ —fluctuations of r, and (3) fluctuations in both c and r. The case $(\tau = 1, \xi = 1)$ corresponds to one with no fluctuations (Eq. 24.3). These fluctuations can destroy the fine-scale detail of the transitions and

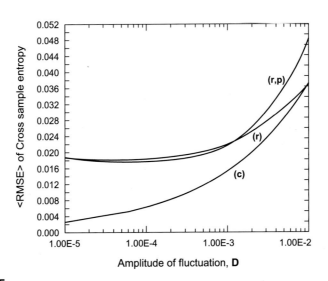

FIGURE 24.5

RMSE of the Cross-SampEn for coupled maps (Eq. 24.3) for fluctuation as a function of the amplitude **D** of fluctuations ranging from 0.00001 to 0.01. The letters next to curves indicate fluctuations in: (c) concentration, (r) logistic parameter, and both (r,c).

the quasiperiodic regions. Fig. 24.3a shows the Lyapunov exponent for coupled maps with no fluctuations (Eq. 24.3) and cross-sample entropy (Cross-SampEn) (Fig. 24.3b) as a function of concentration. c ranging from 0.375 to 0.395, while Fig. 24.4a−c depicts this exponent for the largest amplitude of the fluctuation $\mathbf{D} = 0.01$ and the same ranging interval for c. Obviously, those fluctuations remarkably change the Lyapunov exponent if they occur in both c and r. That disturbance is highly emphasized for the case when fluctuations occur in both the parameters r and c. We calculated the Cross-SampEn with $m = 5$ and $r = 0.05$ for x_n and y_n time series. Fig. 24.4d−f shows high disorder between them, particularly when fluctuations occur in the logistic parameter r and in both r and c. Apparently, the disorder in the entropy is increases when the amplitude of the fluctuations increases.

We calculated the RMSE (root mean square error) of the Cross-SampEn with the state without fluctuations as a referent one. The cross-sample entropy for coupled maps (Eq. 24.3) for fluctuation as a function of the amplitude $\mathbf{D}$ of fluctuations ranging from 0.00001 to 0.01 is shown in Fig. 24.5. From this figure, it is seen that the highest increase of RMSE of the Cross-SampEn is obtained when the fluctuations occur in both the environment parameters, r and c.

REFERENCES

Balaz, I., Mihailovic, D.T., 2010. Modeling the intercellular exchange of signaling molecules depending on intra- and inter-cellular environmental parameters. Arch. Biol. Sci. Belgrade 62 (4), 947−956.

Balaz, I., Mihailovic, D.T., 2011. A model representing biochemical substances exchange between cells. Part I: model formalization. J. App. Funct. Anal. 6, 70−76.

Barkai, N., Shilo, B.-Z., 2007. Variability and robustness in biomolecular systems. Mol. Cell 28, 755−760.

Billings, L.O., Schwartz, I.B., 2002. Exciting chaos with noise: unexpected dynamics in epidemic outbreaks. J. Math. Biol. 44, 31−48.

Dunker, A.K., Brown, C.J., Lawson, J.D., Iakoucheva, L.M., Obradović, Z., 2002. Intrinsic disorder and protein function. Biochemistry 41, 6573−6582.

Dunker, A.K., Lawson, J.D., Brown, C.J., Williams, R.M., Romero, P., Oh, J.S., Oldfield, C.J., Campen, A.M., Ratliff, C.M., Hipps, K.W., Ausio, J., Nissen, M.S., Reeves, R., Kang, C., Kissinger, C.R., Bailey, R.W., Griswold, M.D., Chiu, W., Garner, E.C., Obradovic, Z., 2001. Intrinsically disordered protein. J. Mol. Graph. Model 19, 26−59.

Elowitz, M.B., Levine, A.J., Siggia, E.D., Swain, P.S., 2002. Stochastic gene expression in a single cell. Science 297, 1183−1186.

Hogg, T., Huberman, B.A., 1984. Generic behavior of coupled oscillators. Phys. Rev. A 29, 275−281.

Jen, E., 2003. Stable or robust? What's the difference? Complexity 8, 12−18.

Kitano, H., 2004. Biological robustness. Nat. Rev. Genet. 5, 826−837.

Kitano, H., 2007. Towards a theory of biological robustness. Mol. Syst. Biol. 3, 137.

Liu, Z., Ma, W., 2005. Noise induced destruction of zero Lyapunov exponent in coupled chaotic systems. Phys. Lett. A 343, 300−305.

Longo, D., Hasty, J., 2006. Dynamics of single-cell gene expression. Mol. Syst. Biol. 4, 64.

Mihailovic, D.T., Balaz, I., 2011. A model representing biochemical substances exchange between cells. Part II: effect of fluctuations of environment parameters to behavior of the model. J. App. Funct. Anal. 6, 77–84.

Mihailovic, D.T., Budinčević, M., Balaz, I., Perišić, D., 2010. Emergence of chaos and synchronization in coupled interactions in environmental interfaces regarded as biophysical complex systems. In: Mihailovic, D.T., Lalić, B. (Eds.), Advances in Environmental Modeling and Measurements. Nova Science Publishers, Inc., New York, pp. 89–100.

Purves, W.K., Sadava, D., Orians, G.H., Heller, H.C., 2004. Life: The Science of Biology, seventh ed. Sinauer Associates, Sunderland, MA, and W. H. Freeman, New York.

Schaffer, W.M., Kendall, B.E., Tidd, C.W., Olsen, L.F., 1993. Transient periodicity and episodic predictability in biological dynamics. IMA J. Math. Appl. Med. 10, 227–247.

Schwartz, B., Morgan, D.S., Billings, L., Lai, Y.-C., 2004. Multi-scale continuum mechanics: from global bifurcations to noise induced high dimensional chaos. Chaos 14, 373–386.

Stoodley, P., Sauer, K., Davies, D.G., Costerton, J.W., 2002. Biofilms as complex differentiated communities. Annu. Rev. Microbiol. 56, 187–209.

Swain, P., Elowitz, M.B., Siggia, E.D., 2002. Intrinsic and extrinsic contributions to stochasticity in gene expression. PNAS 99, 12795–12800.

Thattai, M., van Oudenaarden, A., 2001. Intrinsic noise in gene regulatory networks. PNAS 98, 8614–8619.

Waters, C.M., Bassler, B.L., 2005. Quorum sensing: cell-to-cell communication in bacteria. Annu. Rev. Cell Dev. Biol. 21, 319–346.

Complexity and asymptotic stability in the process of biochemical substance exchange in multicell system

25

25.1 COMPLEXITY OF THE INTERCELLULAR BIOCHEMICAL SUBSTANCE EXCHANGE

Brief model description. To explore complexity and asymptotic stability in the model of dynamics of biochemical substance exchange in a multicell system, we first briefly summarize the main features of the intercellular exchange model (Fig. 24.1), which is described in detail in Section 24.1. Thus, the system (Eq. (24.1)) of coupled difference equations for N_c cells exchanging the biochemical substance can be written in the form of matrix equation

$$\mathbf{A} = (\mathbf{B} + \mathbf{C}) \cdot \mathbf{D} \tag{25.1a}$$

where

$$A = \begin{bmatrix} x_{1,n+1} \\ x_{2,n+1} \\ . \\ x_{k-1,n+1} \\ x_{k,n+1} \\ . \\ x_{N-1,n+1} \\ x_{N,n+1} \end{bmatrix}, D = \begin{bmatrix} x_{1,n} \\ x_{2,n} \\ . \\ x_{k-1,n} \\ x_{k,n} \\ . \\ x_{N-1,n} \\ x_{N,n} \end{bmatrix} C = \begin{bmatrix} 0 & c_1 x_{2,n}^{p_1-1} & 0 & 0 & . & 0 & 0 & 0 \\ 0 & 0 & c_2 x_{3,n}^{p_2-1} & 0 & . & 0 & 0 & 0 \\ . & . & . & . & . & . & . & . \\ . & . & . & . & . & . & . & . \\ 0 & 0 & 0 & 0 & 0 & c_k x_{k+1,n}^{p_k-1} & . & 0 \\ . & . & . & . & . & . & . & . \\ 0 & 0 & 0 & 0 & . & 0 & 0 & c_{N-1} x_{N,n}^{p_{N-1}-1} \\ c_N x_{1,n}^{p_N-1} & 0 & 0 & 0 & . & 0 & 0 & 0 \end{bmatrix},$$

$$\tag{25.1b}$$

347

$$
B = \begin{bmatrix}
(1-c_1)\,r(1-x_{1,n}) & 0 & 0 & 0 & . & 0 & 0 & 0 \\
0 & (1-c_2)\,r(1-x_{2,n}) & 0 & 0 & . & 0 & 0 & 0 \\
. & . & . & . & . & . & . & . \\
0 & 0 & 0 & 0 & (1-c_k)\,r(1-x_{k,n}) & 0 & . & 0 \\
. & . & . & . & . & . & . & . \\
0 & 0 & 0 & 0 & . & 0 & (1-c_{N-1})\,r(1-x_{N-1,n}) & 0 \\
1 & 0 & 0 & 0 & . & 0 & 0 & (1-c_N)\,r(1-x_{N,n})
\end{bmatrix}
$$

with condition $\sum c_i = c$ with $0 \le c \le 1$, while x_i represents the concentration of molecules in cells.

In Section 13.1 we discussed the issue of synchronization of substance exchange between cells. However, an unresolved question is how the complexity and stability of the substance exchange processes are affected by the changes in parameters that represent the influence of the environment, cell coupling, and cell affinity.

Complexity of the biochemical substance exchange between cells. In this section, we will first apply model (Eqs. (25.1a) and (25.1b)) for two cells and the case where both cells are strongly influenced by intracellular concentration of signals, while they can provoke a relatively smaller response.

The issue of system complexity in many disciplines, ranging from cognitive science to evolutionary and developmental biology and particle astrophysics (Crutchfield, 2012; Binney eat al., 1992; Cross and Hohenberg, 1993; Manneville, 1990; Wheeler, 1990 and references herein), has been an increasingly discussed topic during the last three decades. An important part in studying complex systems is the analysis of symbolic sequences. Namely, it is believed that most systems whose complexity we would like to estimate can be reduced to them. However, as we already said, according to Adami and Cerf (2000), the idea that the regularity of the string alone is connected to its complexity is meaningless if this analysis is done in the absence of an environment within which the string is to be interpreted. Thus, the complexity of a string representing the complexity of the mentioned systems can be determined only by analyzing its correlation with a system environment (Adami and Cerf, 2000). In this section it will be a sequence of a system's process, as an indication of its complexity. To provide a better insight in these problems, we will model the process of the biochemical substance exchange using the above-described model.

We analyze the complexity of the map (Eqs. (25.1a) and (25.1b)), using Kolmogorov complexity spectrum highest value, K_m^C of the concentration time series $\{x_i\}$, $i = 1, 2, 3, 4, ..., N$ (see Section 15.2), because in this model concentration of signaling molecules can serve as an indicator of dynamics of the whole process of communication (Mihailović et al., 2011). Fig. 25.1 depicts the change of K_m^C against cell affinity p and logistic parameter r, with: (1) r and p taking values in the interval (3.6–4.0) and (0–1.0), respectively, with an increment of 0.005, (2) initial condition $(x_0, y_0) = (0.3, 0.5)$, and (3) $N_c = 100$ cells assembling in a ring. The numerical simulations were done for values of the coupling parameter c of 0.02 and 0.2, respectively. For each x, 5000 iterations of the map (Eqs. (25.1a) and (25.1b)) are applied, and the first 1000 steps are abandoned. The values of K_m^C were calculated

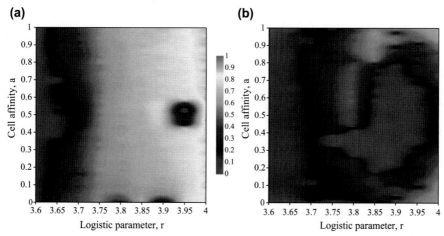

(a)

Cell affinity, a

3.6 3.65 3.7 3.75 3.8 3.85 3.9 3.95 4

Logistic parameter, r

(b)

Cell affinity, a

3.6 3.65 3.7 3.75 3.8 3.85 3.9 3.95 4

Logistic parameter, r

FIGURE 25.1

The dependence of the Kolmogorov complexity spectrum highest value, K_m^C on the cell affinity p and logistic parameter r for the coupling parameter c: (a) 0.02 and (b) 0.2. The calculations were performed for a ring of $N_c = 100$ cells.

Reprinted with permission from Mihailović, D.T., Kostić, V., Balaz, I., Cvetković, L., 2014. Complexity and asymptotic stability in the process of biochemical substance exchange in a coupled ring of cells. Chaos Solitons Fractals 65, 30–43.

for each cell in the ring and then by their averaging, the Kolmogorov complexity spectrum highest value K_m^C of the combined dynamics of cells was obtained. From Fig. 25.1a, it is seen that for $c = 0.02$ (weak coupling), $r > 3.7$, and whole range of the cell affinity p, K_m^C takes high values corresponding to the high complexity in the simulation of process of the biochemical substance exchange between cells in the ring. The only exception is the "island" around the point $(3.95, 0.5)$ in the (r, p) phase space, which can be attributed to the nonlinear features of the map (Eqs. (25.1a) and (25.1b)). In this case, the high complexity indicates the presence of randomness in time series of concentration in the cell. In contrast to this, for higher values of coupling (starting from $c = 0.2$) the process of the biochemical substance exchange (Fig. 25.1a) is much less complex indicating the absence of stochastic processes in comparison with the weak coupling (Fig. 25.1b).

To illustrate complexity of the process of the biochemical substance exchange between cells, we have calculated the Kolmogorov complexity spectra (K^C) for the substance exchange between two cells. Fig. 25.2 depicts K^C of the process of the biochemical substance exchange between two cells (Eqs. (25.1a) and (25.1b)) as a function of normalized concentration in cell, X_i for two values of the coupling parameter c (0.02 and 0.2). We have obtained the normalized values of this series by the transformation $X_i = (x_i - x_{min})/(x_{max} - x_{min})$, where $\{x_i\}$ is a time series of concentration in the cell, and x_{min} and x_{max} are the lowest and highest value of concentration, respectively. From this figure it is seen that the process of the biochemical

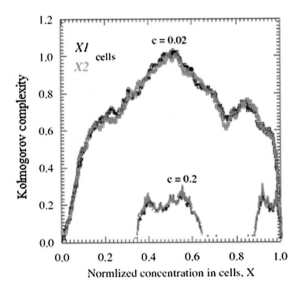

FIGURE 25.2

Kolmogorov complexity spectrum of the process of the biochemical substance exchange between two cells (Eqs. (25.1a) and (25.1b)) as a function of normalized concentration in cell, X_i for the coupling parameter c: 0.02 and 0.2.

Reprinted with permission from Mihailović, D.T., Kostić, V., Balaz, I., Cvetković, L., 2014. Complexity and asymptotic stability in the process of biochemical substance exchange in a coupled ring of cells. Chaos Solitons Fractals 65, 30–43.

substance exchanges, for weak coupling $c = 0.02$, happens for any concentration in the cells. However, for $c = 0.2$ substance exchange is located in the regions of X_i—(0.4, 0.6) and (0.9, 1.0). In the next section, we will test stability of these regions and investigate how they behave in the case of even stronger coupling. So, we will introduce very strong coupling $c = 0.6$.

25.2 ASYMPTOTIC STABILITY OF THE INTERCELLULAR BIOCHEMICAL SUBSTANCE EXCHANGE

The stability of the complex systems is also an important issue that has drawn considerable attention in the analysis of biological and infection-related systems (Adimya et al., 2010; Wang et al., 2012; Yan and Kou, 2012; and herein references). Here, we will analyze asymptotic stability of the dynamical system. Namely, if solutions that are sufficiently close to the equilibrium point remain sufficiently close to that point forever, then the system is said to be Lyapunov stable. If the system is Lyapunov stable and solutions that start sufficiently close to the equilibrium point eventually converge to the equilibrium, then the system is said to be asymptotically stable. The stability analysis for dynamics of different complex systems (Zheng and Wang, 2012; Skryabin, 2000; Choi et al., 2012; Aguiar et al., 2011; among others) can be carried out by using different methods, but the eigenvalue-based methods are commonly used for the asymptotic stability. In this section, for the multicell system presented by a ring of coupled cells, we will provide a set of conditions that describe how stability of the equilibrium of the biochemical substance exchange in time is influenced by the parameters of the model described in the previous section.

Here, following the paper by Mihailović et al. (2014) we address the behavior of the system (Eqs. (25.1a) and (25.1b)) and estimate whether this dynamical system can achieve the asymptotic stability at the possibly existing equilibrium point $\widetilde{\mathbf{X}} = \mathbf{F}(\widetilde{\mathbf{X}})$ for two cells in dependence of coupling parameter c and different values of system parameters r and p. First, let us observe that the parameter constraints of the model assure that $\mathbf{F}: [0, 1] \times [0, 1] \rightarrow [0, 1] \times [0, 1]$. Therefore, since $\mathbf{F}$ is a continuous function on the convex compact subset of Euclidian space $\mathbf{R}^2$, the Brouwer fixed point theorem implies that there exists an equilibrium point. But, since $\mathbf{0} = \mathbf{F}(\mathbf{0})$, $\mathbf{0}$ is an equilibrium point. Since we are not interested in the system without the substance, we need to investigate existence of other (nontrivial) equilibrium points. To that end, notice that $\lim_{\varepsilon \to 0^+} \frac{(1-c)\varphi(\varepsilon)+c\psi(\varepsilon)}{\varepsilon} = +\infty$ holds for all admissible values of parameters c, r, and p. But, this implies that there exists a sufficiently small ε, such that $\mathbf{F}: [\varepsilon, 1] \times [\varepsilon, 1] \rightarrow [\varepsilon, 1] \times [\varepsilon, 1]$. Therefore, again from the Brouwer fixed point theorem, we conclude the existence of a nontrivial equilibrium point $\widetilde{\mathbf{X}} = \mathbf{F}(\widetilde{\mathbf{X}}) \in (0, 1] \times (0, 1]$. Since, in general, our map $\mathbf{F}: (0, 1) \times (0, 1) \rightarrow (0, 1) \times (0, 1)$ is nonlinear; we will discuss the asymptotic stability behavior of the system (Eqs. (25.1a) and (25.1b)) at the given equilibrium by the linearization technique. Namely, it is well known that the asymptotic stability of a dynamical system at the equilibrium point is described by the spectral behavior of its linearization around that equilibrium (Krabs and Pickl, 2010). Therefore, we will compute the Jacobian of the map $\mathbf{F}$ at the point $\widetilde{\mathbf{X}} = (\widetilde{x}_n, \widetilde{y}_n)$ as

$$J_F(\widetilde{\mathbf{X}}) := D\mathbf{F}\big|_{\mathbf{X}=\widetilde{\mathbf{X}}} = \begin{bmatrix} (1-c)r(1-2\widetilde{x}) & cp(\widetilde{y})^{p-1} \\ c(1-p)(\widetilde{x})^{-p} & (1-c)r(1-2\widetilde{y}) \end{bmatrix}, \qquad (25.2)$$

with eigenvalues

$$\lambda_1 = (1-c)r(1-\alpha(\widetilde{x}, \widetilde{y})) + \sqrt{(1-c)^2 r^2(1-\alpha(\widetilde{x}, \widetilde{y}))^2 + c^2(1-c)^2 \frac{\gamma(\widetilde{x}, \widetilde{y})}{\beta(\widetilde{x}, \widetilde{y})}}, \qquad (25.3a)$$

$$\lambda_2 = (1-c)r(1-\alpha(\widetilde{x}, \widetilde{y})) - \sqrt{(1-c)^2 r^2(1-\alpha(\widetilde{x}, \widetilde{y}))^2 + c^2(1-c)^2 \frac{\gamma(\widetilde{x}, \widetilde{y})}{\beta(\widetilde{x}, \widetilde{y})}}, \qquad (25.3b)$$

where $\alpha(\widetilde{x}, \widetilde{y}) := \widetilde{x} + \widetilde{y}$ is the total concentration in both cells ($\widetilde{x}$ and $\widetilde{y}$) at equilibrium state, $\beta(\widetilde{x}, \widetilde{y}) := (\widetilde{x}^p \widetilde{y}^{1-p})/(p(1-p))$ is the weighted generalized geometric mean of concentrations of substance between two cells and $\gamma(\widetilde{x}, \widetilde{y}) := (1 - 2\widetilde{x})(1 - 2\widetilde{y})$ whose sign indicates in which two quadrants, of the domain corresponding to the axis through the point (0.5, 0.5), the provided equilibrium state belongs to.

Since the matrix in (Eq. (25.2)) has only two eigenvalues, using the Theorem from the Appendix, then the discrete nonlinear dynamical system (Eq. (24.1)) at the equilibrium point $\widetilde{\mathbf{X}} = (\widetilde{x}_n, \widetilde{y}_n)$ is

- asymptotically stable for

$$\max\{|\lambda_1|, |\lambda_2|\} < 1, \tag{25.4a}$$

and

- not asymptotically stable if

$$\min\{|\lambda_1|, |\lambda_2|\} > 1, \tag{25.4b}$$

while otherwise we cannot get the answer through analysis of eigenvalues of the Jacobian. Therefore, Eqs. (25.4a) and (25.4b) completely express the stability and instability that is obtained through the linearization. If these conditions are not satisfied, stability purely depends on nonlinearity of the map $\mathbf{F}$ given by (Eq. (24.1)), around the point of equilibrium $\widetilde{\mathbf{X}}$. In this case the regions of stability and instability can be further explored by eigenvalues of Hessian, i.e., using second order of derivatives.

When for a given set of parameters expressions (Eqs. (25.3a) and (25.3b)) are computed, then regions of the domain, for which equilibrium state of the system (Eq. (24.1)) is asymptotically either stable or unstable, can be plotted. To better grasp the meaning of stability/instability conditions given by Eqs. (25.4a) and (25.4b) in Fig. 25.3 (weak coupling $c = 0.02$) and Fig. 25.4 (very strong coupling $c = 0.6$), we have plotted regions of the asymptotic stability and instability in the domain of the map $\mathbf{F}$ for the following values of parameters: $p = 0.25, 0.5, 0.75$ and $r = 1,2,3,4$. In each plot of these figures, the values of the equilibrium concentration of substance in the first cell ($\widetilde{x}$) are given on the horizontal axis, while the values of the second one ($\widetilde{y}$) are given on the vertical axis. The dark gray areas of the state domain $(0,1) \times (0,1)$ represent asymptotic stability, while light gray areas indicate regions of the instability. Note that the white area corresponds to the region where stability/instability is due to purely nonlinear effects of the map $\mathbf{F}$ given by Eq. (24.1). More precisely, if the equilibrium state which we have obtained belongs to the dark gray region of the domain, that equilibrium state is asymptotically a stable one. On the other hand, belonging to a light gray region of the domain implies that the observed state is asymptotically unstable one.

In Fig. 25.3 for logistic parameter r we use the broad range of its values from 1 to 4, although chaotic fluctuations in this equation occur after $r = 3.57$. From this figure, it is seen that the asymptotic instability occurs over the whole domain (i.e., dark gray color completely covers the square section) for $r = 1$ and all values of

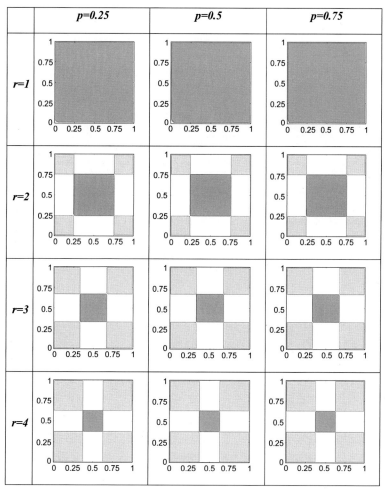

FIGURE 25.3

Regions of the asymptotic stability for the equilibrium points (Eqs. (25.4a) and (25.4b)), in the domain of the map **F** given by (Eq. (24.1)), for weak coupling $c = 0.02$. The values of the equilibrium concentration of substance in the first cell ($\tilde{x}$) and second one ($\tilde{y}$) are given on the horizontal and vertical axes, respectively. Domains of asymptotic stability and instability are indicated by dark and light gray areas, respectively. White area indicates that stability purely depends on nonlinearity of the map F around the point of equilibrium $\tilde{v}$.

Reprinted with permission from Mihailović, D.T., Kostić, V., Balaz, I., Cvetković, L., 2014. Complexity and asymptotic stability in the process of biochemical substance exchange in a coupled ring of cells. Chaos Solitons Fractals 65, 30–43.

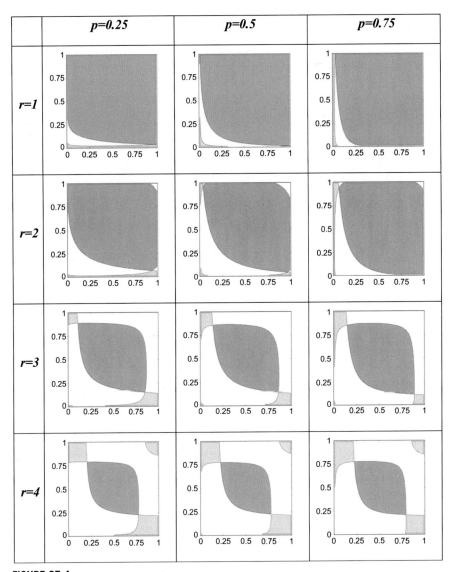

FIGURE 25.4

The same as Fig. 25.3 but for very strong coupling $c = 0.6$.

Reprinted with permission from Mihailović, D.T., Kostić, V., Balaz, I., Cvetković, L., 2014. Complexity and asymptotic stability in the process of biochemical substance exchange in a coupled ring of cells. Chaos Solitons Fractals 65, 30–43.

p (0.25, 0.50, 0.75). With an increase of r, there exist three sections corresponding to asymptotic stability (dark gray), instability (light gray), and stability purely depending on nonlinearity of the map $\mathbf{F}$ (white). Note that the stability square section becomes smaller with increase of r. This is expected for the weak coupling since we approach to region with the chaotic fluctuations. Now we can explain the occurrence of an "island" around the point (3.95, 0.5) in the (r,p) phase space in Fig. 25.1, which depicts the dependence of the Kolmogorov complexity spectrum, K_m^C, on the cell affinity p and logistic parameter r for the coupling parameter $c = 0.02$. This "island" has nearly the same shape and position as the dark shadow square section in Fig. 25.3 ($r = 4$ and $p = 0.5$), which indicates the asymptotic stability of the system (Eq. (24.1)). Namely, all around this section the process of the biochemical substance exchange is highly stochastic because it is not synchronized (Balaž and Mihailović, 2010). In contrast to this, inside of this section the process of the biochemical substance exchange is much less complex indicating the absence of stochastic processes and its synchronization.

From panels in Fig. 25.4, it is seen that for $r = 1$ there exists a large dark gray section that indicates the asymptotic stability, which is expected for the very strong coupling. Namely, in that case the dynamical system (Eq. (24.1)) is not in the chaotic regime and it shows small values of the complexity, since the Kolmogorov complexity (KLL) is low already for $c = 0.2$ (Fig. 25.2). When value of r increases, the dark shadow section becomes smaller with larger sections of the asymptotic instability and stability purely depending on nonlinearity of the map $\mathbf{F}$, particularly when approaching to the chaotic regime.

Since the complexity of the processes of exchange of biophysical substances between cells is chaotic ($r = 4$), in Fig. 25.5 we have plotted the stability regions for different values of $c = 0.1, 0.3, 0.5, 0.7, 0.9$ and $p = 0.25, 0.5, 0.75$, where the meaning of axes is the same as in Figs. 25.3 and 25.4. We note that the section with the asymptotic stability takes much more place in $(\widetilde{x},\widetilde{y})$ space for higher values of r. Finally, from all figures, it is seen that parameter r influences a much more asymptotic instability than the coupling parameter c does.

Here we consider regions of the asymptotic stability for the equilibrium points in the domain of the map $\mathbf{F}$ given by (Eq. (24.1)), through analysis of extreme eigenvalues of the Jacobian, i.e., max $\{|\lambda_1|, |\lambda_2|\}$ and min $\{|\lambda_1|, |\lambda_2|\}$ defined by Eqs. (25.4a) and (25.4b), in dependence of the equilibrium concentration of substance in the cells $(\widetilde{x}, \widetilde{y})$ in a setting that $r = 4$, $p = 0.5$, $c = 0.02$ (weak coupling) and $c = 0.6$ (very strong coupling). The results of computations are depicted in Fig. 25.6. From Fig. 25.6a it is seen that for the weak coupling ($c = 0.02$), the minimal values of the max $\{|\lambda_1|, |\lambda_2|\}$ are reached in the vicinity of the point (0.5, 0.5). Therefore, for such equilibrium points, there exists a strongly pronounced asymptotic stability (the green rectangular section). As expected, for the very strong coupling ($c = 0.6$), this section of the asymptotic instability is larger (Fig. 25.6c). Further inspection of Fig. 25.6, i.e., Fig. 25.6b and d give us information about the asymptotic instability. Namely, for the equilibrium point with cell concentrations close to either higher (toward 1) or lower values (toward 0), values of the

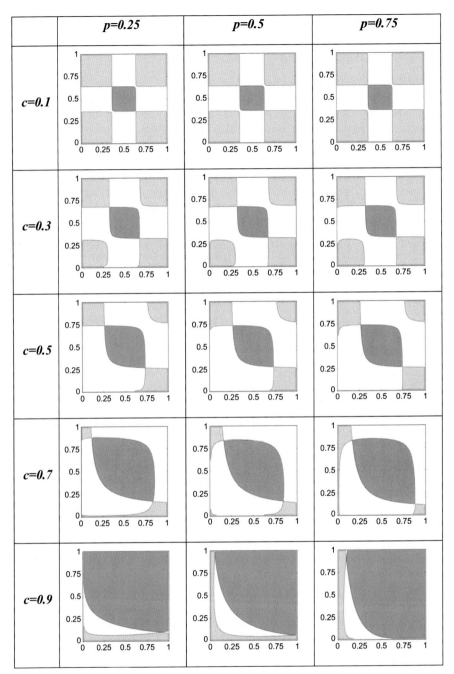

FIGURE 25.5

Regions of the asymptotic stability for the equilibrium points (Eqs. (25.4a) and (25.4b)), in the domain of the map **F** given by (Eq. (24.1)), for $r = 4$ and different values of coupling parameter c.

Reprinted with permission from Mihailović, D.T., Kostić, V., Balaz, I., Cvetković, L., 2014. Complexity and asymptotic stability in the process of biochemical substance exchange in a coupled ring of cells. Chaos Solitons Fractals 65, 30–43.

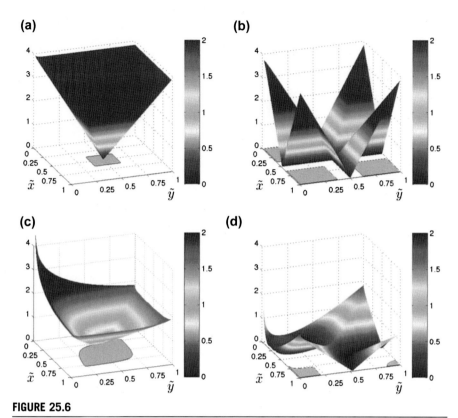

FIGURE 25.6

3D diagram of the extreme eigenvalues of the Jacobian, max $\{|\lambda_1|, |\lambda_2|\}$ ((a) and (c)) min $\{|\lambda_1|, |\lambda_2|\}$; ((b) and (d)), against equilibrium concentration of substance in the cells ($\tilde{x}$, $\tilde{y}$) in setting: $r = 4$, $p = 0.5$, $c = 0.02$ (weak coupling in (a) and (b)) and $c = 0.6$ (strong coupling in (c) and (d)). Sections in ($\tilde{x},\tilde{y}$) plane are due to the regions with stability and instability.

Reprinted with permission from Mihailović, D.T., Kostić, V., Balaz, I., Cvetković, L., 2014. Complexity and asymptotic stability in the process of biochemical substance exchange in a coupled ring of cells. Chaos Solitons Fractals 65, 30–43.

min $\{|\lambda_1|, |\lambda_2|\}$ indicate the existence of a strong instability (green rectangular sections in corners of Fig. 25.6b). Those sections are smaller in case of very strong coupling (Fig. 25.6d).

25.3 BIOCHEMICAL SUBSTANCE EXCHANGE IN A MULTICELL SYSTEM

Stability at an equilibrium state. In this section we consider substance exchange in a multicell system represented by a ring of coupled cells schematically shown in Fig. 25.7.

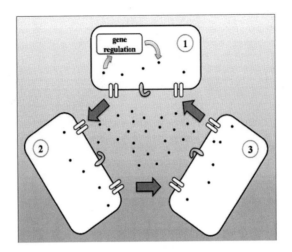

FIGURE 25.7

Schematic diagram of a model of substance exchange in a system represented by a ring of coupled cells.

Reprinted with permission from Mihailović, D.T., Kostić, V., Balaz, I., Cvetković, L., 2014. Complexity and asymptotic stability in the process of biochemical substance exchange in a coupled ring of cells. Chaos Solitons Fractals 65, 30–43.

In our approach, a cell moves locally in its environment without making long pathways. As a generalization of the two-cell system, according to Mihailović and Balaž (2012) and Mihailović et al. (2013), the dynamics of substance exchange in such a multicell system of N_c cells exchanging biochemical substance can be represented by the discrete nonlinear time-invariant dynamical system

$$\mathbf{X}^{(n+1)} = \mathbf{F}\left(\mathbf{X}^{(n)}\right) := C \, \Phi\left(\mathbf{X}^{(n)}\right) + (I - C) \, Z \, \Psi\left(\mathbf{X}^{(n)}\right), \qquad (25.5)$$

where

- $x_k^{(n)}$ is the concentration of the substance in k-th cell in a discrete time step n, $k = 1, 2, ..., N$, $n = 0, 1, 2, ...$ and $\mathbf{X}^{(n)} = (x_1^{(n)}, x_2^{(n)}, ..., x_N^{(n)})$ is the appropriate vector,
- $C = diag(c_1, c_2, ..., c_N)$ is the diagonal matrix of the coupling coefficients for each cell,
- $\Phi(x^{(n)}) := diag\left(\varphi(x_1^{(n)}), \varphi(x_2^{(n)}), ..., \varphi(x_N^{(n)})\right)$ is the diagonal matrix of intracellular behavior modeled by the logistic map $\varphi:(0,1) \to (0,1) \varphi(x) := rx(1-x)$,
- $\Psi(x^{(n)}) := diag\left((x_2^{(n)})^{p_1}, (x_3^{(n)})^{p_2}, ..., (x_N^{(n)})^{p_{N-1}}, (x_1^{(n)})^{p_N}\right)$ is the diagonal matrix of the flow of the substance to each cell, where the cell's affinities fulfill the constraint $p_1 + p_2 + ...p_N = 1$, and
- $Z \in \{0, 1\}^{N \times N}$ is the upper cyclic permutation matrix, i.e., $Z = (e_N, e_1, e_2, ..., e_{N-1})$, where $e_1, e_2, ..., e_{N-1}$ are the standard basis vectors of $\mathbb{R}^N$.

It is interesting to note that in this model of a multicell system for N_c cells in a ring ($N_c = 100$ in our simulations), we permit that coupling coefficients can differ from cell to cell. Therefore, the two-cell model that we have discussed in the previous section is a special case for $N_c = 2$ and $c_1 = c_2 = c$.

As in the Section 25.1, we are interested in the asymptotic stability of the dynamical system (Eq. (25.5)) at the provided equilibrium point $\widetilde{\mathbf{X}} = \mathbf{F}(\widetilde{\mathbf{X}})$. Again, since $\lim\limits_{\mathbf{X} \to 0} \frac{\|\mathbf{F}(\mathbf{X})\|_\infty}{\|\mathbf{X}\|_\infty} = +\infty$ holds for all admissible values of parameters r, $c_1,...,c_N$, and $p_1,...,p_N$, there exists a sufficiently small ε, such that $\mathbf{F} \colon [\varepsilon,\ 1]^N \to [\varepsilon,\ 1]^N$, and consequently, there exists a nontrivial equilibrium point $\widetilde{\mathbf{X}} = \mathbf{F}(\widetilde{\mathbf{X}}) \in (0,\ 1] \times (0,\ 1]$. As before, we determine the asymptotic stability behavior of nonlinear system (Eq. (25.2)) at equilibrium $\widetilde{\mathbf{X}}$ by linearization. The Jacobian of the map $\mathbf{F}$ at the point $\widetilde{\mathbf{X}}$ is

$$J_F(\widetilde{\mathbf{X}}) := DF\big|_{\mathbf{X}=\widetilde{\mathbf{X}}} = C\,\widehat{\Phi}\,(\widetilde{\mathbf{X}}) + (I-C)\,Z\,\widehat{\Psi}\,(\widetilde{\mathbf{X}}), \tag{25.6}$$

where

$$\widehat{\Phi}(\widetilde{\mathbf{x}}) := diag(\varphi'(\widetilde{x}_1),\ \varphi'(\widetilde{x}_2),\ ...,\ \varphi'(\widetilde{x}_N)),$$

$$\varphi'(x) = r\ (1-2x)\ \text{and}$$

$$\widehat{\Psi}(\widetilde{\mathbf{x}}) := diag\left(p_1(\widetilde{x}_2)^{p_1-1},\ p_2(\widetilde{x}_3)^{p_2-1},\ ...,\ p_{N-1}(\widetilde{x}_N)^{p_{N-1}-1},\ p_N(\widetilde{x}_1)^{p_N-1}\right).$$

In this case, if we use the Theorem from the Appendix, then we obtain that the stability of the discrete dynamical system (Eq. (25.5)) at the point $\widetilde{\mathbf{X}}$, which is determined by the spectral radius ρ of the matrix $J_F(\widetilde{\mathbf{X}})$, i.e., $\rho(J_F(\widetilde{\mathbf{X}}))$. Thus, the stability of this system is given by

- asymptotically stable if $\rho(J_F(\widetilde{\mathbf{X}})) < 1$
- not asymptotically stable if $\rho((J_F(\widetilde{\mathbf{X}}))^{-1}) > 1$ and
- can be stable or unstable otherwise, due to purely nonlinear effects.

For the specific choice of parameters, the above criterion gives quite a good answer on the question of the stability through computation of the spectra of the Jacobian. Otherwise, finding an analytic expression for the spectral radius $\rho\,(J_F(\widetilde{\mathbf{X}}))$ in dependence on the values of the system parameters is generally impossible. Therefore, to have a better insight into the dependence of stability on the parameter changes, we will return to the matrix infinity norm. Namely, it is well known that

$$\rho\,(J_F(\widetilde{\mathbf{X}})) \le \|J_F(\widetilde{\mathbf{X}})\|_\infty = \max_{i=1,2,...,N} \sum_{j=1}^N \left|[J_F(\widetilde{\mathbf{X}})]_{i,j}\right|. \tag{25.7}$$

Having this in mind, we obtain only the necessary conditions for the asymptotic stability of the system (Eq. (25.2)) by computing the infinity norm and analyzing the following inequality:

$$\|J_F(\widetilde{\mathbf{X}})\|_\infty = \max_{i=1,2,...,N} \left((1-c_i)\,r(1-2\widetilde{x}_i) + c_i p_i(\widetilde{x}_{i+1})^{p_i-1}\right) < 1. \tag{25.8}$$

At this point, let us note that the specific choice of infinity norm, to bound the spectral radius of a matrix, is not generally the best possible choice. In fact, this is equivalent to approximations of matrix eigenvalues by Geršgorin's circles (Varga, 2004). But, due to its analytic simplicity, it is a good choice in many cases, as a starting point. The better estimations of stability region could be done by the use of other matrix norms or localization areas (Cvetković et al., 2004, 2011; Cvetković and Kostić, 2006; Varga et al., 2008), which will be in the focus of our future work regarding this subject.

Domains for equilibrium points that permit stability for every coupling. We start by observing that inequality (Eq. (25.8)) holds if and only if for every $i \in \{1, 2, ..., N_c\}$

$$(1 - c_i) \, r \, |1 - 2\tilde{x}_i| + c_i p_i \, (\tilde{x}_{i+1})^{p_i - 1} < 1. \tag{25.9}$$

But, since the left-hand side of (Eq. (25.9)) is a convex combination of $r \, |1 - 2\tilde{x}_i|$ and $p_i \, (\tilde{x}_{i+1})^{p_i - 1}$ with parameter $0 < c_i < 1$, we conclude that $r \, |1 - 2\tilde{x}_i| < 1$ and $p_i \, (\tilde{x}_{i+1})^{p_i - 1} < 1$ together imply (Eq. (25.8)) independently of the values of $0 < c_i < 1$, $i \in \{1, 2, ..., N_c\}$.

Thus, since $r \, |1 - 2\tilde{x}_i| < 1$ and $p_i \, (\tilde{x}_{i+1})^{p_i - 1} < 1$ is equivalent to

$$\tilde{x}_i \in \left(\frac{r - 1}{2r}, \frac{r + 1}{2r} \right) \quad \text{and} \quad (p_{i-1})^{\frac{1}{1 - p_{i-1}}} < \tilde{x}_i, \tag{25.10}$$

we obtain the region in the N_c-dimensional space of substance concentrations in the coupled ring of cells

$$\mathbf{S} := \left\{ \tilde{\mathbf{X}} \in (0, 1)^N : \max \left\{ \frac{r - 1}{2r}, (p_{i-1})^{\frac{1}{1 - p_{i-1}}} \right\} < \tilde{x}_i < \frac{r + 1}{2r} \right\}, \tag{25.11}$$

such that for every coupling ($0 < c_i < 1$, $i \in \{1, 2, ..., N_c\}$), if an equilibrium point $\tilde{\mathbf{X}} \in \mathbf{S}$, then this equilibrium point is asymptotically stable. In other words, independently of the coupling parameters of the individual cells in the multicell system, region S, which is always included in a hypercube around the central point the domain (0.5, 0.5, ..., 0.5) with the edge of the length $0.5r^{-1}$ is the place where asymptotic stability of the point of equilibrium is always assured. Looking at Fig. 25.5, we can see the effect of the same behavior in the two-cell system, i.e., while the coupling parameter was changing the square region (0.375, 0.625) × (0.375, 0.625) somehow remains included in the stability region. Now, using an infinity norm bound, in fact, we can explain the existence of this region even for the system of N_c cells coupled in a ring formation. Namely, if we assume that the values of the logistic parameter r are ranged from 3.785 to 4, then, after a closer look at (Eq. (25.11)), it is seen that, independent of the coupling parameters and the cell affinities, the following inclusions hold

$$(0.375, \, 0.625)^N \subseteq \mathbf{S} \subseteq (0.3679, \, 0.6321)^N. \tag{25.12}$$

Namely, if an equilibrium concentration in each cell is between 0.375 and 0.625 then asymptotic stability of (Eq. (25.5)) is assured without any additional constraints on the parameters (c_i, r, p_i), i.e., for each $c_1, c_2, ..., c_N \in (0, 1), p_1, p_2, ..., p_N \in (0, 1)$, and $r \in (3.785, 4)$. In addition, we consider a case when $p_i \leq 0.8$, $i \in \{1, 2, ..., N\}$. Similarly, as in the above analysis of (Eq. (25.11)) we conclude that this constraint extends the stability interval for individual cell's equilibrium concentration to

$$(0.375, \ 0.625)^N \subseteq \mathbf{S} \subseteq (0.3333, \ 0.6667)^N. \tag{25.13}$$

Here, the constraint $p_i \leq 0.8$, $i \in \{1, 2, ..., N_c\}$ encounters practically in the majority of the cases of the multicell systems since $p_i + p_2 + ... + p_N = 1$.

Finally, we underline a conclusion, which can be derived from the above discussion. Namely, the asymptotic stability of the dynamic system represented by a ring of coupled cells given by (Eq. (25.5)) is assured without any additional constraints for $r \in (3.785, 4)$, i.e., the interval, which includes the size of the r interval for the "island" of the low complexity for a two-cell system exchanging the biochemical substance (Fig. 25.1a). This is interesting since in this interval, the coupled maps (Eqs. (25.1a) and (25.1b)) show a chaotic behavior. It means that in those conditions, there exists a space of parameters (c_i, r, p_i), for which the process of biochemical substance exchange in a coupled ring of cells is stable.

Appendix (Stability of the discrete nonlinear dynamical system through linearization). Linearization technique is generally used in obtaining asymptotic stability of discrete dynamical systems. Here we provide one version of this well-known result in the form of the following Theorem and its proof to better clarify the analysis in this section (Mihailović et al., 2014). Slight difference from this result can be found, among others in Krabs and Pickl (2010) and Michel et al. (2008).

Theorem. *Given a nonlinear continuously Fréchet differentiable map* $\mathbf{F}: (0, 1)^N \rightarrow (0, 1)^N$ *and the fixed point* $\widetilde{\mathbf{X}} = \mathbf{F}(\widetilde{\mathbf{X}})$ *of the discrete nonlinear dynamical system*

$$\mathbf{X}^{(n+1)} = \mathbf{F}\left(\mathbf{X}^{(n)}\right), \ n = 0, \ 1, \ 2, \ ..., \tag{A1}$$

Let $J_F(\widetilde{\mathbf{X}}) := DF|_{\mathbf{X}=\widetilde{\mathbf{X}}}$ denote the Jacobian matrix at the point $\widetilde{\mathbf{X}}$. Then, the following two implications hold:

1. if $\rho\left(J_F(\widetilde{\mathbf{X}})\right) < 1$, then the system (Eq. (A1)) is asymptotically stable at the point $\widetilde{\mathbf{X}}$,
2. if all eigenvalues of $J_F(\widetilde{\mathbf{X}})$ are larger in absolute value than one, i.e., $J_F(\widetilde{\mathbf{X}})$ is invertible and $\rho\left((J_F(\widetilde{\mathbf{X}}))^{-1}\right) > 1$, then the system (Eq. (A1)) is not asymptotically stable at the point $\widetilde{\mathbf{X}}$.

Proof. First, supposing that the map $\mathbf{F}$ is continuously Fréchet differentiable on the domain, we obtain that the Jacobian matrix exists at each point of the domain, that it is a continuous map, and that the Fréchet derivative is given by:

$$\mathbf{F'}_x(h) = J_F(\mathbf{X})h, \quad h \in (0, 1)^N.$$

Now, assume that $\rho\left(J_F(\widetilde{\mathbf{X}})\right) < 1$, then, see Varga (2004), there exists a matrix norm $\|\ \|$ on the space $(0, 1)^{N, N} \subset \mathbb{R}^{N, N}$ induced by some vector norm $\|\ \|_v$, such that $\rho\left(J_F(\widetilde{\mathbf{X}})\right) < \|J_F(\widetilde{\mathbf{X}})\| < 1$. Therefore, we obtain $\|\mathbf{F'}_{\widetilde{\mathbf{X}}}\| = \|J_F(\widetilde{\mathbf{X}})\| = \sup\{\|J_F(\widetilde{\mathbf{X}})h\|_v : \|h\|_v = 1\} \langle 1$.

Now, since every norm is continuous, we have that the map $\mathbf{X} \to \|\mathbf{F'}_{\mathbf{X}}\|$ is a continuous map, too. Therefore, there exists $\varepsilon > 0$ and $M \in [0, 1)$ such that

$$\|\mathbf{F'}_{\mathbf{X}}\| < M, \text{ for all } \mathbf{X} \in \mathbf{B}\left(\widetilde{\mathbf{X}}, \varepsilon\right) := \left\{\mathbf{y} \in (0, 1)^N : \|\widetilde{\mathbf{X}} - \mathbf{Y}\| < \varepsilon\right\}.$$

Further, using the mean value theorem we have that $\|\mathbf{F}(\mathbf{X}) - \mathbf{F}(\mathbf{Y})\|_v \leq \|\mathbf{F'}_{\mathbf{Z}}\|\|\mathbf{X} - \mathbf{Y}\|_v$, for all $\mathbf{X}, \mathbf{Y} \in \mathbf{B}(\widetilde{\mathbf{X}}, \varepsilon)$, and some $\mathbf{z}$ which is a convex combination of $\mathbf{X}$ and $\mathbf{Y}$, and, therefore $\mathbf{z} \in \mathbf{B}\left(\widetilde{\mathbf{x}}, \varepsilon\right)$. But, this implies that

$$\|\mathbf{F}(\mathbf{X}) - \widetilde{\mathbf{X}}\|_v = \|\mathbf{F}(\mathbf{X}) - \mathbf{F}(\widetilde{\mathbf{X}})\|_v \leq M\|\mathbf{X} - \widetilde{\mathbf{X}}\|_v < \varepsilon \text{ for all } \mathbf{X} \in \mathbf{B}(\widetilde{\mathbf{X}}, \varepsilon), \text{ i.e.,}$$

$$\mathbf{F}[\mathbf{B}(\widetilde{\mathbf{X}}, \varepsilon)] \subseteq \mathbf{B}(\widetilde{\mathbf{X}}, \varepsilon).$$

Therefore, if the starting point $\mathbf{X}^{(0)}$ of the discrete dynamical system (Eq. (A1)) is in the ε-neighborhood of the fixed point $\widetilde{\mathbf{X}}$, then for $\mathbf{X}^{(n)}$, the state of (Eq. (A1)) at the nth discrete time step, we have that

$$\left\|\mathbf{X}^{(n)} - \widetilde{\mathbf{X}}\right\|_v = \left\|\mathbf{F}\left(\mathbf{X}^{(n-1)}\right) - \widetilde{\mathbf{X}}\right\|_v \leq M\left\|\mathbf{X}^{(n-1)} - \widetilde{\mathbf{X}}\right\|_v = \ldots \leq M^n\left\|\mathbf{X}^{(0)} - \widetilde{\mathbf{X}}\right\|_v < M^n \varepsilon.$$

and, as a consequence, when $n \to \infty$, $\mathbf{X}^{(n)} \to \widetilde{\mathbf{X}}$, and the asymptotic stability is obtained.

To prove item (b), we observe that if $\rho\left([J_F(\widetilde{\mathbf{X}})]^{-1}\right) < 1$, according to the same results of Cvetković and Kostić (2006), then there exists vector norm $\|\ \|_v$ and its induced matrix norm $\|\ \|$, such that $\rho\left([J_F(\widetilde{\mathbf{X}})]^{-1}\right) < \left\|[J_F(\widetilde{\mathbf{X}})]^{-1}\right\| = \left\|[\mathbf{F'}_{\widetilde{\mathbf{X}}}]^{-1}\right\| < 1$. Thus, we have obtained that $\left\|[\mathbf{F'}_{\widetilde{\mathbf{X}}}]^{-1}\right\|^{-1} > 1$.

But, then, for every $\mathbf{X}, \mathbf{Y} \in \mathbf{B}(\widetilde{\mathbf{X}}, \varepsilon)$,

$$\|\mathbf{X} - \mathbf{Y}\|_v = \left\|[\mathbf{F'}_{\widetilde{\mathbf{X}}}]^{-1}(\mathbf{F'}_{\widetilde{\mathbf{X}}}(\mathbf{X})) - [\mathbf{F'}_{\widetilde{\mathbf{X}}}]^{-1}(\mathbf{F'}_{\widetilde{\mathbf{X}}}(\mathbf{Y}))\right\|_v \leq \left\|[\mathbf{F'}_{\widetilde{\mathbf{X}}}]^{-1}\right\|\left\|\mathbf{F'}_{\widetilde{\mathbf{X}}}(\mathbf{X} - \mathbf{Y})\right\|_v,$$

which implies that $\|\mathbf{F'}_{\widetilde{\mathbf{X}}}(\mathbf{X} - \mathbf{Y})\|_v \geq q\|\mathbf{X} - \mathbf{Y}\|_v$, for $q = \left\|[\mathbf{F'}_{\widetilde{\mathbf{X}}}]^{-1}\right\|^{-1} > 1$.

Since the map $\mathbf{F}$ is Fréchet differentiable, we have that

$$\mathbf{F}(\mathbf{X}) - \mathbf{F}(\widetilde{\mathbf{X}}) = \mathbf{F'}_{\widetilde{\mathbf{X}}}(\mathbf{X} - \widetilde{\mathbf{X}}) + \theta\left(\|\mathbf{X} - \widetilde{\mathbf{X}}\|_v\right) \text{ for all } \mathbf{X} \in (0, 1)^N,$$

and, thus, we can choose $\delta > 0$ such that $\widehat{q} = q - \delta > 1$, and $\varepsilon > 0$ such that for every $\mathbf{X} \in \mathbf{B}(\widetilde{\mathbf{X}}, \varepsilon)/\{\widetilde{\mathbf{X}}\}$, $\left\|\theta\left(\|\mathbf{X} - \widetilde{\mathbf{X}}\|_v\right)\right\|_v \leq \delta\|\mathbf{X} - \widetilde{\mathbf{X}}\|_v$, and obtain that

$$\|\mathbf{F}(\mathbf{X}) - \widetilde{\mathbf{X}}\|_v = \|\mathbf{F}(\mathbf{X}) - \mathbf{F}(\widetilde{\mathbf{X}})\|_v \geq \|\mathbf{F'}_{\widetilde{\mathbf{X}}}(\mathbf{X} - \widetilde{\mathbf{X}})\|_v - \left\|\theta\left(\|\mathbf{X} - \widetilde{\mathbf{X}}\|_v\right)\right\|_v \geq$$

$$\widehat{q}\|\mathbf{X} - \widetilde{\mathbf{X}}\|_v - \delta\|\mathbf{X} - \widetilde{\mathbf{X}}\|_v = q\|\mathbf{X} - \widetilde{\mathbf{X}}\|_v,$$

for all $\mathbf{x} \in \mathbf{B}(\widetilde{\mathbf{X}}, \varepsilon)/\{\widetilde{\mathbf{X}}\}$.

Now, let $\mathbf{X}^{(0)}$ be an arbitrary starting point of the discrete dynamical system (Eq. (A1)) in the ε-neighborhood of the fixed point $\widetilde{\mathbf{X}}$, such that all the states of (Eq. (A1)) up to the $n - 1$th discrete time step belong to that ε-neighborhood of $\widetilde{\mathbf{X}}$. Then, we have that

$$\left\|\mathbf{X}^{(n)} - \widetilde{\mathbf{X}}\right\|_v = \left\|\mathbf{F}\left(\mathbf{X}^{(n-1)}\right) - \widetilde{\mathbf{X}}\right\|_v \geq q\left\|\mathbf{X}^{(n-1)} - \widetilde{\mathbf{X}}\right\|_v \geq \dots \geq q^n\left\|\mathbf{X}^{(0)} - \widetilde{\mathbf{X}}\right\|_v,$$

which implies that there must exist (sufficiently large) discrete time steps n such that $\mathbf{X}^{(n)} \notin \mathbf{B}\left(\widetilde{\mathbf{X}}, \varepsilon\right)$. Therefore, in the chosen ε-neighborhood of the fixed point $\widetilde{\mathbf{X}}$, there cannot exist a sequence of states of (Eq. (A1)) that will converge to $\widetilde{\mathbf{X}}$, i.e., (Eq. (A1)) cannot be asymptotically stable at the point $\widetilde{\mathbf{X}}$.

REFERENCES

Adami, C., Cerf, N.J., 2000. Physical complexity of symbolic sequences. Physica D 137, 62−69.

Adimya, M., Crauste, F., Marquet, C., 2010. Asymptotic behavior and stability switch for a mature−immature model of cell differentiation. Nonlinear Anal. Real. 11, 2913−2929.

Aguiar, M., Ashwin, P., Dias, A., Field, M., 2011. Dynamics of coupled cell networks: synchrony, heteroclinic cycles and inflation. J. Nonlinear Sci. 21, 271−323.

Balaz, I., Mihailovic, D.T., 2010. Modeling the intercellular exchange of signaling molecules depending on intra- and inter-cellular environmental parameters. Arch. Biol. Sci. Belgrade 62 (4), 947−956.

Binney, J.J., Dowrick, N.J., Fisher, A.J., Newman, M.E.J., 1992. The Theory of Critical Phenomena. Oxford University Press, Oxford.

Choi, Y.-P., Ha, S.-Y., Jung, S., Kim, Y., 2012. Asymptotic formation and orbital stability of phase-locked states for the Kuramoto model. Physica D 241, 735−754.

Cross, M.C., Hohenberg, P.C., 1993. Pattern formation outside of equilibrium. Rev. Mod. Phys. 65, 851−1112.

Crutchfield, J.P., 2012. Between order and chaos. Nat. Phys. 8, 17−24.

Cvetković, Lj, Bru, R., Kostić, V., Pedroche, F., 2011. A simple generalization of Gersgorin's theorem. Adv. Comput. Math. 35, 271−280.

Cvetković, Lj, Kostić, V., 2006. New subclasses of block H-matrices with applications to parallel decomposition-type relaxation methods. Num. Alg. 42, 325−334.

Cvetković, Lj, Kostić, V., Varga, R.S., 2004. A new Geršgorin-type eigenvalue inclusion set. ETNA (Elec. Trans. Num. An.) 18, 73−80.

Krabs, W., Pickl, S., 2010. Dynamical Systems − Stability, Controllability and Chaotic Behavior. Springer.

Manneville, P., 1990. Dissipative Structures and Weak Turbulence. Academic Press, Boston.

Michel, A.N., Hou, L., Liu, D., 2008. Stability of Dynamical Systems: Continuous, Discontinuous and Discrete Systems. Birkhauser, Boston.

Mihailović, D.T., Balaž, I., 2012. Synchronization in biochemical substance exchange between two cells. Mod. Phys. Lett. B 26, 1150031−1.

Mihailović, D.T., Balaž, I., Arsenić, I., 2013. A numerical study of synchronization in the process of biochemical substance exchange in a diffusively coupled ring of cells. Cent. Eur. J. Phys. 11, 440−447.

Mihailović, D.T., Budincevic, M., Balaž, I., Mihailović, A., 2011. Stability of intercellular exchange of biochemical substances affected by variability of environmental parameters. Mod. Phys. Lett. B 25, 2407−2417.

Mihailović, D.T., Kostić, V., Balaz, I., Cvetković, L., 2014. Complexity and asymptotic stability in the process of biochemical substance exchange in a coupled ring of cells. Chaos Solitons Fractals 65, 30−43.

Skryabin, D.V., 2000. Stability of multi-parameter solitons: asymptotic approach. Physica D 139, 186−193.

Varga, R.S., 2004. Geršgorin and His Circles. Springer Series in Computational Mathematics, vol. 36. Springer-Verlag, Berlin, Heilderberg.

Varga, S., Cvetković, L., Kostić, V., 2008. Approximation of the minimal Geršgorin set of a square complex matrix. ETNA (Elec. Trans. on Num. An.) 30, 398−405.

Wang, J., Huang, G., Takeuchi, Y., 2012. Global asymptotic stability for HIV-1 dynamics with two distributed delays. Math. Med. Biol. 29, 283−300.

Wheeler, J.A., 1990. In: Zurek, W. (Ed.), The Proceedings of the Workshop on Complexity, Entropy, and the Physics of Information, Held May-June, 1989, Entropy, Complexity, and the Physics of Information, vol. VIII. Addison-Wesley, Santa Fe, New Mexico. SFI Studies in the Sciences of Complexity.

Yan, Y., Kou, C., 2012. Stability analysis for a fractional differential model of HIV infection of CD4+ T-cells with time delay. Math. Comput. Simulat. 82, 1572−1585.

Zheng, Y.G., Wang, Z.H., 2012. Stability analysis of nonlinear dynamic systems with slowly and periodically varying delay. Commun. Nonlinear Sci. 17, 3999−4009.

Use of pseudospectra in analyzing the influence of intercellular nanotubes on cell-to-cell communication integrity

26

26.1 BIOLOGICAL IMPORTANCE OF TUNNELING NANOTUBES

Communication between cells is highly important for their survival. Using communication, cells can coordinate their collective actions, modify metabolism, and adapt to environmental changes (Levin, 2006; Pikovsky et al., 2001; Arenas et al., 2008; Chen et al., 2003). Cells can communicate by exchanging signaling molecules using several strategies. They can utilize the direct interaction between cell-surface molecules, they can synthetize and release specialized signaling molecules into the environment, or they can directly transport molecules through gap junctions (GJs). As specialized structures whose purpose is to put cells into tight contact, GJs are of special importance. From the perspective of functional role, GJs have evolved independently three times, if one includes plasmodesmata in plants, which are structurally very distinct from GJs but mediate similar functions (Nicholson, 2003). In the vertebrates and invertebrates, GJs are similar in both structure and function. The only major difference is that in vertebrates GJs are composed of proteins from the connexin family (Eiberger et al., 2001; Willecke et al., 2002) while in the invertebrates composing proteins are innexins, which is an unrelated but similar family to connexins (Phelan and Starrich, 2001). In mammals, GJ channels are composed of two hemi-channels, connexons, each composed of six subunits, connexins. Connexins are *trans*-membrane proteins, encoded by a large gene family that in humans have at least 20 isoforms. In the pair, each connexon is provided by one of the two neighboring cells. They are tightly connected to each other in the extracellular space, to form a double-membrane intercellular channel (Unger et al., 1999). Sometimes, the channels can form aggregates (termed plaques) that are highly dynamic, both spatially and temporally. Since connexins are coded by such a large gene family, it is clear that their synthesis and degradation are very precisely regulated, to secure proper functioning of GJs. The GJ-mediated intercellular communication is crucial for

Developments in Environmental Modelling, Volume 29, ISSN 0167-8892, http://dx.doi.org/10.1016/B978-0-444-63918-9.00026-0

coordination of development, tissue function, and cell homeostasis. Opening and closing of the GJ channels is a main way of regulating GJ-mediated communication. However, on a longer time-scale, all processes involved into the formation of GJ channels (delivery, assembly, removal) are additional independent mechanism to control GJ-mediated communication (Segretain and Falk, 2004).

In addition to stable communication, recent findings show that GJs may play a significant role in unstable, transient cell—cell contacts and that GJ hemi-channels by themselves may function in intracellular/extracellular signaling (Segretain and Falk, 2004). However, recent discoveries demonstrate that additional structures can play an important role in transient, dynamically more adaptable communication—tunneling nanotubes (TNTs).

TNTs were first reported as highly sensitive structures that are formed de novo between cells, connecting them into a complex network (Rustom et al., 2004). Later research demonstrated that they are highly dynamic and highly variable structures with heterogeneous molecular composition. The average length of TNTs can vary from 6 μm between PC12 cells, up to 44 μm between retinal pigment epithelial (ARPE-19) cells, while maximal observed length is in the range of 120 μm (Austefjord et al., 2014). Even within the same cell line, length can dynamically adapt as cells migrate. Exact mechanism of how length change is regulated is unknown. Diameter of TNTs is also highly variable: typically in range of 50—200 nm.

TNTs are sensitive to light leading to visible vibrations and rupture. They are also sensitive to shearing stress and chemical fixation. Most of them contain F-actin through the whole length of the TNTs. In some cell lines, microtubules are also detected, but how their presence is regulated is yet to be investigated. TNTs have no contact to the substrate and they hang in the medium. They are probably formed as membrane continuation because (1) in transmission electron micrographs, their membrane appeared to be continuous with the membranes of connected cells, (2) in scanning electron micrographs, they exhibit seamless transition to the surface of both connected cells, and (3) they permit a lateral diffusion of fluorescent membrane proteins between the plasma membranes of both connected cells (Rustom et al., 2004). Due to limited characterization so far, and their ability to perform numerous functions, proper classification of TNTs is still incomplete.

Formation of TNTs is rapid, usually within a few minutes. Two distinct mechanisms have been proposed (Gerdes and Carvalho, 2008). According to the first, filopodia-like protrusions grow by actin polymerization, directly toward a neighboring cell. Such a process implies existence of chemotactic signaling that guides the formation of TNTs. Some indirect results indicate that such a process indeed exists (Ramirez-Weber and Kornberg, 1999; Sherer et al., 2007). The second proposed mechanism proposes that TNTs emerge when previously attached cells depart from one another (Gerdes et al., 2007; Önfelt et al., 2004). This mechanism is based on observation that formation of membrane bridges between T-cells requires cell interaction before dislodging (Sowinski et al., 2008). Average lifetime of TNTs ranges from several minutes up to 1 h.

In the original finding, TNTs were proposed as a novel way for long-range cell-to-cell interaction primarily used for intercellular transfer of organelles (Rustom et al., 2004). Later research showed that the role of TNTs is much broader. Similar to GJs, these nanotubes can transfer diverse signals between the cells. Also they are detected in prokaryotes where they enable interspecies communication and share of antibiotic resistance (Dubey and Ben-Yehuda, 2011; Ficht, 2011). It has been shown that in both cases, these TNTs can facilitate cell-to-cell communication and intercellular transfer of cytoplasmic molecules, organelles, and viruses (Belting and Wittrup, 2008; Davis and Sowinski, 2008; Bukoreshtliev et al., 2009). Existence of clusters of TNTs enables formation of complex cellular networks with both local and long-distance interactions based on membrane continuity between TNT-connected cells (Fig. 26.1). Empirical evidence indicate that they can have an important role in many pathophysiological processes, like in activation of natural killer cells, regulation of osteoclastogenesis, or in the tumor formation and growth (Chauveau et al., 2010; Takahashi et al., 2013; Lou et al., 2012). In prokaryotes they can play the important part in transferring virulence from pathogenic to nonpathogenic bacteria (Dubey and Ben-Yehuda, 2011). These findings indicate that TNTs significantly contribute in a multitude of physiological processes, in both prokaryotic and eukaryotic cells.

It has been shown that TNTs have an important role in the migration and differentiation of neurons. Astrocytes are crucial for the maintenance of the microenvironment of mature neurons by clearing neurotransmitters from the

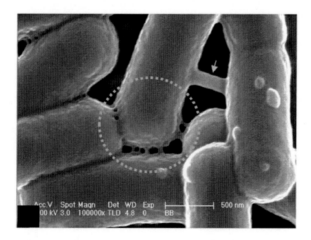

FIGURE 26.1

Example of intercellular TNTs between neighboring, prokaryotic cells. A field of cells with a cluster of smaller TNTs (highlighted by a *dashed circle*) and a more pronounced larger tube (indicated by an *arrow*).

Reprinted with permission from Dubey, G., Ben-Yehuda, S., 2011. Intercellular nanotubes mediate bacterial communication. Cell 144, 590–600.

synapses (Bergles et al., 1999). Also, astrocytes can modify the formation and function of synapses (Allman et al., 2011; Volterra and Meldolesi, 2005). However, little is known about how these interactions are initiated. Wang et al. (2012) demonstrated that immature hippocampal neurons generated short protrusions toward astrocytes resulting in TNTs formation, with an average lifetime of 15 min. Their findings suggest that within a limited maturation period developing neurons establish electrical coupling and exchange of calcium signals with astrocytes via TNTs. This novel cell—cell communication pathway between cells of the central nervous system provides new concepts in our understanding of neuronal migration and differentiation.

In addition to exchange of molecules, TNTs also convey electrical signals between distant cells (Wang et al., 2010). Such long-distance electrical coupling is assisted by GJs, while strength of electrical coupling depended on the length and number of TNT connections. Additionally, their results suggest that there are at least two different types of TNTs: those that interpose connexins and thus participate in electrical coupling, and those that lack connexins and do not display electrical coupling. More important is that the electrical signals transferred from one cell to another are sufficient to induce a transient calcium elevation in the recipient cell by activating low voltage-gated Ca^{2+} channels. Together with other research that demonstrated TNT-dependent intercellular calcium signaling via calcium diffusion through TNTs (Watkins and Salter, 2005; Hase et al., 2009), these findings suggest that different mechanisms of intercellular calcium signaling exist. It opens up the possibility that TNTs participate in physiologically relevant cell functions. In particular, the collective behavior of solitary or loosely attached migratory cells that follow the same tracks during diverse developmental processes could be coordinated by a long-distance signaling network (Wang et al., 2010). In experiments with the rat kidney cells $\sim 80\%$ of the TNTs between those cells mediate electrical coupling (Wang et al., 2010), but $\sim 50\%$ of TNTs between the cells from the same line are involved in organelle transfer (Gurke et al., 2008). This finding suggests that a number of TNTs are likely to be involved in both processes and raises the question as to the mechanism by which organelle transfer occurs between TNT-connected cells. Suggested mechanisms involve exocytic and endocytic events at the membrane interface or a transient fusion of the TNT membrane with the target-cell membrane.

Besides the activation of Ca^{2+} channels, TNT-mediated electrical coupling may affect other processes in the network of cells. They can modulate activity of small-molecule transporters (Levin, 2007) and the activity of different enzymes (Levin, 2007; Murata et al., 2005; Zhang et al., 2004). Also, it has been shown that during the wound healing, synthesis of TNTs increase, probably to synchronize the observed F-actin remodelling by activation of downstream signaling cascades (Wang et al., 2010). These results indicate that the transfer of electrical signals via TNTs and the subsequent activation of physiologically relevant biophysical signals may provide a unique mechanism for long-distance cellular signaling.

As a result of the heterogeneous structure, TNTs can exhibit different functional roles. For example, nanotubes connecting human macrophages can be divided in two groups with different functional properties: thick nanotubes containing both F-actin and microtubules, where intracellular vesicles are transferred through the nanotubes, and thin nanotubes, containing only F-actin, that facilitate transport of bacteria along their surface (Önfelt et al., 2006). Within thick nanotubes vesicles move in a step-wise manner, and the presence of microtubules is necessary for their transport. Vesicles move through the full length of nanotubes, from the connection with one cell into the cytoplasm of the other connected cell. It has been observed that in addition to endosomes and lysosomes, mitochondria can also enter thick nanotubes. On the other hand, bacteria move along the surface of thin TNTs. This process is termed surfing because they use a constitutive flow of the nanotube surface for movement. Surface transport along thin nanotubes is dependent on ATP but independent of microtubules. In addition to the role in transfer of bacteria, both viruses (Sherer et al., 2007; Sowinski et al., 2008) and prions (Gousset et al., 2009) also can exploit TNTs for invading eukaryotic cells.

Since TNTs are relatively newly discovered structures, whose regulation and dynamics is still largely unknown, there have not been many attempts to model their behavior. To our knowledge, only one model has been published so far (Suhail et al., 2013). They propose a mathematical model to explain passive protein transfer between cells via formation of TNTs. The model makes a critical assumption about distinct characteristics of transport of membrane versus cytosolic components through TNTs. It is assumed that there is a difference in transfer speed depending on the size of the molecules that diffuse through TNTs. Large cytosolic components, as intracellular proteins, can diffuse much slower that the membrane components. Therefore, there are two rate-limiting steps. For large components, it is their diffusion speed through TNTs, while for the smaller components, it is the rate of their access to TNTs on the side of the donor cell. As a consequence and due to the transient nature of TNTs, only smaller components would reach a steady state distribution over the length of TNT. Larger, cytosolic components will form a diffusion-based gradient.

First-order approximation in the model is that the abundance of the TNT lengths falls linearly with length. Therefore, it is assumed that the length r of TNTs follows a distribution:

$$p(r) \propto \begin{cases} l - r, & \text{if } r < l, \\ 0, & \text{otherwise.} \end{cases} \tag{26.1}$$

where l is the maximum length based on experimental observations. According to the model, greater mean diffusion length increases the observed levels of the transferred molecules adjacent to the donor cells and it also sharpens the fall in the concentration of the molecule as we move farther from the donor cells. The model explains that while transfer of cytoplasmic proteins may occur between cells, it would be in relatively smaller amounts in comparison to membrane proteins or smaller biochemical components present in the cytosol. Simulations of protein transfer for membrane proteins show that the membrane proteins can be transferred into the acceptor cells within the distance up to the maximum length of the TNTs, while the decline of the protein levels is approximately linear with the distance from the boundary of the donor cells. Also, the amount of transferred molecules depends on the mean diffusion length.

Despite their functional multitude, several recent results point toward a close functional link between TNTs, GJs, and stress response. First, their formation is induced by stress (Wang et al., 2011) but direction of their formation depends on cell types. For example, in neurons, cells that undergo stress will generate more TNTs, while in endothelial cells, stressed cells will synthesize signals that will induce formation of TNTs in nonstressed cells (Marzo et al., 2012). Second, TNTs can extend through GJs (Wang et al., 2010) modifying GJ functionality. Finally, TNTs are dynamic structures whose formation and decomposition is very sensitive to both intracellular and extracellular factors, with lifetimes ranging from several minutes up to 4–5 h (Bukoreshtliev et al., 2009; Lou et al., 2012). Their short lifetime suggests that they can promote unstable, transient cell-to-cell communication, in contrast to more stable communication mediated by GJs (Goodenough and Paul, 2009). Moreover, this transient influence on communication is promoted by stress, when integrity of intercellular communication is of special interest. Therefore, we believe that it is of importance to systematically explore how the perturbations in communication, induced by the existence of clusters of TNTs, can influence stability of intercellular communication.

26.2 COMPUTING THE THRESHOLD OF THE INFLUENCE OF INTERCELLULAR NANOTUBES ON CELL-TO-CELL COMMUNICATION INTEGRITY

In this section we explore how the substance exchange through TNTs affects the functional stability of a multicellular system, following the paper by Mihailović et al. (2016). We suppose that GJs form the main communication line, while formation of TNTs can modify dynamics of communication, keeping GJs intact. Also, in the model, we will consider only TNTs as transient structures. We are aware that both GJs and TNTs are dynamic structures. Formation of GJs is tightly regulated and their number can significantly change over time. However, in this section we are only interested in how TNTs can influence the already established stable communication. Therefore, the time scale in our model is shorter than the time needed for GJs to be synthesized or degraded. From these starting points we focus on two issues: (1) whether transient clusters of TNTs can either stabilize or destabilize intercellular communication governed by GJs? and (2) how to determine the threshold at which influence of TNTs destabilize GJ-mediated communication? Therefore, we introduce a model of the substance exchange in a multicellular system, represented by ordinary differential equations, where cell-to-cell communication is mediated by both the GJs and TNTs while metabolic processes in the cell follow Michaelis—Menten dynamics. In this model, the GJ function governs the time evolution of the intercellular network while the TNTs function simulates the exchange mediated by the TNTs including a scaling parameter for that mediation. So, we consider the influence of TNTs as a functional perturbation of the main communication mediated by GJs. To determine the threshold for which the multicellular system remains stable, despite TNTs influence, we compute the *distance to instability* (Higham, 1989), using nonconvex optimization algorithm from Kostic et al. (2015), and we derive numerically cheap lower bounds based on pseudospectral localizations (Kostic et al., 2016).

General model dynamics and pseudospectra. To investigate how TNTs affect the stability of the intercellular communication network, we model the network dynamics as

$$\dot{x}(t) = \Psi(x(t)) := \Phi(x(t)) + \xi\Theta(x(t)), \tag{26.2}$$

where $x = (x_1, x_2, ..., x_n)$. Here $x_i(t) \in [0, 1]$ is the relative concentration of molecules and ions in the cell $i \in N := [1, 2, ..., n]$, $\Phi : \mathbf{R}^n \to \mathbf{R}^n$ is a GJ function that governs time evolution of the intercellular network, while $\Theta : \mathbf{R}^n \to \mathbf{R}^n$ models the exchange mediated by TNTs and $\xi > 0$ is a scaling parameter for that mediation. Since many questions remain unanswered about how cargo is transported through TNTs, we consider their effect on the model dynamics as an uncertainty described by Θ. The system (26.2) is generally a nonlinear one whose stability is typically investigated at the equilibrium state $x^\star \in \mathbf{R}^n$ as the local asymptotic stability, where $x^\star$ is a state vector such that $\Psi(x^\star) = 0$. An equilibrium state $x^\star$ is locally asymptotically stable if there exists $\varepsilon > 0$ such that for every $x(0)$ that is in ε neighborhood of the equilibrium $x^\star$ (i.e., $\|x^\star - x(0)\| < \mu$), it holds that $\lim_{t \to \infty} \|x(t) - x^\star\| = 0$. This local stability property is characterized by the spectra of the Jacobian matrix $A = \left[\frac{\partial \psi_i}{\partial x_j}(x^\star)\right]$ of (26.2) in the state $x = x^\star$, which can be written as $A = \widehat{A} + \xi\Delta$, where $\widehat{A} = \left[\frac{\partial \Phi_i}{\partial x_j}(x^\star)\right]$ corresponds to the exchange through GJs that will be named the measurable communication. The term $\widehat{\Delta} = \left[\frac{\partial \Theta_i}{\partial x_j}(x^\star)\right]$ is determined by the transport through TNTs that is generally unknown. Thus, we call it the unmeasurable communication. Furthermore, in order that $\xi > 0$ to reflect the scale of TNT mediation, we assume that uncertainty parameters are unit scaled in the chosen matrix norm, i.e., $\|\widehat{\Delta}\| = 1$.

To investigate how TNTs can influence stability of communication, we will analyze the sensitivity of the spectrum of the measurable communication matrix $\widehat{A}$ upon perturbation $\xi\Delta$. More precisely, we are interested, in general, in which scale of TNTs influence ($\xi > 0$) is capable of changing the asymptotic stability/instability of the network dynamics (26.2) from what would be expected if TNTs were not present ($\xi = 0$). Computing the spectra $\Lambda(\widehat{A})$ of the measurable part of the Jacobian, corresponding to GJs interactions and intracellular metabolic processes, we can determine the expected stability $\Lambda(\widehat{A}) \subseteq \mathbb{C}^-$ or instability $\Lambda(\widehat{A}) \nsubseteq \mathbb{C}^-$ of the substance fluxes. However, the unmeasurable part of cellular communication can change this spectral property and lead to the different dynamics of the network. Thus, in general, we distinguish two cases.

First, if GJ network dynamics is stable, i.e., $\Lambda(\widehat{A}) \subseteq \mathbb{C}^-$, then we are interested in computing the critical value of $\xi > 0$ such that the full network (GJ and TNTs) becomes unstable, i.e., $\Lambda(A) \nsubseteq \mathbb{C}^-$. If we denote $\nu_{\widehat{A}}^- := \max\{\xi > 0 :$ $\Lambda(\widehat{A} + \xi\Delta) \subseteq \mathbb{C}^-, \Delta \in \mathbb{C}^{n,n}, \|\Delta\| = 1\}$, this threshold value $\nu_{\widehat{A}}^-$ is known as the *distance to instability* (Higham, 1989), whose computation requires solving a nonconvex optimization problem, and, thus, the use of numerical algorithms; for details see (Trefethen and Embre, 2005; Byers, 1988; He and Watson, 1999; Freitag and Spence, 2011; Gugliemi et al., 2015). The main tool often used in determining the distance to instability, and thus the threshold value of TNTs scale $\xi > 0$, is the concept of matrix pseudospectra (Trefethen and Embre, 2005).

Given a matrix $\widehat{A} \in \mathbb{C}^{n,n}$ and $\varepsilon > 0$, the ε-pseudospectrum of a matrix $\widehat{A}$, denoted by $\Lambda_\varepsilon(\widehat{A})$, is composed of all eigenvalues of matrices which are "$\varepsilon-close$}" to $\widehat{A}$: $\lambda \in \Lambda_\varepsilon(\widehat{A})$ if and only if there exists $x \in \mathbb{C}^n\backslash 0$ and $\widehat{\Delta} \in \mathbb{C}^{n,n}$ such that $\|\widehat{\Delta}\| \leq \varepsilon$ and $(\widehat{A} + \widehat{\Delta})x = \lambda x$, i.e., $\Lambda_\varepsilon(\widehat{A}) = \bigcup_{\|\widehat{\Delta}\| \leq \varepsilon} \Lambda(\widehat{A} + \widehat{\Delta})$. Consequently, as noted above, we use ε-pseudospectrum to establish spectral properties that are robust under matrix perturbations bounded in a given norm $\|\cdot\|$ by the parameter $\varepsilon > 0$. To conclude, $\Lambda(A) \subset \Lambda_\xi(\widehat{A}) \subseteq \mathbb{C}^-$ if and only if $\xi < \nu_{\widehat{A}}^-$, and, therefore, if the TNTs influence is dominated by the distance to instability of the measurable matrix $\widehat{A}$, we can safely conclude that the mediation of TNTs does not change the stability of (26.2) from what would be expected by observing GJ processes, while, for the values $\xi \geq \nu_{\widehat{A}}^-$, we cannot do that.

Second, if GJ network dynamics is unstable, i.e., $\Lambda(\widehat{A}) \nsubseteq \mathbb{C}^-$, then we are interested in computing the critical value of $\xi > 0$ such that the full network (GJ and TNTs) can become stable, i.e., $\Lambda(A) \subseteq \mathbb{C}^-$. So, if we denote $\nu_{\widehat{A}}^+ := \min\{\xi > 0 : \Lambda(\widehat{A} + \xi\Delta) \subseteq \mathbb{C}^-, \Delta \in \mathbb{C}^{n,n}, \|\Delta\| = 1\}$, this threshold value $\nu_{\widehat{A}}^+$ is known as the *distance to stability* of the measurable matrix $\widehat{A}$ (Higham, 1989) and it is the critical value of the TNTs influence for which the system (26.2) can have the stable behavior despite the expected unstable one deduced from the observations of GJs. Contrary to the previous case, pseudospectra methods cannot be used to compute such $\nu_{\widehat{A}}^+$, which makes it a much harder computational problem: for detail treatment see Orbandexivry et al. (2013).

Therefore, the integrity of intercellular communication (either stable or unstable one) under the influence of TNTs is not only dependant on GJ network's *resilience*, i.e., real part of the least negative eigenvalue of the measurable Jacobian matrix, as often assumed in the literature, but also, highly, on the GJ network structure which can, due to nonnormality (Trefethen and Empre, 2005), amplify the importance of TNT mediation.

26.3 ANALYSIS OF A SIMPLE DETERMINISTIC MODEL OF INTERCELLULAR COMMUNICATION

Mathematical background. We introduce a simple deterministic model for substance exchange in a multicellular system which is mediated by GJs and TNTs. Both, GJs and TNTs, allow various molecules and ions to pass freely between cells through the channels by the diffusion-like process. However, diffusion in cells and between them, known as the anomalous diffusion, can differ from the "classical" one due to spatial inhomogeneity (Cherstvy and Metzler, 2013; Mullineaux et al., 2008). In a situation like this, it is suitable to consider the kinetics of substance exchange between cells in terms of an *exchange coefficient* g_{ij} with a dimensional unit of inverse time. In the simplest case, communication from the cell i to cell j is proportional to the concentrations between the cells. Therefore, we can define the substance exchange between cell j to cell i as $g_{ij}(x_j(t) - x_i(t)) + \xi\delta_{ij}x_j(t)$, where $\xi > 0$ is a small value that determines the strength of influence of TNTs on communication modeled by the uncertainty parameter δ_{ij}. Since exchanged substances play a role in the metabolic processes inside the cell and are released into the environment, we introduce the parameters $\alpha_i > 0$ that describe the rate by which the substance is metabolized by the cell $i \in N$ in time t. Since most of the metabolic processes follow Michaelis–Menten dynamics, we introduce $\beta_i > 0$ as the half-time saturation coefficient for the cell i. Accordingly, we express the intercellular communication as:

$$\dot{x}_i(t) = -\frac{\alpha_i x_i(t)}{\beta_i + x_i(t)} + \sum_{j \neq i} g_{ij}(x_j(t) - x_i(t)) + \xi \sum_{j \in N} \delta_{ij}x_j(t), \quad (i \in N). \tag{26.3}$$

While this model is rather restrictive in terms of the real dynamics of a multicellular system, and more complex dynamical systems can be used instead, we will use it to emphasize the use of the pseudospectra in understanding influence of TNTs which is the main focus of this section, which could become otherwise more vague due to the technical complexity of mathematical analysis.

In order to show how the pseudospectra can be used to determine the threshold of TNT mediation that can affect stability of GJ network, we use the model given by (26.3) and restrict to the case of zero equilibrium $x^\star = 0$, when the initial state $x(0) \neq 0$ reflects starting distribution of the substance in the network. Consequently,

$$\widehat{\Delta} = [\delta_{ij}] \quad \text{with} \quad \|\widehat{\Delta}\| = 1, \quad \text{while} \quad \widehat{A} = \left[\frac{\partial \Phi_i}{\partial x_j}(0)\right] = [\widehat{a}_{ij}] \quad \text{where} \quad \text{for} \quad i, j \in N$$

$\widehat{a}_{ij} = -\left(\frac{\alpha_i}{\beta_i} + \sum_{j \neq i} g_{ij}\right)$ if $i = j$ or $\widehat{a}_{ij} = g_{ij}$ if $i \neq j$.

Depending on the system's property, we wish to examine the norm in which distances measured is chosen. Here we discus three such cases:

- using the 1-norm $\|x\|_1 = \max_{i \in N} |x_i|$ and $\|A\|_1 = \max_{j \in N} \sum_{i \in N} |a_{ij}|$ measures the total substance concentration in the network,

- using the Euclidean norm $\|x\|_2 = \sqrt{\sum_{i \in N} |x_i|^2}$ and $\|A\|_2 = \sqrt{\max_{i \in N} |\lambda_i(A^*A)|}$ measures the network's total squared deviation from the equilibrium,

- using the infinity norm $\|x\|_\infty = \max_{i \in N} |x_i|$ and $\|A\|_\infty = \max_{i \in N} \sum_{j \in N} |a_{ij}|$ measures the network's maximal per cell substance concentration.

In that setting, for the communication mediated exclusively by GJs, i.e., $\xi = 0$, using pseudospectral localization (Kostić et al., 2015), we can conclude that zero equilibrium point of (26.3) is exponentially stable one. Moreover, the following holds:

Theorem 1. *Let $x(t)$ be the flow of (26.3) for an arbitrary, sufficiently small, initial condition $x(0)$. Then, there exists $M > 0$ such that $\|x(t)\|_p \leq M e^{-(\omega_p + \xi)t} \|x(0)\|_p$, for $t \geq 0$, holds for every $\xi \geq 0$ and every $p \in \{1, 2, \infty\}$, where*

$$\omega_p = \min_{i \in N} \left\{ \frac{\alpha_i}{\beta_i} + \frac{1}{p} \sum_{j \neq i} (g_{ij} - g_{ji}) \right\}.$$

Proof. Let $x(t)$ be the flow of (26.3) for the sufficiently small initial value $x(0)$. Then, it is well known that there exists $M > 0$ such that $\|x(t)\|_\infty \leq M e^{\mu t} \|x(0)\|_\infty$, for all $t \geq 0$, where $\mu = \max\{\lambda : \lambda \in \Lambda(A)\}$ (Hinrichsen and Pritchard, 2005). But, then, $\mu \in \Lambda(\widehat{A} + \xi \Delta) \subseteq \Lambda_\xi(\widehat{A})$. On the other hand, Theorem 1 of Kostic et al. (2016) states that $\Lambda_\xi(\widehat{A}) \subseteq \Gamma^\xi(\widehat{A})$, where $\Gamma^\xi(\cdot)$ denotes the ξ-pseudo Geršgorin set, i.e.,

$$\Gamma^\xi(\widehat{A}) = \bigcup_{i \in N} \left\{ z \in \mathbb{C} : |z - \widehat{a}_{ii}| \leq \sum_{j \neq i} |\widehat{a}_{ij}| + \xi \right\}.$$

But, then, there exists an index $i \in N$ such that $|\mu - \widehat{a}_{ii}| \leq \sum_{j \neq i} |\widehat{a}_{ij}| + \xi$, and, therefore, $\mu \leq \widehat{a}_{ii} + \sum_{j \neq i} |\widehat{a}_{ij}| + \xi$, i.e., $\mu \leq$

$\max_{i \in N} \left\{ -\frac{\alpha_i}{\beta_i} - \sum_{j \neq i} g_{ij} + \sum_{j \neq i} g_{ij} + \xi \right\} = -\omega_\infty + \xi$, which completes the proof for the case $p = \infty$.

When $p = 1$, we can do the similar reasoning applied to the A^T, since $\Lambda(A^T) = \Lambda(A)$. In such a way we conclude that $\mu \in \Gamma^\xi(\widehat{A}^T)$, i.e.,

$$\mu \leq \max_{i \in N} \left\{ -\frac{\alpha_i}{\beta_i} - \sum_{j \neq i} g_{ij} + \sum_{j \neq i} g_{ji} + \xi \right\} = -\omega_1 + \xi.$$

Finally, to analyze the Euclidean norm case, $p = 2$, observe that (Kostić et al., 2016)

$$\mu \in \Lambda_\xi\left(\widehat{A}\right) \text{ if and only if } \sigma_{\min}\left(\widehat{A} - \mu I\right) = \|\left(\widehat{A} - \mu I\right)^{-1}\|_2^{-1} \le \xi.$$

Thus, applying the bound of the minimal singular value obtained in Johnson (1989) we have that

$$\min_{i \in N}\left\{|\widehat{a}_{ii} - \mu| - \frac{1}{2}\sum_{j \ne i}(\widehat{a}_{ij} + \widehat{a}_{ji})\right\} \le \sigma_{\min}\left(\widehat{A} - \mu I\right) \le \xi,$$

and, having that $\mu \le 0$, consequently,

$$\min_{i \in N}\left\{\mu + \frac{\alpha_i}{\beta_i} + \sum_{j \ne i} g_{ij} - \frac{1}{2}\sum_{j \ne i}(g_{ij} + g_{ji})\right\} \le \xi.$$

Finally, last inequality yields $\mu \le -\omega_2 + \xi$, which completes the proof.

The previous result provides computationally cheap lower bounds of the distance to instability of $\widehat{A}$ in norm $\|\cdot\|_p$. Since for the multicellular system (26.3) we always have that $\omega_\infty > 0$, this means that its communication integrity is maintained when the TNTs influence is scaled bellow ω_∞. Contrary to that, constants ω_1 and ω_2 can be, independently, positive or negative, depending of the structure of the GJ network, and, therefore, we may not always have this conclusion in cases $p = 1$ and $p = 2$ and more computations are needed. As noted above, determining the precise threshold value $\nu_{\widehat{A}}^-$ is a computationally demanding task, and all of the existing algorithms treat a case when $p = 2$. The basis for such numerical computations is based on the following fact (here formulated in the setting of GJ and TNT multicellular networks).

Theorem 2. Let $\widehat{A}$ be the measurable part of the Jacobian matrix of the multicellular system (26.3). Then, $\Lambda_\xi(\widehat{A}) \subset \backslash \mathbb{C}_-$ if and only if $\xi < \nu_{\widehat{A}}^- = \min_{t \in \mathbb{R}} \sigma_{\min}(A - itI)$, where $\sigma_{\min}$ denotes minimal singular value of a matrix and i is an imaginary unit.

Furthermore, the worst-case TNT configuration matrix $\widehat{\Delta}$ for the critical case $\xi_{crit} = \nu_{\widehat{A}}^-$ is then given by $\widehat{\Delta} = \widehat{u}\widehat{v}^*$, where $\widehat{u}$ and $\widehat{v}$ are, respectively, left and right singular vector corresponding to the singular value ξ_{crit}.

For the proof, see Kostić et al. (2016) and Trefethen and Embre (2005).

Numerical simulation. To illustrate the use of the introduced concepts, we consider a few realistic scenarios based on the simple deterministic model of a 100-cell network, as could be the case in highly packed clusters of cells in prokaryotic biofilms or in eukaryotic tissues. In all the test cases, we compute the lower bound of threshold of TNTs influence in infinity norm (ω_∞), the exact value in Euclidean norm ($\nu_{\widehat{A}}^-$), and construct pseudospectral portrait with transient plot.

For GJ communication we use (1) the spatially distributed Newman-Gastner network with weight parameter set to 0.001 (Gastner and Newman, 2006) and (2) the Erdös-Rényi modular network (Erdös and Rényi, 1959) with 10 clusters connected with 0.03 overall probability of the attachment and proportion of links within

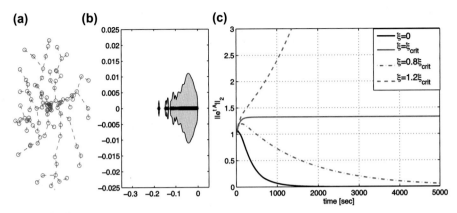

FIGURE 26.2

Distance to instability computed for 100-cell GJ network constructed as the Newman-Gastner spatial network. (a) Graphical image of that network; (b) pseudospectral portrait of its Jacobian matrix $\widehat{A}$ (*asterisks* mark eigenvalues of $\widehat{A}(\xi = 0)$, while the *shadowed area* represents all the possible locations of eigenvalues for the full network Jacobian A, when GJs and TNTs are included ($\xi = \xi_{crit}$ where $\xi_{crit} = \nu_{\widehat{A}}$ is a threshold value)); (c) transient growth of substance concentration within cells from the initial state measured in the Euclidian norm $\|e^{tA}\|_2$, due to nonnormality of the GJs Jacobian matrix (McCoy, 2013) for the following cases: $\xi = 0$ (*solid black*), $\xi = \xi_{crit}$ (*solid gray*), $\xi = 0.8\xi_{crit}$ (*solid dashed-dotted*), and $\xi = 1.2\xi_{crit}$ (*solid dashed*). Note that peak of concentration and duration of concentration decay depends on the scenario used as seen in Figs. 26.2c–26.4c. The computed values are $\xi_{crit} = 2.38 \cdot 10^{-3}$ and $\omega_{\infty} = 4.47 \cdot 10^{-5}$ (Mihailović et al., 2016).

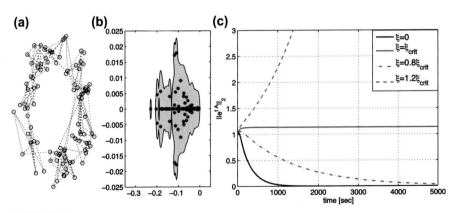

FIGURE 26.3

The same as in Fig. 26.2 but for Erdös-Rényi modular network. The computed values are $\xi_{crit} = 3.15 \cdot 10^{-3}$ and $\omega_{\infty} = 6.30 \cdot 10^{-5}$.

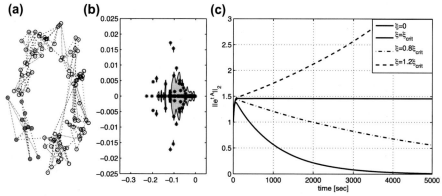

FIGURE 26.4

The same as in Fig. 26.2 but for Erdös-Rényi modular network simulating a pathological state of the system (*colored circles* (gray in print versions) show nodes with altered capacity to receive signals). The computed values are $\xi_{crit} = 6.55 \times 10^{-4}$ and $\omega_\infty = 6.30 \times 10^{-5}$.

modules set to 90%, graphically depicted in Figs. 26.2a, 26.3a, and 26.4a. In the spatially distributed network, proximity of cells is not explicitly defined but is a function of their communication range. Therefore, small changes of distances due to cell motility converge into one network configuration. For values of physiological parameters in simulations, we assume that all cells in the population are of the same type and therefore have the same time scales for metabolizing the substance. Therefore, for each $i = 1, \ldots, 100$, we take β_i randomly on a uniformly distributed interval [0.9, 1.1]. On the other hand, the saturation constants $\{\alpha_i\}$ (having the same order of value as the exchange coefficients), that are of the same order as the exchange coefficients, can differ more significantly, and are chosen randomly from [0, 0.01] with the uniform distribution. Finally, following Mullineaux et al. (2008), for exchange coefficients g_{ij}, we use random values from the interval [0, 0.05] with the uniform distribution.

We compute the critical pseudospectra of the GJs Jacobian matrix $\widehat{A}$ of the network (Figs. 26.2b, 26.3b, and 26.4b). Here, the term critical pseudospectra stand for the fact that $\varepsilon = \nu_{\widehat{A}}$ which is the computed threshold of TNTs influence. The shadowed area indicates how the system is sensitive to changes in cell communication determined by TNTs. To demonstrate how formation of TNTs can affect network dynamics, we compute the first-order approximated behavior of (26.3), measured in Euclidean norm, i.e., $\|e^{tA}\|_2$ (Figs. 26.2c, 26.3c, and 26.4c), for the following cases.

1. (idealized case) when cell-to-cell communication takes place only through GJs ($\xi = 0$). Then the system, after passing through short transient interval, is reaching the stability either faster (Figs. 26.2c and 26.3c) or slower (Fig. 26.4c). The corresponding curves are depicted by the solid black lines.

2. (critical case), when $\xi = \xi_{crit}$ and $\widehat{\Delta}$ is the worst arrangement of TNTs that move eigenvalues of A to the imaginary line (*marginal instability*). Such $\widehat{\Delta}$ is constructed via suitable singular vectors (Kostić et al., 2016). From Figs. 26.2c, 26.3c, and 26.4c (solid gray line), it is seen that the substance exchange in the system is in the state of an oscillating mode, waiting to start toward either stability or instability.

3. (case of stability) when $\xi < \xi_{crit}$ (here $\xi = 0.8\xi_{crit}$). In this case the system maintains the communication integrity (gray dashed-dotted line in Figs. 26.2c, 26.3c, and 26.4c).

4. (case of instability), when $\xi > \xi_{crit}$ (here $\xi = 1.2\xi_{crit}$). Correspondingly, the system is communicationally disintegrated (gray dashed line in Figs. 26.2c, 26.3c, and 26.4c).

The situation of the disrupted cell-to-cell communication can cause numerous diseases. For example, inborn cardiac diseases are among the most frequent congenital anomalies and are caused by mutations in genes that form GJs (Salameh et al., 2013). Therefore, it is crucial to investigate stability of intercellular communication and determine possible thresholds for disruption of cell-to-cell communication integrity. To investigate a possible influence of TNTs in the case of disrupted communication, we simulate pathological situations by modifying the Erdös-Rényi modular network as follows. In this network arrangement, only one module (colored circles (gray in printed version) in Fig. 26.4a) exhibits a cascade degradation of their capacity to receive the substance under exchange, while their capacity to send it in the fixed network flux direction (corresponding to the node enumeration) remains the same. More precisely, in the original realization of Erdös-Rényi modular network we take $0.1g_{ij}$, instead of g_{ij} for $i = 1, \ldots, 100$ and $j = 1, \ldots, 10$. In the example we create, when only one module is disrupted, pseudospectral portrait (Fig. 26.4b) shows that sensitivity of the whole network to communication changes is significantly increased, compared to the nonpathological state (Fig. 26.3b). Also, critical level of oscillations in the Euclidian norm deviates more from the equilibrium state indicating that formation of TNTs can disrupt the system more easily.

As a summary, an overview of what has been in this section is presented. The main novelty in considering the subject, we were talking about, lies in the fact that the uncertainty of TNTs influence to the overall cellular communication can be treated as the matrix nearness problems, i.e., either as distance to instability (treated in this section) or distance to stability (reverse problem that will be treated in the further researches). We have presented how this concept can provide meaningful insights using the simple deterministic model of cellular networks with asymptotically stable GJs cell-to-cell communication, where TNTs can destabilize the system. The problem is analyzed in terms of maximal individual deviation ($\|\cdot\|_\infty$) and total square deviation (Euclidian norm) of the cells substance concentration. In the simulations, the threshold of such TNTs influence is computed using recently developed pseudospectra methods for two standard structures of cellular networks (spatial Newman-Gastner and modular Erdös-Rényi) modelling healthy and pathological states of the system. The reasoning presented here is a first step toward understanding of the influence of TNTs as uncertainty of the system using matrix analysis and computational methods. Therefore, many open questions remain, on both mathematical and experimental sides. For example, analyzing the tie between matrix perturbations as stochastic processes, where the pseudospectra combined with the Bregman divergences (Dhillon and Tropp, 2007) can help to reliably estimate the mathematical expectation of the threshold of destabilizing/stabilizing intercellular communication. On the empirical side, this model implies that the pattern of cellular arrangement influences the stability of intercellular communication. This insight has the potential to enhance understanding of the role of cellular architecture on basic intracellular processes via communication. Thus, the model should be extended with experimentally verified data which requires the experimental studies of biochemical mechanisms of dynamics of TNTs in different environments. Also, it could be successfully applied in the FRAP technique (fluorescence recovering after photobleaching) to observe and quantify rapid diffusion of calcein or other molecules between the cytoplasm of the cells.

REFERENCES

Allaman, I., Belanger, M., Magistretti, P.J., 2011. Astrocyte-neuron metabolic relationships: for better and for worse. Trends Neurosci. 34, 76–87.

Arenas, A., Diaz-Guilera, A., Kurths, J., Moreno, Y., Zhou, C., 2008. Synchronization in complex networks. Phys. Rep. 469, 93–153.

Austefjord, M.W., Gerdes, H.-H., Wang, X., 2014. Tunneling nanotubes: diversity in morphology and structure. Commun. Integr. Biol. 7, e27934, 1–5.

Belting, M., Wittrup, A., 2008. Nanotubes, exosomes, and nucleic acid-binding peptides provide novel mechanisms of intercellular communication in eukaryotic cells: implications in health and disease. J. Cell. Biol. 183, 1187–1191.

Bergles, D.E., Diamond, J.S., Jahr, C.E., 1999. Clearance of glutamate inside the synapse and beyond. Curr. Opin. Neurobiol. 9, 293–298.

Bukoreshtliev, N.V., Wang, X., Hodneland, E., Gurke, S., Barroso, J.F., Gerdes, H.H., 2009. Selective block of tunneling nanotube (TNT) formation inhibits intercellular organelle transfer between PC12 cells. FEBS Lett. 583, 1481–1488.

Byers, R., 1988. A bisection method for measuring the distance of a stable matrix to the unstable matrices. SIAM J. Sci. Statist. Comput. 9, 875–881.

Chauveau, A., Aucher, A., Eissmann, P., Vivier, E., Davis, D.M., 2010. Membrane nanotubes facilitate long-distance interactions between natural killer cells and target cells. Proc. Natl. Acad. Sci. U.S.A. 107, 5545–5550.

Chen, Y., Rangarajan, G., Ding, M., 2003. General stability analysis of synchronized dynamics in coupled systems. Phys. Rev. E 67, 026209.

Cherstvy, A.G., Metzler, R., 2013. Population splitting, trapping, and non-ergodicity in heterogeneous diffusion processes. Phys. Chem. Chem. Phys. 15, 20220–20235.

Davis, D.M., Sowinski, S., 2008. Membrane nanotubes: dynamic long-distance connections between animal cells. Rev. Mol. Cell. Biol. 9, 431—436.

Dhillon, I.S., Tropp, J.A., 2007. Matrix nearness problems with Bregman divergences. SIAM J. Matrix Anal. Appl. 29, 1120—1146.

Dubey, G., Ben-Yehuda, S., 2011. Intercellular nanotubes mediate bacterial communication. Cell 144, 590—600.

Eiberger, J., Degen, J., Roumaldi, A., Deutsch, U., Willecke, K., Sohl, G., 2001. Connexin genes in the mouse and human genome. Cell Commun. Adhes. 8, 163—165.

Erdös, P., Rényi, A., 1959. On random graphs I. Publ. Math Debr. 6, 290—297.

Ficht, T., 2011. Bacterial exchange via nanotubes: lessons learned from the history of molecular biology. Front. Microbiol. 2, 179.

Freitag, M.A., Spence, A., 2011. A Newton-based method for the calculation of the distance to instability. Linear Algebra Appl. 435, 3189—3205.

Gastner, M.T., Newman, M.E.J., 2006. The spatial structure of networks. Eur. Phys. J. B 49, 247—252.

Gerdes, H.-H., Bukoreshtliev, N.V., Barroso, J.F., 2007. Tunneling nanotubes: a new route for the exchange of components between animal cells. FEBS Lett. 581, 2194—2201.

Gerdes, H.-H., Carvalho, R.N., 2008. Intercellular transfer mediated by tunneling nanotubes. Curr. Opin. Cell Biol. 20, 470—475.

Goodenough, D.A., Paul, D.L., 2009. Gap junctions. Cold Spring Harb. Perspect. Biol. 1, a002576.

Gousset, K., Schiff, E., Langevin, C., Marijanovic, Z., Caputo, A., et al., 2009. Prions hijack tunnelling nanotubes for intercellular spread. Nat. Cell Biol. 11, 328—336.

Guglielmi, N., Kressner, D., Lubich, C., 2015. Low rank differential equations for Hamiltonian matrix nearness problems. Numer. Math. 129, 279—319.

Gurke, S., Barroso, J.F., Hodneland, E., Bukoreshtliev, N.V., Schlicker, O., Gerdes, H.H., 2008. Tunneling nanotube (TNT)-like structures facilitate a constitutive, actomyosin-dependent exchange of endocytic organelles between normal rat kidney cells. Exp. Cell Res. 314, 3669—3683.

Hase, K., Kimura, S., Takatsu, H., Ohmae, M., Kawano, S., Kitamura, H., Ito, M., Watarai, H., Hazelett, C.C., 2009. M-Sec promotes membrane nanotube formation by interacting with Ral and the exocyst complex. Nat. Cell Biol. 11, 1427—1432.

He, C., Watson, G.A., 1999. An algorithm for computing the distance to instability. SIAM J. Matrix Anal. Appl. 20, 101—116.

Higham, N.J., 1989. Matrix nearness problems and applications. In: Gover, M.J.C., Barnett, S. (Eds.), Applications of Matrix Theory. University Press, pp. 1—27.

Hinrichsen, D., Pritchard, A.J., 2005. Mathematical Systems Theory I, Modelling, State Space Analysis, Stability and Robustness. Springer-Verlag, Berlin.

Johnson, C.R., 1989. A Gersgorin-type lower bound for the smallest singular value. Linear Algebra Appl. 112, 1—7.

Kostić, V., Cvetković, L., Cvetković, D., 2016. Pseudospectra localizations and their applications. Numer. Linear Algebra Appl. 23, 356—372.

Kostić, V., Miedlar, A., Stolwijk, J., 2015. On matrix nearness problems: distance to delocalization. SIAM J. Matrix Anal. Appl. 36, 435—460.

Levin, M., 2007. Large-scale biophysics: ion flows and regeneration. Trends Cell Biol. 17, 261—270.

Levin, S.A., 2006. Fundamental questions in biology. PLoS Biol. 4, e300.

Lou, E., Fujisawa, S., Barlas, A., Romin, Y., Manova-Todorova, K., Moore, M.A., Subramanian, S., 2012. Tunneling nanotubes: a new paradigm for studying intercellular communication and therapeutics in cancer. Commun. Integr. Biol. 5, 399–403.

Marzo, L., Gousset, K., Zurzolo, C., 2012. Multifaceted roles of tunneling nanotubes in intercellular communication. Front. Physiol. 3, 1–14.

McCoy, J.H., 2013. Amplification without instability: applying fluid dynamical insights in chemistry and biology. New J. Phys. 15, 113036.

Mihailović, D.T., Kostić, V.R., Balaž, I., Kapor, D., 2016. Computing the threshold of the influence of intercellular nanotubes on cell-to-cell communication integrity. Chaos Solitons Fractals 91, 174–179.

Mullineaux, C.W., Mariscal, V., Nenninger, A., Khanum, H., Herrero, A., Flores, E., Adams, D.G., 2008. Mechanism of intercellular molecular exchange in heterocyst-forming cyanobacteria. EMBO J. 27, 1299–1308.

Murata, Y., Iwasaki, H., Sasaki, M., Inaba, K., Okamura, Y., 2005. Phosphoinositide phosphatase activity coupled to an intrinsic voltage sensor. Nature 435, 1239–1243.

Nicholson, B.J., 2003. Gap junctions – from cell to molecule. J. Cell Sci. 116, 4479–4481.

Önfelt, B., Nedvetzki, S., Benninger, R.K., Purbhoo, M.A., Sowinski, S., Hume, A.N., Seabra, M.C., Neil, M.A., 2006. Structurally distinct membrane nanotubes between human macrophages support long-distance vesicular traffic or surfing of bacteria. J. Immunol. 177, 8476–8483.

Önfelt, B., Nedvetzki, S., Yanagi, K., Davis, D.M., 2004. Cutting edge: membrane nanotubes connect immune cells. J. Immunol. 173, 1511–1513.

Orbandexivry, F.-X., Nesterov, Y., Dooren, P.V., 2013. Nearest stable system using successive convex approximations. Automatica 49, 1195–1203.

Phelan, P., Starrich, T., 2001. Innexins get into the gap. Bioessays 23, 388–396.

Pikovsky, A., Rosenblum, M., Kurths, J., 2001. Synchronization: A Universal Concept in Nonlinear Sciences. Cambridge University Press, London.

Ramirez-Weber, F.-A., Kornberg, T.B., 1999. Cytonemes: cellular processes that project to the principal signaling center in Drosophila imaginal discs. Cell 97, 599–607.

Rustom, A., Saffrich, R., Markovic, I., Walther, P., Gerdes, H.H., 2004. Nanotubular highways for intercellular organelle transport. Science 303, 1007–1010.

Salameh, A., Blanke, K., Daehnert, I., 2013. Role of connexins in human congenital heart disease: the chicken and egg problem. Front. Pharmacol. 4, 70.

Segretain, D., Falk, M.M., 2004. Regulation of connexin biosynthesis, assembly, gap junction formation, and removal. Biochim. Biophys. Acta 1662, 3–21.

Sherer, N.M., Lehmann, M.J., Jimenez-Soto, L.F., Horensavitz, C., Pypaert, M., Mothes, W., 2007. Retroviruses can establish filopodial bridges for efficient cell-to-cell transmission. Nat. Cell Biol. 9, 310–315.

Sowinski, S., Jolly, C., Berninghausen, O., Purbhoo, M.A., Chauveau, A., Köhler, K., Oddos, S., Eissmann, P., Brodsky, F.M., Hopkins, C., Onfelt, B., Sattentau, Q., Davis, D.M., 2008. Membrane nanotubes physically connect T cells over long distances presenting a novel route for HIV-1 transmission. Nat. Cell Biol. 10, 211–219.

Suhail, Y., Kshitiz, Lee, J., Walker, M., Kim, D.H., Brennan, M.D., Bader, J.S., Levchenko, A., 2013. Modeling intercellular transfer of biomolecules through tunneling nanotubes. Bull. Math. Biol. 75, 1400–1416.

Takahashi, A., Kukita, A., Li, Y.J., Zhang, J.Q., Nomiyama, H., Yamaza, T., Ayukawa, Y., Koyano, K., Kukita, T., 2013. Tunneling nanotube formation is essential for the regulation of osteoclastogenesis. J. Cell. Biochem. 114, 1238–1247.

Trefethen, L.N., Embre, M., 2005. Spectra and Pseudospectra: The Behavior of Nonnormal Matrices and Operators. Princeton University Press, Princeton.

Unger, V.M., Kumar, N.M., Gilula, N.B., Yeager, M., 1999. Three-dimensional structure of a recombinant gap junction membrane channel. Science 283, 1176–1180.

Volterra, A., Meldolesi, J., 2005. Astrocytes, from brain glue to communication elements: the revolution continues. Nat. Rev. Neurosci. 6, 626–640.

Wang, X., Bukoreshtliev, N.V., Gerdes, H.-H., 2012. Developing neurons form transient nano-tubes facilitating electrical coupling and calcium signaling with distant astrocytes. PLoS One 7 (10), e47429.

Wang, X., Veruki, M.L., Bukoreshtliev, N.V., Hartveit, E., Gerdes, H.H., 2010. Animal cells connected by nanotubes can be electrically coupled through interposed gapjunction channels. Proc. Natl. Acad. Sci. U.S.A. 107, 17194–17199.

Wang, Y., Cui, J., Sun, X., Zhang, Y., 2011. Tunneling-nanotube development in astrocytes depends on p53 activation. Cell. Death Differ. 18, 732–742.

Watkins, S.C., Salter, R.D., 2005. Functional connectivity between immune cells mediated by tunneling nanotubules. Immunity 23, 309–318.

Willecke, K., Elberger, J., Degen, J., Eckardt, D., Romualdi, A., Guldenangel, M., Deutsch, U., Sohl, G., 2002. Structural and functional diversity of connexin genes in the mouse and human genome. Biol. Chem. 383, 725–737.

Zhang, C., et al., 2004. Calcium- and dynamin-independent endocytosis in dorsal root ganglion neurons. Neuron 42, 225–236.

Index

Printed in the United States
By Bookmasters